Lecture Notes of the Institute for Computer Sciences, Social Informatics and Telecommunications Engineering 548

The LNICST series publishes ICST's conferences, symposia and workshops.

LNICST reports state-of-the-art results in areas related to the scope of the Institute. The type of material published includes

- Proceedings (published in time for the respective event)
- Other edited monographs (such as project reports or invited volumes)

LNICST topics span the following areas:

- General Computer Science
- E-Economy
- E-Medicine
- Knowledge Management
- Multimedia
- Operations, Management and Policy
- Social Informatics
- Systems

Lin Yun · Jiang Han · Yu Han

Editors

Advanced Hybrid Information Processing

7th EAI International Conference, ADHIP 2023
Harbin, China, September 22–24, 2023
Proceedings, Part II

 Springer

Editors
Lin Yun
Harbin Engineering University
Harbin, China

Jiang Han
Harbin Engineering University
Harbin, China

Yu Han
Harbin Engineering University
Harbin, China

ISSN 1867-8211   ISSN 1867-822X (electronic)
Lecture Notes of the Institute for Computer Sciences, Social Informatics
and Telecommunications Engineering
ISBN 978-3-031-50545-4   ISBN 978-3-031-50546-1 (eBook)
https://doi.org/10.1007/978-3-031-50546-1

This Springer imprint is published by the registered company Springer Nature Switzerland AG
The registered company address is: Gewerbestrasse 11, 6330 Cham, Switzerland

Paper in this product is recyclable.

Preface

We are delighted to introduce the proceedings of the 7th edition of the European Alliance for Innovation (EAI) International Conference on Advanced Hybrid Information Processing (ADHIP 2023). This conference brought together researchers, developers and practitioners around the world who are leveraging and developing advanced information processing technology. This conference aimed to provide an opportunity for researchers to publish their important theoretical and technological studies of advanced methods in social hybrid data processing, and their novel applications within this domain.

The technical program of ADHIP 2023 consisted of 108 full papers. The topics of the conference were novel technology for social information processing and real applications to social data. Aside from the high-quality technical paper presentations, the technical program also featured three keynote speeches. The three keynote speakers were Cesar Briso from Technical University of Madrid, Spain, Yong Wang from Harbin Institute of Technology, China, and Yun Lin from Harbin Engineering University, China.

Coordination with the steering chairs, Imrich Chlamtac, Shuai Liu and Yun Lin was essential for the success of the conference. We sincerely appreciate their constant support and guidance. It was also a great pleasure to work with such an excellent organizing committee team for their hard work in organizing and supporting the conference. In particular, the Technical Program Committee, led by our TPC Co-Chairs, Yun Lin, Ruizhi Liu and Shan Gao completed the peer-review process of technical papers and made a high-quality technical program. We are also grateful to the Conference Manager, Ivana Bujdakova, for her support and to all the authors who submitted their papers to the ADHIP 2023 conference.

We strongly believe that the ADHIP conference provides a good forum for all researchers, developers and practitioners to discuss all technology and application aspects that are relevant to information processing technology. We also expect that the future ADHIP conferences will be as successful and stimulating as indicated by the contributions presented in this volume.

Yun Lin

Organization

Organizing Committee

General Chair

Yun Lin — Harbin Engineering University, China

General Co-chairs

Zheng Dou — Harbin Engineering University, China
Yan Zhang — University of Oslo, Norway
Shui Yu — University of Technology Sydney, Australia
Joey Tianyi Zhou — Institute of High-Performance Computing, A*STAR, Singapore
Hikmet Sari — Nanjing University of Posts and Telecommunications, China
Bin Lin — Dalian Maritime University, China

TPC Chair and Co-chairs

Yun Lin — Harbin Engineering University, China
Guangjie Han — Hohai University, China
Ruolin Zhou — University of Massachusetts Dartmouth, USA
Chao Li — RIKEN-AIP, Japan
Guan Gui — Nanjing University of Posts and Telecommunications, China
Ruizhi Liu — Harbin Engineering University, China

Sponsorship and Exhibit Chairs

Yiming Yan — Harbin Engineering University, China
Ali Kashif — Manchester Metropolitan University, UK
Liang Zhao — Shenyang Aerospace University, China

Local Chairs

Jiang Hang   Harbin Engineering University, China
Yu Han   Harbin Engineering University, China
Haoran Zha   Harbin Engineering University, China

Workshops Chairs

Nan Su   Harbin Engineering University, China
Peihan Qi   Xidian University, China
Jianhua Tang   Nanyang Technological University, Singapore
Congan Xu   Naval Aviation University, China
Shan Gao   Harbin Engineering University, China

Publicity and Social Media Chairs

Jiangzhi Fu   Harbin Engineering University, China
Lei Chen   Georgia Southern University, USA
Zhenyu Na   Dalian Maritime University, China

Publications Chairs

Weina Fu   Hunan Normal University, China
Sicheng Zhang   Harbin Engineering University, China
Wenjia Li   New York Institute of Technology, USA

Web Chairs

Yiming Yan   Harbin Engineering University, China
Zheng Ma   University of Southern Denmark, Denmark
Jian Wang   Fudan University, China

Posters and PhD Track Chairs

Lingchao Li   Shanghai Dianji University, China
Jibo Shi   Harbin Engineering University, China
Yulong Ying   Shanghai University of Electric Power, China

Panels Chairs

Danda Rawat	Howard University, USA
Yuan Liu	Tongji University, China
Yan Sun	Harbin Engineering University, China

Demos Chairs

Ao Li	Harbin University of Science and Technology, China
Guyue Li	Southeast University, China
Changbo Hou	Harbin Engineering University, China

Tutorials Chairs

Yu Wang	Nanjing University of Posts and Telecommunications, China
Yi Zhao	Tsinghua University, China
Qi Lin	Harbin Engineering University, China

Technical Program Committee

Zheng Dou	Harbin Engineering University, China
Yan Zhang	University of Oslo, Norway
Shui Yu	University of Technology Sydney, Australia
Joey Tianyi Zhou	A*STAR, Singapore
Hikmet Sari	Nanjing University of Posts and Telecommunications, China
Bin Lin	Dalian Maritime University, China
Yun Lin	Harbin Engineering University, China
Guangjie Han	Hohai University, China
Ruolin Zhou	University of Massachusetts Dartmouth, USA
Chao Li	RIKEN-AIP, Japan
Guan Gui	Nanjing University of Posts and Telecommunications, China
Zheng Ma	University of Southern Denmark, Denmark
Jian Wang	Fudan University, China
Lei Chen	Georgia Southern University, USA
Zhenyu Na	Dalian Maritime University, China
Peihan Qi	Xidian University, China
Jianhua Tang	Nanyang Technological University, Singapore

Congan Xu   Naval Aviation University, China
Ali Kashif   Manchester Metropolitan University, UK
Liang Zhao   Shenyang Aerospace University, China
Weina Fu   Hunan Normal University, China
Danda Rawat   Howard University, USA
Yuan Liu   Tongji University, China
Yi Zhao   Tsinghua University, China
Ao Li   Harbin University of Science and Technology,
                                 China
Guyue Li   Southeast University, China
Lingchao Li   Shanghai Dianji University, China
Yulong Ying   Shanghai University of Electric Power, China

Contents – Part II

**Wireless Networks for Social Information Processing, Image
Information Processing**

Mobile Education, Mobile Monitoring, Behavior Understanding and Object Tracking

A Method for Integrating Sports Information Resources Based on Fuzzy Clustering Algorithm

Xiaoxian Xu[1](✉) and Qiao Wu[2]

[1] Physical Education Department, Xi'an Shiyou University, Xi'an 710065, China
13289273723@163.com
[2] Changchun University of Architecture and Engineering, Changchun 130119, China

Abstract. To improve the accuracy of sports information resource integration, a fuzzy clustering algorithm based method for sports information resource integration is studied. First, classify the sports information resources. According to the classification results of resources, build the sports information resource model. Use different sports concepts as nodes and their relationships as edges to build the concept network model. Based on the concept network model, denoise the sports information data. Based on the denoised sports information data, use the fuzzy clustering algorithm to cluster the sports information cluster analysis, Obtain relevant clusters of data, and then adjust the clustering algorithm parameters accordingly through statistical analysis of the clustering results to obtain accurate and effective integration results of sports information resources. The experimental results show that the accuracy of sports information integration in this method is the highest at 98%, the recall rate is the highest at 96%, the F1 is the highest at 0.97, and the longest time is 3.77 s, indicating the practicality of this method.

Keywords: Sports information resources · Fuzzy clustering algorithm · Conceptual network model

1 Introduction

In contemporary society, sports competitions have gradually developed from a simple sports event to providing entertainment and commercial value for people [1, 2]. Sports competitions [3, 4] are no longer just pure competitions between individuals and teams, but there are many factors and issues that need to be considered. Sports information resources are generated to address these issues. It is also closely related to sports, training, management, and business [5]. Firstly, in terms of sports, sports information resources can help players and coaches better understand their opponents' tactics and lineup, in order to respond faster and more accurately. In terms of training, sports information resources can provide more data and statistical results, which can be compared with past experience and judgments, helping coaches better improve and adjust training plans to improve the level of athletes. In terms of management, sports information resources can provide more useful information for event organizers and management institutions. For

© ICST Institute for Computer Sciences, Social Informatics and Telecommunications Engineering 2024
Published by Springer Nature Switzerland AG 2024. All Rights Reserved
L. Yun et al. (Eds.): ADHIP 2023, LNICST 548, pp. 3–19, 2024.
https://doi.org/10.1007/978-3-031-50546-1_1

example, TV broadcast data, event ticket booking, etc. [6]. By analyzing this data, managers can develop better marketing strategies, increase advertising revenue, and better operate competitions.In terms of business, with the increasing popularity of sports competitions, more and more enterprises and brands have noticed the commercial value of sports events and have invested in the precise marketing strategy of sports events. At this time, sports information resources have become particularly important [7]. Through the use of sports information resources, major brands can obtain more user data and traffic, improve brand value and brand influence, and enhance the stickiness with consumers. In short, sports information resources play a very important role in promoting the development of sports industry. The accuracy and timeliness of sports information resources are directly related to the smooth development of sports events, the healthy development of sports industry and the audience's feeling of watching sports events.

Therefore, the integration of sports information resources is a very important part of the current development of the sports industry. With the continuous development and popularization of computer technology and information network technology, sports information resources now widely cover athlete personal information, technical and tactical data, competition videos, ticket sales, marketing, news reports, and other aspects of information [8]. However, these different types of information are usually stored in different databases, using different systems and platforms, and cannot be shared or interacted with each other. Therefore, integrating various sports information resources is very important for promoting the development of the sports industry and improving the level and quality of sports. Effective integration of sports information resources can improve data validity: Data from different sources often have problems such as duplication, missing, and errors. After integration, it can effectively eliminate duplication, fill in gaps, verify, and improve the reliability and effectiveness of the data. Ability to better analyze and predict: The integrated data is richer and more comprehensive, enabling better data mining and analysis, and making predictions and analyses on the performance of sports athletes, event results, etc. Can provide support for management and decision-making: The integrated data can provide basic data for personnel engaged in sports management and decision-making, analyzing trends and needs, formulating better plans and decisions, and better meeting the needs of athletes and spectators. Can improve efficiency and reduce costs: The integration of sports information resources can realize the automatic collection, processing and sharing of data, reduce the duplication of work, improve work efficiency, and reduce the cost of data management and processing. Can promote marketing: Integrated data can provide strong support for the marketing of sports industry, promote sports events and brands, and provide better commercialization opportunities for sports enterprises. In short, the integration of sports information resources is very important for comprehensively improving the management and competitive level of sports, promoting the healthy development of sports industry and realizing the sustainable development goals. Some relevant researchers have studied the integration methods of sports information resources, for example, literature [9] proposes the integration study of college sports teaching and ethnic traditional sports resources, literature [10] proposes the optimization and integration of Chinese ice and snow sports resources based on the perspective of H-O-S theory, but the accuracy of the above methods is low, which affects the effect of sports information integration.

The fuzzy clustering algorithm can process fuzzy data and handle fuzzy and uncertain data without the need to divide the data into clear categories. At the same time, it has a certain degree of flexibility, and the clustering results obtained by fuzzy clustering algorithms can be fuzzy, rather than unique hard classification results. This makes the clustering results more flexible and can better reflect the similarity between data. And the anti noise ability is strong: the fuzzy clustering algorithm has a strong ability to resist outlier and noise. Suitable for data analysis and processing in various fields. This article studies the integration method of sports information resources based on fuzzy clustering algorithm.

2 Research on Sports Information Resource Integration Based on Fuzzy Clustering Algorithm

2.1 Classification of Sports Information Resources

Sports information resources refer to various data and information related to the field of sports, which can be used to support various decisions and activities, including but not limited to athlete personal data, competition results, event arrangements, schedule, player data, hot topics, historical records and trend analysis. We can analyze the characteristics of excellent athletes, factors that win in competitions, historical development trends of sports events, et al. [11]. Sports information resources can be classified according to different dimensions, and the following are some common classification methods:

1. Competitive events: including football, basketball, volleyball, track and field, swimming and other kinds of sports.
2. Event type: including professional league, national championship, World Championship, Olympic Games, World Cup and other types of sports events.
3. Data types: including game results, player data, team data, tactical analysis, technical statistics and other different types of data.
4. Data sources: including official data, third-party data, social media data and other types of data sources.
5. Time dimension: including historical data, real-time data, forecast data and other data of different time dimensions.
6. Business applications: including sports analysis, data analysis, media reports, on-site operation of sports events, event promotion and other business applications.

Different classification methods can provide demand oriented data extraction and integration for different business needs, meeting the data needs in different scenarios. The above classification process has completed the classification of sports information resources, laying a solid foundation for the subsequent construction of sports information resource models.

2.2 Construction of Sports Information Resource Model

Based on the above classification results of sports information resources, the model of sports information resources is constructed. The conceptual network model is a computer model that simulates the structure of human knowledge. The purpose of constructing sports information resource model is to prepare for the introduction of fuzzy clustering algorithm and the construction of sports information resource integration method. The better the sports information resource model, the better the integration effect. The core idea of the model is to express knowledge as concepts, and to construct knowledge network by expressing the relationship between concepts. Conceptual network model, also known as conceptual map model, is a graphical model that represents concepts and their relationships. It can be mainly used for knowledge management and knowledge representation. It can help people organize complex knowledge and information, and facilitate people to understand and use these knowledge and information [12]. It can also be used in the fields of natural language processing, information retrieval, text classification and machine translation. In this paper, the concept network model is adopted to construct the sports information resource model.

The conceptual network consists of nodes and arcs. Construct a conceptual network model by using different sports concepts as nodes and their relationships as edges. The network includes two types of nodes: concept nodes and document nodes. The arc connecting nodes expresses the correlation between nodes and quantifies the strength of the relationship using weights [13]. Set the concept node set as:

$$C = (c_1, c_2, \cdots, c_n) \tag{1}$$

The document node set is:

$$D = (d_1, d_2, \cdots, d_n) \tag{2}$$

$c_i \xrightarrow{\mu} c_j$ represents the correlation weight between concept nodes c_i and c_j as μ, which can also be expressed as $f(c_i, c_j) = \mu$. $d_i \xrightarrow{\eta} c_j$ represents the correlation weight between d_i and concept c_j as η, which can also be expressed as $f(d_i, c_j) = \eta$.

Rule 1: If there are nodes c_i, c_j and c_k, the weight of the correlation relation between $f(c_i, c_k) = \alpha, f(c_k, c_j) = \beta$, c_i and c_j is $F(c_i, c_k) = \min(\alpha, \beta)$.

Rule 2: If multiple paths are connected between node c_i and c_j, the correlation between c_i and c_j is the maximum path weight. The conceptual network model is shown in Fig. 1 below.

As shown in Fig. 1, the basic elements of the conceptual network model are mainly divided into two categories: nodes and edges.

A node represents a concept or entity, usually represented by text or icons. Nodes may include the following content: name, description, attributes, and labels. The specific element content is shown in Table 1.

Edges represent relationships or connections between nodes. Edges can also have attributes. There are several common types of edges: associative edges, inherited edges, combined edges, and aggregated edges. The specific element content is shown in Table 2.

Building a sports information resource model based on the above element content, the specific process is as follows: Document d is preprocessed (word segmentation, removal of stop words) and represented as keyword set $T = \{t_1, t_2, \cdots, t_n\}$. The frequency of each keyword appearing in the main text, title, keyword, hyperlink, and hyperlink

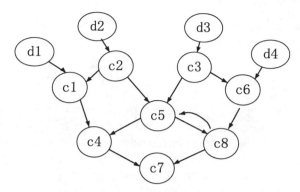

Fig. 1. Schematic diagram of conceptual network model

Table 1. Main elements of conceptual network model nodes

Element	Element content
Name	The name of the concept or entity that the node represents
Description	Description of a concept or entity corresponding to a node, including text, pictures, or videos
Attribute	Attributes of the concept or entity corresponding to the node, such as brand, color, size, et al.
Label	Nodes can use multiple labels to identify and classify the concepts or categories to which they belong

Table 2. Main elements of conceptual network model

Element	Element content
Associated edge	Indicates a relationship between two nodes, such as "basketball player" and "basketball game"
Inherited edge	Indicates that one node is a specialization of another node. For example, "China National Football team" is a subclass of "Asian Cup Football match"
Combined edge	Indicates that one node is a part of another node, for example, "Player v is a component of the" Chinese Women's Volleyball team"
Convergent edge	Indicates that one node contains other nodes, but other nodes can belong to different instances of other nodes. For example, "football team" includes "football player"

description in the statistical word set is represented as $tf_{other}(t_i)$, $tf_{title}(t_i)$, $tf_{key}(t_i)$, $tf_{url}(t_i)$, and $tf_{Anchore}(t_i)$. The frequency calculation formula for keyword t_i is:

$$tf(t_i) = tf_{other}(t_i) + A_1 \cdot f_{title}(t_i) + A_2 \cdot f_{key}(t_i) + A_3 \cdot f_{url}(t_i) + A_4 \cdot f_{anchor}(t_i) \qquad (3)$$

where, A_1, A_2, A_3 and A_4 are adjustment coefficients. Then the formula for calculating the weight of feature words in the document is:

$$w_t \rightarrow d(t_i, d) = tf(t_i) \times \log\left(\frac{N}{df(t_i)} + 0.5\right) \qquad (4)$$

Among them, $df(t_i)$ represents the number of documents containing keyword t_i, and N represents the total number of documents. Words are the manifestation of concepts, and a concept node may contain multiple corresponding words. Let the words corresponding to concept node c_i form a set $T_i = (t_1, t_2, \cdots, t_m)$, represented in vector form $c_i = \langle w_{i1}, w_{i2}, \cdots, w_{im} \rangle$, where w_{ij} represents the weight of keyword t_j in concept node c_i. Calculate the correlation between document d and concept c_i as follows:

$$rel_{d \rightarrow c}(d, c_i) = \frac{\sum_{j=1}^{m} w_{t \rightarrow d}(t_j, d) + w_{t \rightarrow c}(t_j, c_i)}{Max(Tn(d), Tn(c_i)) * 2} \qquad (5)$$

In the formula, $w_{t \rightarrow c}(t_j, c_j)$ represents the weight of keyword t_j in concept node c_i, $Tn(d)$ represents the total weight of all keywords in document d, and $Tn(c_i)$ represents the total weight of keywords contained in concept node c_i.

There are semantic associations between words contained in the same document. This correlation is formally manifested as the co-occurrence of words [14]. We use these phenomena to find correlations between concepts. Some samples are selected to constitute sample set S, $S = (S_1, S_2, \cdots, S_M)$ and M as the number of documents. Set the concept node set $C = (c_1, c_2, \cdots, c_n)$, and calculate the correlation between the documents in the sample set and the concept node. For concept node c_i, its correlation with the document can be expressed in vector form: $c_i = \langle e_{1i}, e_{2i}, \cdots, e_{Mi} \rangle$, e_{ji} represents the correlation between document d_j and concept node c_i. The formula for calculating the correlation between conceptual nodes c_i and c_j is:

$$rel_{c \rightarrow c}(c_i, c_i) = \frac{\sum_{k=1}^{M} e_{ki} e_{kj}}{\sqrt{\sum_{k=1}^{M} e_{ki}^2 \sum_{k=1}^{M} e_{kj}^2}} \qquad (6)$$

The generation of concept nodes in conceptual networks can be achieved through two methods: clustering and gradual addition. When using clustering methods, each keyword in the initial stage corresponds to an independent concept node. Calculate the correlation of concept nodes, and according to the set threshold, concept nodes with correlation exceeding the threshold are merged into new nodes.

Gradually add methods to implement concept mapping using HowNet. CNKI is a common sense knowledge base based on concepts and their characteristics, revealing the relationships between concepts and the characteristics they possess. CNKI describes various relationships between concepts and their attributes. By querying CNKI, we can obtain the concepts corresponding to feature words. We map the keywords in the document to the corresponding concepts. Due to the existence of multiple semantics in some feature words, there may be a one-to-many mapping situation, so auxiliary processes need to be taken to confirm the correct conceptual mapping relationship.

Define co-occurrence: If feature words x and y appear in the same sentence in document d_i, they are considered co-occurrence, and the expression of interword co-occurrence rate CO is as follows:

$$CO(x, y) = \frac{freq(x, y)}{freq(x) + freq(y)} \tag{7}$$

Among them, $freq(x, y)$ is the number of sentences where the feature word x and y co appear, and $freq(x)$ is the number of sentences where the feature word x appears. Document feature word set $T = \{t_1, t_2, \cdots\}$, co-occurrence feature word set for feature word $t_i \in T$ with multiple mapping relationships:

$$T_{CO}(t_i) = \left\{ t_j \,\middle|\, CO(t_i, t_j) \geq \psi \right\} \tag{8}$$

ψ is the default threshold. The concept of t_i is $c_1, c_2, \ldots c_k$. The feature word t_i belongs to the concept c_i. The possibility is calculated as:

$$p_{ti}(c_i) = \sum_{t_k \in T_{CO}(t_i)} CO(t_i, t_k) \delta_{rel(c_i, C(t_k))} \tag{9}$$

Among them,

$$d_{rel(c_i, c_j)} = \begin{cases} 0.7 & c_i \text{ and } c_j \text{ are co-occurrence concepts} \\ 0.5 & c_i \text{ and } c_j \text{ are agents or beneficiaries} \\ 0.2 & c_i \text{ and } c_j \text{ are synonymous} \\ 0.1 & \text{other} \end{cases} \tag{10}$$

Among them, $C(x)$ represents the concept of feature word x. Choose the concept with the highest degree of membership as its feature word.

For concept node c, the page set is:

$$D(c) = \{d \,|\, d \in D, rel_{d \to c}(d, c) > \lambda\} \tag{11}$$

λ indicates the threshold. According to the above page set, calculate the weight of the key words corresponding to the concept node:

$$w_{t \to c}(t_i, c) = \frac{\sum_{d_j \in D(c)} w_{t \to d}(t_i, d_j)}{m} \tag{12}$$

In the formula, m is the number of documents in document set $D(c)$. Normalize the $w_{t \to c}(t_i, c)$ value.

Weights can reflect the importance or relevance of a keyword in a specific text or field. Finally, we can calculate the weight of each concept node as the sum of the keyword weights contained in its label. This can obtain an indicator that reflects the importance or correlation of nodes in the conceptual network. By labeling and weight calculation of nodes mentioned above, we can obtain a more comprehensive and accurate conceptual network model, and can more effectively apply sports information resource models when processing large amounts of data and conducting complex reasoning. Based on the above process, the construction of a sports information resource model was completed, and support was provided for the subsequent preprocessing of sports information resources. In the process of building the sports information resource model, this paper labels the information nodes, combines the weights, and optimizes the correlation between concept nodes to build the feature word set of sports information resources. In order to further optimize the resource model, the weights of the keywords corresponding to the concept nodes are further determined. A more perfect sports information resource model is constructed.

2.3 Preprocessing of Sports Information Resources

There is often some noise in sports information resources, which affects the integration of sports information resources, so it is necessary to preprocess and reduce the noise of resources. Based on the model of sports information resources, the representation form of sports information resources is determined. According to the representation form of sports information resources, the denoising process of sports information data is carried out using the principle of wavelet transform [15, 16]. In general, the data after wavelet transform consists of two parts. Are respectively the low frequency part corresponding to the effective signal and the high frequency part corresponding to the noise. Therefore, when using wavelet transform to denoise data, an appropriate threshold should be selected to control the denoising accuracy. The expression of threshold function is as follows:

$$\hat{w}_{i,k} = \begin{cases} w_{j,k}, & |w_{j,k}| \geq \lambda \\ 0, & |w_{j,k}| < \lambda' \end{cases} \tag{13}$$

where, $w_{j,k}$ is the wavelet coefficient after threshold processing, $\lambda = \delta_n \sqrt{2 \ln N}$ and δ_n are the standard deviation of noise, and N is the number of discrete points.

Let the observation data be $f(t)$, which can be expressed as the combination of the original data $s(t)$ and noise $n(t)$, namely $f(t) = s(t) + n(t)$. The objective of wavelet denoising is to estimate the original signal $s(t)$ and remove the noise $n(t)$, which can be achieved by the following formula:

$$\hat{s}(t) = \sum_{j=1}^{J} \sum_{k=0}^{2^j-1} wthresh\left(\left|W_{j,k}(f(t))\right| - \lambda_j W_{j,k}(f(t))\right) \tag{14}$$

Based on the above content, complete the preprocessing of sports information data, laying the foundation for the subsequent integration of sports information resources.

2.4 Sports Information Resource Integration Based on Fuzzy Clustering Algorithm

In order to improve the integration effect of sports information resources, fuzzy clustering algorithm [17–19] is introduced. Based on the above preprocessed sports information data, the fuzzy clustering algorithm is used to cluster the sports information cluster analysis to obtain the relevant clusters of the data. Then, through the statistical analysis of the clustering results, the clustering algorithm parameters are adjusted accordingly to obtain more accurate and effective clustering results.

Due to the numerous factors that affect the integration of sports information resources, and the significant correlation between each factor and the integration of sports information resources, different aspects of the impact are comprehensively considered, and the representativeness and comparability in the resource classification process are followed to determine the clustering objects and related indicators. Based on the classification results of sports information resources mentioned above, select a clustering center, which is the basis for resource classification. Assume that the given data set is $X = (x_1, x_2, \ldots, x_N)$, that is, the data set contains N samples, where $x_j (j = 1, 2, \ldots, N)$ is a D-dimensional data point. If the data set X is considered to be classified into class C, $V = (v_1, v_2, \ldots, v_c)$ represents the clustering center. According to membership matrix u_{ij}, each data point is assigned to C clustering centers. u_{ij} represents the probability that the j data point belongs to the i clustering center, where $u_{ij} \in [0, 1]$ can reflect the probability of the data point belonging to the corresponding category according to the probability. The membership function also needs to satisfy $\sum_{i=1}^{c} u_{ij} = 1$, that is, the sum of probabilities from a pixel point to each cluster center is always equal to 1. After the membership matrix is determined, each pixel can be divided into the category with the largest corresponding membership value according to the principle of maximum similarity, as shown in Fig. 2.

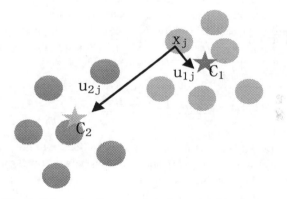

Fig. 2. Schematic diagram of pixel membership division

Among them, the membership value is inversely proportional to the distance, and the distance between data points x_j and c_2 is larger, resulting in smaller membership values. Therefore, x_j is classified into the category where c_1 belongs. The similarity of similar data points in the final clustering results is higher, while the similarity of data points in different classes is lower. Based on the fuzzy C-means clustering algorithm, an alternating iteration scheme is used to minimize the objective function J to obtain the optimal segmentation result. Among them, the objective function represents the weighted sum of Euclidean distances from the data point to the corresponding clustering center.

The definition formula of objective function J is as follows:

$$J(U, V) = \sum_{i=1}^{C} \sum_{j=1}^{N} u_{ij}^{m} \|x_j - v_i\|^2 \tag{15}$$

where, m is a membership weighted value, which can be set artificially. Generally speaking, m is 2 according to experimental experience. Membership degree must meet: $\sum_{i=1}^{C} u_{ij} = 1, \forall j \in \{1, 2, \ldots, N\}$. v_i represents the i cluster center, and $\|x_j - v_i\|^2$ represents the Euclidean distance between data points and the cluster center. A better clustering result is represented by a smaller distance between data points in one category and a larger distance between data points in different categories. So the iterative process is constantly adjusting the membership to minimize the weight and objective function J. Firstly, the membership degree is calculated and then the clustering center is calculated, and then the membership degree is recalculated. Then the membership degree is iterated until the clustering center and the membership matrix do not change too much. The optimal conditions for the implementation of each iteration are as follows:

$$\begin{cases} U^{(t+1)} = \underset{U}{\arg\min} \, J_m^{(FCM)}\{U, V^{(t)}\} \\ U^{(t+1)} = \underset{V}{\arg\min} \, J_m^{(FCM)}\{U^{(t+1)}, V\} \end{cases} \tag{16}$$

Among them, t represents the iteration step. Usually, random initialization is used to obtain $U^{(0)}$ or $V^{(0)}$, and then alternating updates are made.

U and V until convergence. Due to the constraint conditions of the membership function being: $\sum_{i=1}^{C} u_{ij} = 1, \forall j \in \{1, 2, \ldots, N\}$. Therefore, the Lagrangian factor can be introduced to solve the objective function J:

$$L(u_{ij}, v_i) = \sum_{i=1}^{C} \sum_{j=1}^{N} u_{ij}^{m} \|x_j - v_i\|^2 - \lambda \left(\sum_{i=1}^{C} u_{ij} - 1 \right) \tag{17}$$

where the undetermined coefficient λ is the Lagrange factor. In this way, the minimum problem of objective function J with respect to membership function u_{ij} can be transformed into the minimum problem of function $L(u_{ij}, v_i)$. According to the constraints of the membership function [20], it can be obtained:

$$\lambda = m \left[\|x_j - v_i\|^{\frac{2}{m-1}} \right]^{m-1} \tag{18}$$

By incorporating formula (14) into formula (13), the iterative formula for the membership function u_{ij} can be obtained:

$$u_{ij} = \sum_{l=1}^{C} \left(\frac{\|x_j - v_i\|}{\|x_j - v_l\|} \right)^{\frac{-2}{m-1}} \tag{19}$$

The iterative formula of cluster center v_i can be obtained by differentiating v_i:

$$v_i = \frac{\sum_{j=1}^{N} u_{ij}^m x_j}{\sum_{j=1}^{N} u_{ij}^m} \qquad (20)$$

The fuzzy C-means clustering algorithm usually initializes the membership matrix U randomly in the process of algorithm iteration, obtains the cluster center matrix V through formula (20), and then updates U, so as to iterate repeatedly to obtain the solution of the objective function J. In summary, the algorithm flowchart of the fuzzy C-means algorithm is shown in Fig. 3.

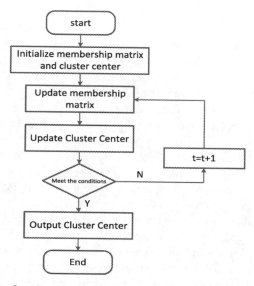

Fig. 3. Algorithm flowchart of fuzzy C-means algorithm

Before iteration, the clustering algorithm initializes two clustering centers, and then updates the membership matrix continuously according to iteration operation, and then updates the clustering centers according to the updated membership function until the iteration stop condition set by the initial parameter is satisfied, the algorithm stops iteration and outputs the clustering result, which is the integration result of sports information resources. It can help to improve the efficiency of sports information resources. At this point, the design of sports information resource integration method based on fuzzy clustering algorithm has been completed, which classifies sports information resources and considers the influence of resource model on resource integration effect. Therefore, the sports information resource model is constructed by introducing weight algorithm, labeling processing and correlation calculation, and considering the existence of certain noise in sports information resources. It seriously affects the integration effect. Therefore, before introducing the fuzzy clustering algorithm, the wavelet transform algorithm is introduced to preprocess the sports information resources and introduce the fuzzy clustering algorithm. Through the combination of these algorithms, the integration method of sports information resources is constructed and the integration effect is improved.

3 Experimental Analysis

3.1 Preparation for Experiment

This article is based on the integration method of sports information resources using fuzzy clustering algorithm for sports information resource integration. To verify the performance of this method, experimental tests were conducted using the integration research of university sports teaching and ethnic traditional sports resources (Method 1) and the optimization and integration of China's ice and snow sports resources based on the H-O-S theory perspective (Method 2) as comparative methods. Retrieve sports information resources through a database for experimental testing. This experiment adopts the sports special database, which has collected all kinds of sports resources since the establishment of People's Sports Publishing House, divided into seven categories: classic textbooks, academic monographs, competition rules and judging laws, sports videos, sports books, sports dictionaries, Olympic channels, and more than 100 series. The library contains more than 600 kinds of teaching materials, 2,269 kinds of academic books, more than 700 kinds of public books, more than 1,000 kinds of videos, more than 90,000 professional entries, and more than 4,000 professional example sentences and phrases. The database content is systematic and thematic. And the database combined with the requirements of the new national curriculum standards, through video, books, monographs, sports dictionaries, competition rules, referee law, and Olympic topics and other forms of all-round display in sports teaching, research and daily application of resources. In order to verify the data security performance and sharing time efficiency of the method proposed in this article, the entire experiment needs to be completed on a powerful server. The server software and hardware parameters are shown in the table below (Table 3).

Table 3. Software and hardware parameters of the server

Parameter	Model number
Server	Inter(R)Core(TM)i7-7700HQ CPU@3.8 GHz
Database	MySQL 5.5
Database management tool	Navicat
Environment	D2RO, D2R Server
Code writing software platform	Python 4.8
Memory	512G
Operating system	Windows10

3.2 Experimental Indicators

The accuracy of sports information resource integration is generally quantitatively evaluated using evaluation indicators such as Precision, Recall, and F1 in the field of information retrieval.

Among them, Precision represents the ratio of the number of relevant resources retrieved to the number of all resources retrieved, expressing the proportion of relevant information content in the search results. The closer it is to 1, the more truly relevant resources in the search results, that is, the more accurate the search results are. The calculation formula is:

$$P = \frac{TP}{(TP + FP)} \tag{21}$$

TP (True Positive) indicates the number of resources retrieved correctly, and FP (False Positive) indicates the number of resources retrieved incorrectly.

Recall refers to the ratio between the number of retrieved resources and the actual number of all relevant resources, that is, the number of retrieved resources. The closer it is to 1, the more truly relevant resources can be found, that is, the more comprehensive the search results. The formula is as follows:

$$Recall = TP/(TP + FN)$$

$$R = \frac{TP}{(TP + FN)} \tag{22}$$

Among them, FN (False Negative) represents the actual number of related resources that exist but cannot be retrieved.

F1 value is the weighted harmonic mean of Precision and Recall, which can quantify the accuracy and integrity of the integration results. The closer to 1, the higher the overall accuracy and recall, and the best comprehensive evaluation effect. Its calculation formula is:

$$F1 = 2 \times P \times R \frac{R}{(P + R)} \tag{23}$$

The F1 value can be regarded as the comprehensive performance of Precision and Recall. The larger it is, the higher the accuracy and completeness of the integrated results will be.

3.3 System Function Test Analysis

Before verifying the performance of the sports information resource integration method based on fuzzy clustering algorithm designed in this paper, it is necessary to ensure that all functions of the method in this paper can operate normally, otherwise the accuracy of the experimental analysis may be affected. Therefore, systematic functional testing and analysis of the method in this paper are carried out, and the analysis results are shown in Table 4.

Table 4. Results of systematic functional test of sports information resource integration method

Serial number	Test item	Operation condition
1	Classification of sports information resources	Normal
2	Information resource label	Normal
3	Multiple semantic feature word mapping	Normal
4	Sports information resources denoising	Normal
5	Clustering of different types of resources	Normal
6	Clustering of resources of the same type	Normal

As can be seen from Table 4, all functions of the method in this paper can oper-ate normally and have the basis for experimental analysis. During the experiment, the operation of each module function of the method will not affect the integration effect of sports information resources.

3.4   Accuracy of Sports Information Resource Integration

To verify the effectiveness of the method proposed in this article, experimental tests were conducted using the accuracy of information resource integration as the experimental indicator. The test results are shown below.

As shown in Fig. 4, after applying the method proposed in this article, the accuracy, recall, and F1 of sports information resource integration are higher than those of Method 1 and Method 2. Among them, the maximum accuracy of the sports information integration method in this article is 98%, the maximum recall rate is 96%, and the maximum F1 is 0.97. It can be seen that the sports information integration effect of this method is the best.

3.5   Efficiency of Sports Information Resource Integration

To further validate the practicality of the method proposed in this article, experimental testing was conducted using the efficiency of information resource integration as the experimental indicator. The test results are shown in Fig. 5.

As shown in Fig. 5, the integration efficiency of sports information resources after the application of the proposed method is higher than that of methods 1 and 2. Among them, the longest time of sports information integration of the method in this paper is 3.77 s, the longest time of the method in method 1 is 6.34 s, and the longest time of the method in method 2 is 8.21 s. It can be seen that the method in this paper has the shortest time of sports information integration, the highest efficiency and practicability.

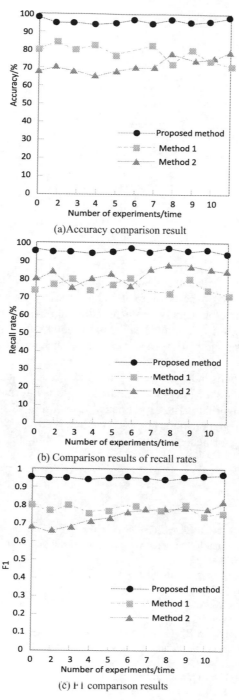

(a)Accuracy comparison result

(b) Comparison results of recall rates

(c) F1 comparison results

Fig. 4. Integration accuracy of sports information resources

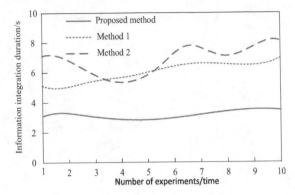

Fig. 5. Integration efficiency of sports information resources

4 Conclusion

With the continuous development and popularization of computer technology and information network technology, sports information resources have now widely covered athletes' personal information, game video, marketing and other aspects of information. However, these different types of information are usually stored in different databases, using different systems and platforms, and cannot be shared or interacted with. Therefore, the integration of various sports information resources is very important to promote the development of sports industry and improve the level and quality of sports. In this regard, this paper studies the integration method of sports information resources based on fuzzy clustering algorithm. The experimental results show that the accuracy rate of sports information integration of the proposed method is 98%, the recall rate is 96%, the F1 maximum is 0.97, and the longest time is 3.77 s, indicating that the proposed method is practical.

References

1. Escamilla-Fajardo, P., Alguacil, M., García-Pascual, F.: Business model adaptation in spanish sports clubs abetaording to the perceived context impact on the social cause perfonnance. Sustainability **13**(6), 1–11 (2021)
2. Wei, L., Xing, L.: Sociological analysis of Chinese sports viewers with differences in social capital. Complexity **2021**(6), 1–7 (2021)
3. Subathra, P., Elango, M., Arumugam, S.: Influence of mental training on aggression and sports competition anxiety among volleyball players. Gorteria **34**(1), 377–382 (2021)
4. Ju, L., Huang, L., Tsai, S.B.: Online data migration model and ID3 algorithm in sports competition action data mining application. Wirel. Commun. Mobile Comput. **2021**(7), 1–11 (2021)
5. Wang, J.: Research on the integration path of school and city sports resources based on network platform. Contemp. Sports Technol. **11**(18), 233–235 (2021)
6. Yi, S.: The value evolution and resource integration path of school physical education under the "Double Reduction Policy". Sci. Technol. Stationery Sport. Goods (23), 130–132 (2022)

7. Tian, H.: Resource integration and inheritance innovation of traditional sports culture in local universities-a case study of south lion talent training in Jiaying University. J. Jiaying Univ. **40**(6), 82–87 (2022)

8. Chen, H., Nan, M.: Research on the archive resources integration mode of large-scale sports events from the perspective of digital humanities. Shanxi Arch. (01), 27–39 (2022)

9. Deng, X.: Research on the integration of college physical education teaching and ethnic traditional sports resources. China Adult Educ. (04), 51–53 (2021)

10. Wang, Z., Li, K.: The optimization and integration of snow and ice sports resources in China based on H-O-S theory. China Winter Sports **44**(3), 49–53 (2022)

11. Zheng, C.: The exploration and research of school physical education informationization teaching in DT Era. Contemp. Sports Technol. (6), 130–133 (2021)

12. Jiang, H., Fu, Z.: Research on the construction and application of a school school joint sports resource sharing platform based on regional digitalization. Chin. Sci. Technol. J. Database (Full Text Edition) Educ. Sci. (11), 199–202 (2022)

13. Wang, Y.: Algorithm design of air logistics resource optimal allocation based on fuzzy clustering. Electron. Des. Eng. **30**(15), 20–24 (2022)

14. Wang, Y., Li, X.: Hybrid recommendation algorithm combining wolf colony algorithm and fuzzy clustering. Comput. Eng. Appl. **58**(5), 104–111 (2022)

15. Meng, L., Liu, L., Peng, Y., et al.: A data hiding scheme based on U-Net and wavelet transform. Knowl.-Based Syst. **223**(8), 22–36 (2021)

16. Izat, S., Ruslan, M., Peter, C., et al.: Prediction of gamma ray data from pre-stack seismic reflection partial angle stacks using continuous wavelet transform and convolutional neural network approach. J. Appl. Geophys. **197**(6), 523–533 (2022)

17. Li, K.X., Zhao, J., Liu, T., et al.: A fuzzy spectral clustering algorithm for hyperspectral image classification. IET Image Process. **15**(12), 2810–2817 (2021)

18. Zhu, Y., Han, Y.: Marine environment monitoring based on virtual reality and fuzzy C-means clustering algorithm. Mob. Inf. Syst. **2021**(7), 1–11 (2021)

19. Chen, Z.: Using big data fuzzy K-means clustering and information fusion algorithm in english teaching ability evaluation. Complexity **2021**(5), 1–9 (2021)

20. Wang, X., Hao, L.: Event-triggered fault detection for Takagi-Sugeno fuzzy systems via an improved matching membership function approach. Inf. Sci. **593**(6), 35–48 (2022)

Research on Energy Consumption Data Monitoring of Smart Parks Based on IoT Technology

Hao Zhu[✉]

Business School, Shanghai Sanda University, Shanghai 201209, China
zhu_scmglobal@163.com

Abstract. Intelligent park energy consumption data monitoring is the basis of effective energy management. By monitoring energy consumption data, you can understand the energy consumption of each equipment, system and region in the park, and analyze and evaluate energy use. This helps identify energy consumption problems, optimize energy use, and develop sound energy management practices. Therefore, a smart park energy consumption data monitoring method based on Internet of Things technology is proposed. The perception layer of the Internet of Things technology is used to control the energy of the smart park through the collection, transmission and control of monitoring data. The energy consumption data of each layer of the large-scale smart park is collected through the sensor network. The host computer uses USB interface to obtain data from the gateway. Based on this, the energy consumption data is preprocessed by using power error data correction and missing data fitting compensation steps. By using Gaussian function to analyze the characteristics of energy consumption sample data of the smart park, a multiple linear regression model is constructed to complete the monitoring of energy consumption data of the smart park. The experimental results show that the smart park energy consumption sequence under the proposed method is more stable in fit degree, more accurate in prediction and shorter in response time.

Keywords: Internet of Things Technology · Smart Park · Zigbee · Grey Short - and Long-Term Memory Network

1 Introduction

Smart Park [1] is a modern park based on information technology, fully utilizing technologies such as the Internet of Things, cloud computing, big data, artificial intelligence, and blockchain to achieve intelligent facilities management, enterprise services, urban governance, and other aspects of the park. Through the application of Internet of Things technology, smart parks have achieved functions such as intelligent perception, interconnection, and autonomous decision-making of items, thereby improving the operational efficiency and enterprise service level within the park. For example, by deploying intelligent monitoring equipment to monitor and analyze the use of various resources (water,

L. Yun et al. (Eds.): ADHIP 2023, LNICST 548, pp. 20–37, 2024.
https://doi.org/10.1007/978-3-031-50546-1_2

electricity, gas, etc.) in the park in real time, energy efficiency has been improved and waste has been reduced. In addition, smart parks can also utilize big data and artificial intelligence technology for data analysis and prediction, optimize park planning and operational decision-making, and achieve information management of various internal links in the park. At the same time, smart parks can also work closely with urban management departments to achieve intelligent urban governance. In short, smart parks are a new direction for the development of modern parks and an important means of promoting the economic development of parks and enhancing the level of urban intelligence.

At present, many scholars have gradually proposed new energy consumption methods. For example, Ji et al. [2] proposed an energy consumption prediction method based on convolutional neural networks. Convolutional kernels can continuously extract time series features to obtain accurate results. However, in practical applications, reasonable output weights are not set, resulting in slow network training speed and weak generalization ability. Xiao et al. [3] proposed a support vector machine method for predicting energy consumption, which verifies the input variables of the model through univariate validation. Although the support vector machine considers the temporal correlation of prediction, when the sample data is small, the energy consumption prediction error of the method is relatively large.Kladas A et al. [4] in order to realize efficient photovoltaic research, proposed an energy data detection method, which combined the Ramer-Douglas-Peucker algorithm with the Timescaledb compression method to reduce the space of time series data to ensure the maximum saving of disk space. The monitoring speed is slow. Nascimento GFS et al. [5] adopted non-invasive load monitoring (NILM) technology to compress building data sets to reduce energy consumption, and used factorial hidden Markov model to complete data detection. However, in the actual application process, the calculation amount is large and the detection efficiency is low.

To improve the drawbacks of the above methods, this article proposes a smart park energy consumption data monitoring method based on Internet of Things technology.The main innovations of this research are reflected in the following aspects:

(1) Use the perception layer of the Internet of Things technology to collect energy consumption data of each layer of the smart park through the sensor network. This paper adopts advanced wireless sensor technology to realize real-time monitoring and collection of energy consumption data of smart park.
(2) The steps of power error data correction and missing data fitting compensation are proposed, and the collected data is preprocessed. The innovation of this step lies in the ability to effectively deal with errors and omissions in the data, improving the accuracy and integrity of the data.
(3) Gaussian function was used to analyze the characteristics of the energy consumption sample data of the smart park. This analysis method can reveal the underlying patterns and regularities in the data, providing a basis for subsequent modeling and prediction.
(4) Construct multiple linear regression model for energy consumption monitoring. This paper uses multiple linear regression model to monitor the energy consumption data of smart parks. This model can comprehensively consider the influence of many factors on energy consumption and provide more comprehensive and accurate energy consumption monitoring results.

2 Smart Park Energy Consumption Data Monitoring

2.1 Energy Control and Information Transmission in Smart Parks Based on the Internet of Things

This article utilizes IoT technology to achieve energy management and control in smart parks. The Internet of Things mainly plays a monitoring role in this system, with the perception layer being the most critical. It generally includes sensors, sink nodes, upper computers, etc., which use the cooperative relationship of nodes to transmit the acquired information through the routing of other nodes, and finally send it to the sink node, and then use the external network to send it to the control center, so as to complete the collection, transmission and control of monitoring data.

Develop an overall concept for the energy management and control system, dividing it into three layers: acquisition, transmission, and application through a layered approach. The overall architecture is shown in Fig. 1.

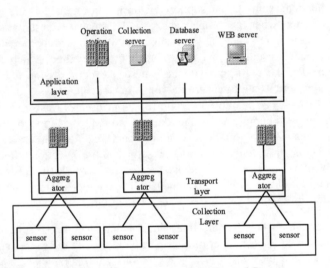

Fig. 1. Overall Architecture of the Control System

(1) Collection layer

The system acquisition part mainly uses different types of sensors, such as smart meter [6] and temperature sensing devices, through which various energy consumption information and environmental information are tested.

Power supply information collection: including data such as electricity quantity, voltage, power, and harmonics, providing a basis for the system's electricity analysis.

Water supply information collection: divided into water temperature, flow rate, and pH value, which can enable users to timely understand water use information and water quality.

Boiler information collection: mainly collects steam temperature, fuel consumption, etc., and can provide real-time data for system efficiency evaluation.

(2) Transport layer

The system transmission section includes on-site level and management level, with the former communicating through Zig Bee [7] and the latter using Ethernet for communication.

Among them, Zig Bee is suitable for short distance wireless communication due to its advantages such as low cost and good security. Its two biggest characteristics are: good self-organization performance, no need for manual operation; Strong self-healing ability, when a node fails, the network can repair itself to ensure normal communication, and this process does not require manual operation.

(3) Application layer

It has multiple functions such as information collection, analysis, organization, and prediction. Through information collection, real-time understanding of energy utilization status, and utilizing functions such as analysis and organization, the energy utilization status is organized to identify problems; Finally, predict the use of energy consumption, grasp energy consumption trends, achieve effective control, and reduce energy waste.

2.2 Monitoring Information Transmission Based on the Internet of Things

The monitoring method in this article consists of three parts: sensor nodes, gateways, and hosts. Multiple sensors and ZigBee [8] form a sensor network, whose main task is to collect energy consumption data from various layers of large smart parks. By processing the data of the power system, sensor nodes send data to the gateway, and the upper computer uses USB interface to obtain data from the gateway, forming a visual human-machine interface, which facilitates users to observe the numerical changes of electrical equipment in real time on a PC. Based on the three-layer architecture of the Internet of Things, energy consumption monitoring is achieved through technologies such as wireless sensors and the Internet. It consists of three main modules: perception layer, communication layer, and application layer.

The perception layer is the lowest level of the system and the center of the entire monitoring. The main function of the sensing layer is to collect energy consumption data, and a wireless sensor network is established using ZigBee technology. The perception layer of this unit is located indoors in a large smart park, consisting of various indoor sensors and wireless communication nodes. And analyze, preprocess, and store the wireless sensor nodes connected to it, completing the configuration of ZigBee and the establishment of the wireless Mesh network.

The communication layer consists of two main parts: inbound communication and remote data transmission, with the function of transmitting energy consumption information. Utilizing an embedded gateway, the ZigBee communication protocol for close proximity is converted into remote GPRS to meet the needs of remote data transmission;

The GPRS communication network is a bridge between the application layer information management center and the perception layer. Through the embedded gateway of the communication layer, the collected energy consumption data of the smart park is connected to the wireless GPRS network through the GPRS module. Then, the GPRS base station of the mobile operator is connected to the internet to complete the storage and management of the data, which is finally sent to the monitoring management information center through various switches and routers. The application layer is the energy consumption management center, whose main role is to analyze and integrate energy consumption data. And the parameter information of each energy consumption measurement point is transmitted to the database through the network of the communication layer, which is managed by the administrator.

2.3 Pre Processing of Historical Energy Consumption Data in Smart Parks

This article takes electricity as the energy consumption of smart parks that needs to be predicted. During the process of collecting electricity data, due to statistical errors and other reasons, a large number of errors and missing information may appear in the data used for prediction. Therefore, it is necessary to preprocess the original dataset, which includes the correction of electrical energy error data and the fitting compensation of missing data.

The external factors that affect the overall energy consumption of smart parks include: urban development capacity, smart park area, temperature changes, thermal conditions of smart parks, power of smart park electrical equipment, etc. However, there has been no actual change in the thermal situation of smart parks in many regions over the years, and the power equipment in smart parks has always been an original piece of equipment, without any actual changes.

Therefore, the main influencing factors for the overall power consumption prediction of urban smart parks only need to consider three external factors: urban development capacity, urban smart park area, and temperature changes. Characteristic indicator analysis is needed for these three external factors. The different indicators of these three external factors and the variable indicator R of smart park power consumption are as follows:

$$R = \sum (X - \overline{X})(Y - \overline{Y})/\sqrt{(X - \overline{X})^2(Y - \overline{Y})^2} \tag{1}$$

In the formula, X represents indicator sample 1, $\overline{X}$ represents indicator sample 2, Y represents the mean of indicator sample 1, and $\overline{Y}$ represents the mean of indicator sample 2.

Based on the absolute value of variable indicators, the power consumption of smart parks for various influencing factors was determined, and six indicators were obtained: monthly sequence variables, monthly cycle variables, per capita disposable income of urban and rural people, smart park area, average summer hour, and average winter hour. The relationship between various indicators and the electricity consumption of the smart park is shown in Table 1.

2.4 Energy Consumption Prediction of Smart Parks Under Long-Term and Short-Term Memory Networks

In the traditional recurrent neural network, when any hidden layer memory unit calculates the weight matrix and activation function, the obtained memory data will be fleeting,

Table 1. Relationship between various indicators and electricity consumption in smart parks

Impact variable indicators	The relationship between power consumption and smart park
Smart Park Area	This indicator has a linear relationship with the total electricity consumption of the smart park, increasing in the same proportion
Per capita disposable income	This indicator is highly correlated with the total power consumption of the smart park
Month cycle variable	This indicator can display the cycle length of power consumption in smart parks
Month sequence variable	This indicator can display the growth of power consumption in smart parks
Average Summer Hour	This indicator is clearly biased towards the air conditioning and cooling power consumption in the smart park's electricity consumption
Winter average hour	This indicator is clearly biased towards the heating and heating electricity consumption in the smart park's electricity consumption

while the long-term and short-term memory network makes the memory data disappear instantaneously without being affected by the fusion of memory data and current input data. The short-term memory network can store and use historical data by introducing input gates and output gates into the model, and adjust the output of different time series to bring stable processing to the subsequent actions of the energy consumption prediction model [9, 10].

Although the long-term and short-term memory network [11] has excellent nonlinear fitting ability, its prediction deviation is high when there are few samples. The energy consumption cycle of smart parks is unstable, making it difficult to find similar fitting curves for replacement. By using the grey theory method [12, 13] to cumulatively transform the raw data and reduce input instability, rapid prediction of energy consumption can be achieved. The optimized energy consumption prediction model process is shown in Fig. 2.

In the process of establishing the prediction model of grey long and short term memory network, it is necessary to accurately calculate the relevant parameters in the network. By using the Gaussian function to analyze the characteristics of the energy consumption sample data of the smart park, the optimized prediction model uses the K-means clustering method [14, 15] to calculate the height and central coordinate point of the Gaussian function, complete the reasonable allocation of resources, and make the central coordinate point of the prediction model coincide with the central point of the input sample.

Grey process the output and input data, and normalize the formula:

$$T = T_{\min} + \frac{T_{\max} - T_{\min}}{E_{\max} - E_{\min}}(E - E_{\min}) \tag{2}$$

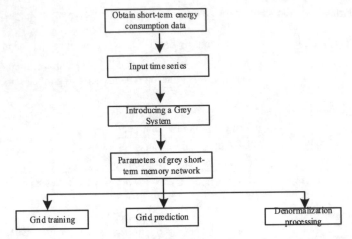

Fig. 2. Flow chart of energy consumption prediction of grey short-term memory network

In the formula, E represents raw data, T represents ashed data, E_{max} represents the maximum value of raw data, E_{min} represents the minimum value of raw data, T_{max} represents the maximum value of ashed data, and T_{min} represents the minimum value of ashed data. The data samples are constructed by using the position sliding method, and the corresponding weights and parameters are obtained through the learning and training of the gray long-term and short-term memory network.

2.5 Energy Consumption Data Monitoring for Smart Park Operation

From a statistical perspective, the possibility of the existence of lower limit data samples for energy consumption monitoring in smart parks was analyzed, and a multiple nonlinear regression model was used to analyze the energy consumption and influencing factors of smart parks. In order to obtain the actual monitoring values of energy consumption in smart parks, reasonable sample energy consumption data for smart parks can be determined more scientifically, which is of great significance for the determination of accurate monitoring models for energy consumption in smart parks.

The multiple linear regression model [16, 17] is one of the most commonly used methods in dealing with variable relationships at present. The expression for a multiple regression model with m variables is as follows:

$$Z = \beta_0 + \beta_1 v_1 + \beta_2 v_2 + ... + \beta_m v_m + e \tag{3}$$

In the formula, A represents the regression coefficient, B represents random error, and v represents the variable influencing factor of the smart park. The variables of the smart park and any one of them need to satisfy a linear relationship, and the formula for solving the linear correlation coefficient is as follows:

$$r = \frac{\sum\limits_{i=1}^{m}(v_i - \bar{v})(u_i - \bar{u})}{\sqrt{\sum\limits_{i=1}^{m}(v_i - \bar{v})^2 \sum\limits_{i=1}^{m}(u_i - \bar{u})^2}} \tag{4}$$

Substitute the energy consumption data of the smart park operation without missing values into the formula, set the linear coefficient between CPU utilization and power as $r_{cpu} = 0.9603$, and the linear coefficient between memory utilization and power as $r_{men} = 0.9537$. The expression of the multiple linear regression energy consumption model is:

$$Q = \beta_0 + \beta_1 U_{cpu} + \beta_2 U_{mem} \tag{5}$$

Due to the mutual influence of different energy consumption during the data monitoring process of smart parks, and the non absolute and complete nonlinear relationship between operating power and the presented utilization rate, in order to improve the accuracy of the smart park energy consumption monitoring model in this article, a nonlinear relationship is introduced for discussion and analysis.

Using a polynomial model as the basic function, a nonlinear energy consumption model is obtained. The expression for monitoring energy consumption in smart parks using the smart park energy consumption model is as follows:

$$Q = \beta_0 + \beta_1 U_{cpu} + \beta_2 U_{cpu}^2 + \beta_3 U_{mem} + \beta_4 U_{mem}^2 \tag{6}$$

The number of energy efficiency equipment in the smart park and the area of the smart park are the main influencing factors of energy consumption monitoring in the smart park. The actual value of $\beta_0, \beta_1, \beta_2, ...\beta_m$ is obtained by polynomial regression analysis.

In order to ensure the accuracy of the monitoring results, the energy consumption data monitoring results of the smart park are calibrated through relative deviation and average relative deviation. The relative deviation can obtain the monitoring accuracy of each smart park impact factor, and the relative deviation can reflect the error of model monitoring. The expression for the relative deviation and average relative deviation [18, 19] of energy consumption data monitoring in smart parks is as follows:

$$Q_{\text{relative deviation}} = \frac{Q_{\text{predictive value}} - Q_{\text{true value}}}{Q_{\text{true value}}} \tag{7}$$

$$Q_{\text{Average relative deviation}} = \frac{\sum\limits_{i=1}^{m}(Q_{\text{deviation}})}{m} \tag{8}$$

During the monitoring process, there may also be some data anomalies that cause changes in energy consumption values and cannot reflect the actual situation, as shown in Table 2.

Table 2. Data Classification of Energy Consumption Supervision Platform

Data type	Data characteristics	Cause of occurrence	Processing method
Real data	Normal data usage	business as usual	retain
Distorted data	Using abnormal data	Abnormal energy usage behavior, equipment failure	Reserved, marked
data type	Mutation data	Measurement transmission failed	Eliminate
	0 data	There is a problem with the transmission or recording device	Eliminate

From Table 1, it can be seen that based on whether the data can reflect the actual energy consumption of buildings, they can be divided into two types: real data and distorted data [20].

(1) Real data
　　1) Normal data: When the monitored energy consuming equipment is operating normally, the recorded data is the measured data and calculated energy consumption results.
　　2) Abnormal energy usage data: The collected electrical energy values and consumption data in the event of abnormal use of energy consuming devices. The abnormal operation of the device is caused by human energy consumption behavior and the failure of the electrical device.
(2) Distorted data
　　1) Catastrophic data: measuring the quality of the equipment itself.
　　　　Adjacent electromagnetic interference [21] can cause a significant increase or decrease in measurement results, but it quickly returns to normal. This result cannot truly reflect the actual situation of energy consumption.
　　2) 0 data: When the measuring instrument or transmission device malfunctions, the collected data will be displayed as 0. If not detected or repaired in a timely manner, it will result in all collected data being 0 for a period of time.

To improve the data quality [22] of energy consumption monitoring platforms, it is necessary to screen and integrate the identified zero data and mutation data. Therefore, an intelligent data supplementation method using energy mode is proposed.

Assuming there is a continuous problem data with 0 and mutation data [23, 24]. The problem data includes m type of energy usage data, j type of energy usage data has h_j problem data points. Before the problem data occurs, the cumulative display value A' on the time meter is displayed, and after the problem data occurs, the cumulative display value B' on the time meter is displayed. By analyzing the time, temperature, and other information of the problem data, the energy consumption types of each problem data are determined, and the mathematical expectation μ and variance Δ of energy consumption are obtained. The supplement of problem data can be represented by the following mathematical problems.

A' and B' represent the energy consumption numbers included in the missing part, while E_j follows the layout of $F(\mu_j, \Delta_j)$; Among them, E_j describes the compensation number for the problem data under the j-type energy consumption situation. The equation used in this problem is not closed, and the $\sum_{j=1}^{m} \left(\frac{E_j - \mu_j}{\Delta_j}\right)^2$ minimum limit is introduced to measure the supplementary quality of the data.

Obtain the calculation result using the pull method, expressed as:

$$E_j = \mu_j + \frac{h_j \Delta_j^2}{\sum_{j=1}^{m} \left(h_j \Delta_j^2\right)} \left(B' - A' - \frac{\sum_{j=1}^{m} h_j \mu_j}{\sum_{j=1}^{m} h_j} \right) \tag{9}$$

Integrate the obtained energy consumption data of each item into one monitoring result. If it is greater than the variance data, the energy consumption is not in a normal state and requires the intervention of the technical department to check whether there is a fault or simply excessive electricity consumption [25], in order to take corresponding measures.In order to improve the readability and understandability of the article, the method flow chart is presented below. By observing the flow chart, you can quickly understand the overall framework of the research method, which helps readers to understand and master the research content faster. The flow chart is shown in Fig. 3:

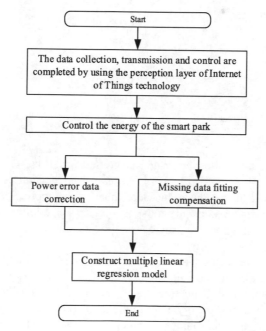

Fig. 3. Smart park energy consumption data monitoring flow chart

3 Experimental Design and Result Testing

Conduct experiments to demonstrate the effectiveness of the proposed method. Select a large shopping mall smart park as the monitoring target, the use of IBM Watson IoT iot platform to obtain real data, with advanced data storage and processing capabilities, can safely store the energy consumption data of the mall's smart park, and provides strong security measures to ensure the confidentiality, integrity and availability of data, support a variety of communication protocols and data formats, Real-time access to various energy consumption data of the mall's smart park,and the specific experimental parameters are shown in Table 2. The 6-story building is from the positive 5th floor to the negative 1st floor, with the negative 1st floor being the underground garage; The first floor is the department store sales center; The second floor is for shoe sales; The third floor is for clothing sales; The fourth floor is for food and beverage sales, and the fifth floor is for electronic product sales and movie theaters. A total of 120 GPRS transmission devices will be installed in the mall, with 20 devices per layer (Table 3).

Table 3. Experimental Parameters

Name	Data
the measure of area	$250000 \ m^2$
position	Located in the North China region
Number of layers	6th floor
Cargo Lift	4 parts
Elevator	7 parts
Getting on and off the elevator	12 parts
Business Hours	9:00–21:00

According to the laws of local climate change, set April, May, October, and November as the transitional seasons; Set January, February, March, and December as the heating season; Set June, July, August, and September as the cooling season. The energy consumption monitoring results during the heating season, cooling season, and transition season using the proposed energy consumption monitoring method are shown in Fig. 4.

From Fig. 4, it can be seen that the energy consumption of lighting, sockets, and power is in a stable state during the heating, cooling, and transition seasons, with particularly small changes in amplitude, indicating the non seasonal relationship between lighting energy consumption, socket energy consumption, and power energy consumption. The annual fluctuation of air conditioning power consumption is relatively severe, indicating the seasonality of air conditioning energy consumption. Due to the fact that most student dormitories and office buildings use non electric heating methods such as gas for heating, the use of air conditioning for refrigeration is greater than that of air conditioning for heating, and the energy consumption monitoring is completely consistent with the actual situation.

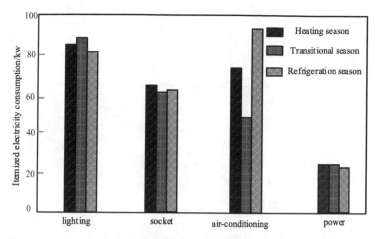

Fig. 4. Energy consumption during heating, cooling, and transition seasons

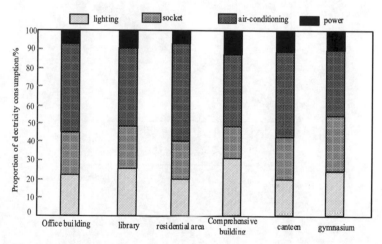

Fig. 5. Proportion of energy consumption by sub projects in the smart park

From Fig. 5, it can be clearly seen that lighting energy consumption, socket energy consumption, and air conditioning energy consumption are the main energy consumption sub items. In various types of smart parks, the proportion of these three items is far greater than power energy consumption. Among them, the highest proportion of energy consumption for air conditioning is in office smart parks, student dormitories, and canteens. This is because these locations have frequent student access, high mobility, and higher indoor air quality compared to other smart parks.

Through the MATLAB platform software, the standardized original processing samples and the ashed samples are used as the input of the long-term and short-term memory network prediction model, of which 125 groups of daily data from January 1, 2021 to May 5, 2021 are used as training data. Comparing the method proposed in this paper with the energy consumption prediction method based on convolutional neural networks proposed in reference [2] and the energy consumption prediction method based on support vector machines proposed in reference [3], Fig. 6 shows the comparison results of the three models.

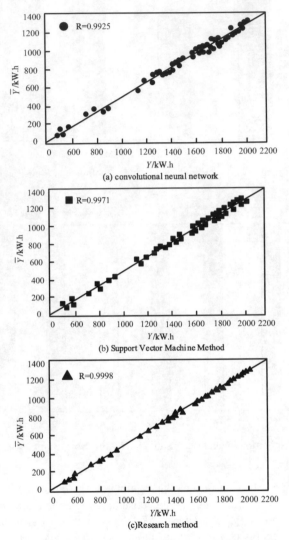

(a) convolutional neural network

(b) Support Vector Machine Method

(c)Research method

Fig. 6. Comparison of three building prediction models

Figure 6 shows the correlation distribution between the predicted value $\overline{Y}$ and the actual value $\overline{Y}$ of all models, where R represents the Pearson correlation function, the Pearson correlation coefficient value predicted by the neural network model is $R = 0.9925$, the Pearson correlation coefficient value predicted by the adaptive neural network model is $R = 0.9971$, and the Pearson correlation coefficient value predicted by the method in this paper is $R = 0.9998$. The correlation coefficient value of the method in this paper is significantly higher than that of other methods, The predicted values of the smart park prediction model proposed in this article have always been within the standard linear range, and the fit of the smart park energy consumption series is more stable, resulting in more accurate prediction results.

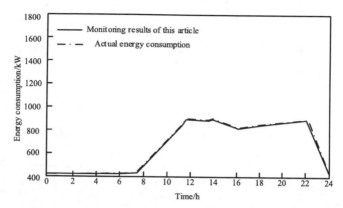

Fig. 7. Comparison of Energy Consumption and Actual Monitoring on the First Floor

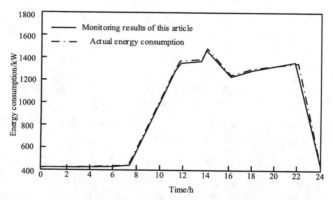

Fig. 8. Comparison of Energy Consumption and Actual Monitoring on the 2nd Floor

From Fig. 7 and Fig. 8, it can be observed that the positive 1st and 2nd floors are basically in a flat state from 0 to 7, and the energy consumption gradually increases after 7 o'clock, reaching a peak at 11 o'clock and 14 o'clock. This indicates that the power consumption of the 1st and 2nd floors is relatively high at this time, and the energy consumption increases.

Observing Fig. 9, it can be seen that the energy consumption of the third floor is relatively high at 12:00 and 19:00, which is during lunch and dinner time, Oil fumes can cause an increase in indoor temperature and a higher demand for air conditioning. As the meal time ends, energy consumption also decreases.

By observing Fig. 10, it can be seen that the energy consumption on the fourth floor is relatively high throughout the day, especially from 12:00 to 14:00 and from 20:00 to

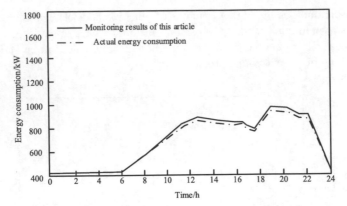

Fig. 9. Comparison of Energy Consumption and Actual Monitoring on the 3rd Floor

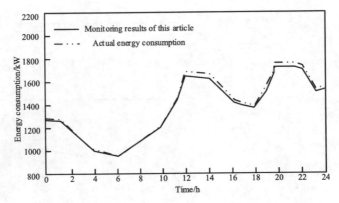

Fig. 10. Comparison of Energy Consumption and Actual Monitoring on the 4th Floor

22:00. At this time, the movie theater on the fifth floor is open 24 h, which is not consistent with the business hours of the mall. In addition, the electronic product sales area is open, resulting in an overall increase in energy consumption. As the opening hours of the mall end, the energy consumption in the electronic product sales area decreases, but at this time, the cinema is still in business hours, and night movies are more popular, so the energy consumption is only slightly reduced. From the experimental results obtained in the above figure, it can be seen that there is an error between the actual energy consumption and the monitoring results in this article, but the numerical value is very small, and overall, the accuracy is high.

Energy consumption monitoring in smart parks requires timely acquisition and processing of energy consumption data to achieve real-time monitoring and management of energy consumption. Therefore, the performance of evaluation methods should have a short response time and be able to quickly respond to and process energy consumption data. The method in this paper, the energy consumption prediction method based on Convolutional neural network proposed in literature [2] and the energy consumption prediction method based on support vector machine proposed in literature [3] are used as comparison methods to test the response time of different methods. The specific test results are shown in Fig. 11.

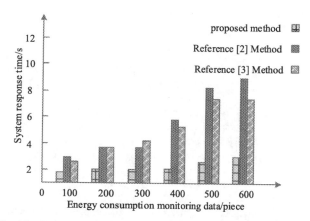

Fig. 11. Comparison of Response Time Testing for Three Methods

Analyzing the data in Fig. 11, it was found that with the increase of energy consumption monitoring data, the response time of the three methods increased to varying degrees. However, the response time of the method in this paper was always lower than that of the other two methods. When monitoring 600 energy consumption data, the method in this paper only took 3 s, while the methods in reference [2] and [3] took 9 s and 7 s, respectively. This verifies that the method proposed in this paper has the fastest response speed, indicating its practicality.

4   Conclusion

The Internet of Things technology can achieve real-time monitoring and analysis of different resources for better planning and utilization, to minimize waste and improve resource utilization efficiency. In this context, a smart park energy consumption data monitoring method based on Internet of Things technology is proposed. Based on the pre processed energy historical data of the smart park, improve the long-term and short-term memory network to predict the energy consumption of the smart park, and build a nonlinear regression model to monitor the energy consumption of the smart park.

However, smart parks often involve multiple departments and enterprises, involving issues such as data privacy and security. Therefore, in the future, it is necessary to consider whether the security and privacy of data can be ensured during the energy consumption data monitoring process.

References

1. Moon, J., Park, S., Rho, S., et al.: Robust building energy consumption forecasting using an online learning approach with R ranger. J. Build. Eng. **47**, 103851 (2022)

2. Ji, T.Y., Wang, T.S.: Building energy consumption prediction based on word embedding and convolutional neural network. J. South China Univ. Technol. (Nat. Sci. Ed.) **49**(06), 40–48 (2021)
3. Xiao, R., Wei, Z.Q., Zhai, X.Q.: Hourly energy consumption forecasting for office buildings based on support vector machin. J. Shanghai Jiaotong Univ. (Chin. Ed.) **55**(03), 331–336 (2021)
4. Kladas, A., Herteleer, B., Cappelle, J.: Scalable data storage for PV monitoring systems (2022)
5. Nascimento, G.F.M., Wurtz, F., Kuo-Peng, P., et al.: Quantifying compressed air leakage through non-intrusive load monitoring techniques in the context of energy audits. Energies **15**, 3213 (2022)
6. Vlker, B., Reinhardt, A., Faustine, A., et al.: Watt's up at home? smart meter data analytics from a consumer-centric perspective. Energies **14**(3), 719 (2021)
7. De la Cruz Severiche Maury, Z., Fernández Vilas, A., Díaz Redondo, R.P.: Low-Cost HEM with arduino and zigbee technologies in the energy sector in Colombia. Energies **15**(10), 3819 (2022)
8. Mustafa, A.S., Al-Heeti, M.M., Hamdi, M.M.: A new approach for smart electric meter based on Zigbee. Bull. Electr. Eng. Inf. **11**(2), 722–730 (2022)
9. Zhu, J., Jiang, Q., Shen, Y., et al.: Application of recurrent neural network to mechanical fault diagnosis: a review. J. Mech. Sci. Technol. **36**(2), 527–542 (2022)
10. Onan, A.: Bidirectional convolutional recurrent neural network architecture with group-wise enhancement mechanism for text sentiment classification. J. King Saud Univ.-Comput. Inf. Sci. **34**(5), 2098–2117 (2022)
11. Chen, Q., Zhang, C., Liu, Y.: Long-term and short-term browsing behavior data mining simulation based on tag mapping. Comput. Simul. **39**(01), 394–398 (2022)
12. Rajagopal, R., Agariya, A.K., Rajendran, C.: Predicting resilience in retailing using grey theory and moving probability based Markov models. J. Retail. Cons. Serv. **62**(2), 102599 (2021)
13. Jia, Y., Li, G., Dong, X., et al.: A novel denoising method for vibration signal of hob spindle based on EEMD and grey theory. Measurement **169**, 108490 (2021)
14. Li, H., Fan, R., Shi, Q., et al.: Class imbalanced fault diagnosis via combining k-means clustering algorithm with generative adversarial networks. J. Adv. Comput. Intell. Intell. Inf. **25**(3), 346–355 (2021)
15. Sun, H., Chen, Y., Lai, J., Wang, Y., Liu, X.: Identifying tourists and locals by k-means clustering method from mobile phone signaling data. J. Transport. Eng. Part A: Syst. **147**(10), 04021070 (2021)
16. Galib, S.S., Islam, S.M.R., Rahman, M.A.: A multiple linear regression model approach for two-class fNIR data classification. Iran J. Comput. Sci. **4**, 45–58 (2021)
17. Tang, S., Li, T., Guo, Y., et al.: Correction of various environmental influences on Doppler wind lidar based on multiple linear regression model. Renew. Energy **184**, 933–947 (2022)
18. Chen, L.P., Yi, G.Y.: Semiparametric methods for left-truncated and right-censored survival data with covariate measurement error. Ann. Inst. Stat. Math. **73**, 481–517 (2021)
19. Yu, H., Wang, X., Ren, B., Zeng, T., Lv, M., Wang, C.: An efficient Bayesian inversion method for seepage parameters using a data-driven error model and an ensemble of surrogates considering the interactions between prediction performance indicators. J. Hydrol. **604**, 127235 (2022)
20. Burgos, C., Cortés, J.-C., Shaikhet, L., et al.: A delayed nonlinear stochastic model for cocaine consumption: stability analysis and simulation using real data. Disc. Contin. Dyn. Syst. Series **14**(4), 1233–1244 (2021)
21. Zazoum, B.: Machine learning enabled prediction of electromagnetic interference shielding effectiveness of poly(vinylidene fluoride)/mxene nanocomposites. Mater. Sci. Forum **1053**, 77–82 (2022)

22. Li, C., Zhang, Y., Sun, Q., et al.: Collaborative caching strategy based on optimization of latency and energy consumption in MEC. Knowl.-Based Syst. **233**, 107523 (2021)
23. Imanparast, M., Kiani, V.: A practical heuristic for maximum coverage in large-scale continuous location problem. University of Guilan (4) (2021)
24. Hou, R., Chen, J., Feng, Y., Liu, S., He, S., Zhou, Z.: Contrastive-weighted self-supervised model for long-tailed data classification with vision transformer augmented. Mech. Syst. Signal Process. **177**, 109174 (2022)
25. Ribeiro, J.C., Cardoso, G., Silva, V.B., et al.: Paraconsistent analysis network for uncertainties treatment in electric power system fault section estimation. Int. J. Electr. Power Energy Syst. **134**, 107317 (2022)

Design of a Multidimensional Teaching Effectiveness Evaluation System Based on Information Integration

Lei Ma[✉], Yanning Zhang, Jingyu Li, and Wei Han

Beijing Polytechnic, Beijing 100016, China
malei235@tom.com

Abstract. The multidimensional teaching effect evaluation system has the problem of high CPU usage in the process of actual use. To solve this problem, a multidimensional teaching effect evaluation system based on information integration is designed. Hardware part: choose RC punch and discharge circuit, convert TTL level into PC communication mode as well; software part: identify the characteristics of multidimensional teaching effect evaluation elements, divide them into evaluation subjects, **optimize** the system software function by using information integration technology, extract and revise the system log intermediate table. Experimental results: The multidimensional teaching effectiveness evaluation system designed this time is lower than the average CPU usage of the other three multidimensional teaching effectiveness evaluation systems: 9.896%, 11.111% and 10.036% respectively, indicating that the performance of the designed multidimensional teaching effectiveness evaluation system is superior after fully integrating information integration technology.

Keywords: Information Integration · Multidimensional Teaching · Teaching Evaluation · Reset Circuit · Teaching Process · Teaching Link

1 Introduction

With the continuous development of information technology, corresponding teaching models and methods of evaluating teaching effectiveness are also undergoing changes [1, 2]. Combining the existing multimedia technology and applying the emerging information integration technology to teaching evaluation is an effective way to improve teaching. Despite the rise of online teaching in the current educational environment, it is still fundamentally classroom based, so quality assessment still plays an important role in research and teaching processes. Different measurement standards represent different value orientations. Undoubtedly, the transformation and development have put forward new requirements for the evaluation of teaching quality. As an important part of the teaching monitoring system and an important guarantee for achieving talent cultivation goals, a multidimensional teaching evaluation system urgently needs to be established. Therefore, studying a multidimensional teaching quality evaluation system suitable for

L. Yun et al. (Eds.): ADHIP 2023, LNICST 548, pp. 38–50, 2024.
https://doi.org/10.1007/978-3-031-50546-1_3

the transformation of applied technology universities, exploring evaluation mechanisms that conform to the development laws of higher education, is of great significance for accelerating the cultivation of applied talents and the transformation of universities, alleviating structural employment conflicts, and providing a more comprehensive interactive way for teaching effectiveness evaluation.

The research on teaching effectiveness evaluation in Western countries began in the early 20th century, mainly through the construction of evaluation scales. In the subsequent development process, classroom observation gradually replaced the scale of evaluation and promoted the standardization of teaching effectiveness evaluation [3]. For example, CHEN Yanhong studied the teaching effect evaluation method of the Internet of Things teaching platform based on long-term and short-term memory networks, and analyzed the current situation of the Internet of Things education platform, including the development and structure of the Internet of Things teaching platform. We have constructed an LSTM model to evaluate the teaching effectiveness of the Internet of Things education platform. Finally, through the study of the model, the teaching effectiveness of the Internet of Things education platform was evaluated. However, in recent years, both Western educators and Chinese scholars have shifted their research focus to the level of developmental teacher evaluation. In addition to clarifying the subject status of teachers, it also acknowledges the value of self-evaluation and mutual evaluation among teachers.

Information integration technology is mainly responsible for mobilizing various data. At the same time, the main way of information integration is to integrate various heterogeneous data sources, and realize data and information sharing while adjusting the consistency of data structure. This paper designs a multi-dimensional teaching effectiveness evaluation system based on information integration. In the hardware part of the system, RC punching and discharging circuits are used, and TTL level is converted to PC communication mode; In the software part of the system, by identifying the characteristics of multidimensional teaching effect evaluation elements, it is divided into evaluation subjects, and uses information integration technology to optimize the system software functions, extract and modify the system log intermediate table. The experimental results show that the average CPU usage of the designed system is lower than that of the comparison system, indicating that the performance of the system in this paper is superior and can effectively improve the quality of teaching evaluation.

2 System Hardware Design

The hardware of this designed multi-dimensional teaching effect evaluation system mainly includes parts such as power supply, serial port, reset circuit and clock signal. The main peripheral circuit interfaces are RS232 serial port and IIC interface circuit, and the digital tube is of 7 segments or more. In order to provide a stable and reliable current and voltage for the whole multidimensional teaching effect evaluation system, the power input method of this system is set to two ways. The two ways are: using the USB interface as a carrier to obtain +5 V power supply; the other is using the power socket as an input port and directly accessing the corresponding AC power supply. The advantage of the former is that it is simple to operate and only requires a computer and

USB cable to be prepared in advance for stable +5 V voltage transmission. In addition to the power supply circuit, the system's clock circuit has to be adjusted accordingly. As some signal interference will inevitably occur in the process of serial communication. The working frequency accuracy of 10.5026 MHz crystal is very high, and after precise adjustment and testing, its frequency stability and accuracy are very high. And the 10.5026 MHz crystal has high reliability, can operate stably in harsh environments, and has good high-temperature resistance. This is particularly important for devices and systems that require high performance in complex environments, long working hours, and require high performance. Therefore, the hardware of the multi-dimensional teaching effectiveness evaluation system designed this time uses a 10.5026 MHz crystal, and the reset circuit of the multidimensional teaching effect evaluation system hardware is set on the basis of the power supply circuit and the clock circuit. According to the needs of the experimental test, the RC punch and discharge circuit is chosen, which only needs to ensure that the system can be quickly reset during the debugging process. The reset circuit diagram of the hardware of the multidimensional teaching effect evaluation system is shown in Fig. 1.

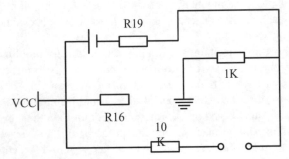

Fig. 1. Reset circuit diagram of the hardware of the multidimensional teaching effectiveness evaluation system

In Fig. 1, the reset circuitry needs to match the user application and operating frequency to select the optimum power operation mode, influenced by the performance of the microcontroller in the hardware. In addition to the above circuitry, the erase function of the audio memory is one of the experiences that affect the use of the system. Often, systems need to be repeatedly debugged under specific conditions during the development and programming process. It is therefore essential to select audio memory that can be erased and written repeatedly. In addition, the two ends of the serial port were connected to the PC port and the peripheral circuit RS232 serial port respectively, while ensuring that the development board of the microcontroller was properly powered up.

In this design of the multi-dimensional teaching effectiveness evaluation system, 8 and more diodes are selected for the light emitting diode circuit. Ensure that the entire circuit can be lit when the output level is low and can be extinguished when the output level is high. Due to the strong conversion performance of the MAX232 level, it is also possible to convert the TTL level to PC communication mode as well after accessing the RS232 serial port. According to the above settings, the operating frequency range in the system is set to 10–50 MHz, ensuring that its operating frequency is kept at around 45 MHz. At the same time, three to five 16-bit timers are set in the system, and one of them is used as the main counter, while the rest of the counters can be used as several 8-bit counters. In order to be compatible with several models of microcontrollers at the same time, the PowerDown mode is introduced in the system and increased to 5 with the guarantee of two basic external interrupts INT.

During the interrupt process, the original wake-up method is used as a technical guarantee to set up branches that can be triggered by external interrupts at low level. At the same time, in order to improve the anti-interference capability of the communication serial port, the ALE clock signal is adjusted to disable the output, which also suppresses power jitter to a certain extent. When the external clock frequency is reduced, the system microcontroller will be in 6T mode for a long time. Based on the above description, the steps to design the hardware of the multi-dimensional teaching effect evaluation system are completed.

3 System Software Design

3.1 Identifying the Characteristics of Multidimensional Teaching Effectiveness Evaluation Elements

Multidimensional teaching effectiveness evaluation is based on multiple dimensions and uses multiple criteria to measure the performance of teachers throughout the teaching process. Multidimensional teaching effectiveness evaluation is a comprehensive and objective evaluation of the teaching implementation process and student learning effectiveness, in order to understand the quality of teaching effectiveness, the intensity of teaching activities, and the learning status of students, in order to timely identify problems, adjust teaching strategies, improve teaching methods, and improve teaching quality. It is also a model for testing the quality of teaching and learning that incorporates planning and objectives. As a common means of evaluating teaching quality, it can be divided into three main subjects, namely teachers, learners and administrators [4, 5]. The basic elements of the teaching and learning process have been broadly encompassed in terms of the teaching dimension of the teacher, the learning dimension of the student and the management dimension of the school. Combined with the above descriptions, the elements of evaluation of multidimensional teaching effectiveness can be summarised in the following four points, as shown in Fig. 2.

As can be seen from Fig. 2, the evaluation elements of multidimensional teaching effectiveness mainly include four levels: level checking, diagnostic management, feedback and guidance, and regulation of progress. Level check: Level check is the foundation of multidimensional teaching effectiveness evaluation, mainly including the inspection and evaluation of students' knowledge, skills, and attitudes, in order to have a basic understanding of students' actual level and potential problems. The elements of proficiency check include: mastery of basic knowledge, mastery of skills, attitude, and values. Diagnosis management: Diagnosis management mainly focuses on the problems that students encounter during the learning process, using various methods to conduct comprehensive analysis, identify the root causes, and promptly adopt targeted teaching methods and management measures, so that students can truly master the knowledge and

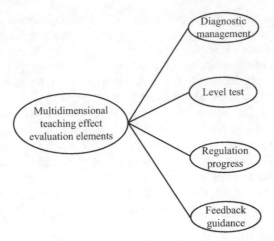

Fig. 2. Elements of evaluation of multidimensional teaching effectiveness

skills they have learned and improve. Diagnostic management elements include: analyzing student differences, scientific diagnosis, problem positioning, and problem-solving solutions. Feedback guidance: Feedback guidance mainly aims to provide targeted guidance and suggestions to students through timely transmission of student performance and performance feedback, helping them summarize their experiences, identify deficiencies, correct problems in a timely manner, and make rapid progress and improve learning outcomes. The elements of feedback guidance include: performance feedback, performance analysis, providing personalized guidance, encouragement, and motivation. Progress adjustment: Progress adjustment mainly refers to the scientific planning and adjustment of the progress in the teaching process to adapt to students' learning progress and characteristics, and ensure that the teaching achieves the expected results. The elements of progress adjustment include: teaching schedule planning, synchronous adjustment of teaching and learning, adjustment of teaching methods and means, and adjustment of teaching duration. Based on these four levels of evaluation, the overall quality of teaching can be judged to be at a certain level, and the subsequent teaching plan can be adjusted according to the evaluation results.

As a rule, the students are the focus of the training and their academic performance is a visual indication of the level of teaching at a given stage and allows the teacher to see how they compare with other teachers in the overall teaching environment. Although there may be some subjective elements in the evaluation process, the overall results of student evaluations are credible in light of the data from the research data. From the teacher's perspective, the multidimensional teaching effectiveness evaluation system includes personal evaluation, evaluation by others and mutual evaluation. From the learner's point of view, this includes logging into the system, making remarks, marking and evaluating and managing the evaluation. Evaluation includes aspects such as teaching attitude, teaching management and the use of modern multimedia. From the level of school management, this includes teaching effectiveness evaluation maintenance, comment management maintenance, and speech warning control.

Combined with the principle of the role of information integration, the system should fully understand the links between the various teaching effectiveness evaluations and identify commonalities and differences in order to achieve data integration and information integration of teaching effectiveness. From a multi-dimensional standpoint, the evaluation of teaching effectiveness should not only be limited to students' report cards, but also consider aspects such as teaching attitudes, students' interest in learning and teacher-student relationships. In today's increasingly diverse teaching formats, teaching effectiveness evaluation should also be adjusted in the direction of diversity and democratisation [6].

According to different evaluation criteria, a variety of evaluation types can be summarised for different scenarios. From the point of view of the teaching process, multi-dimensional evaluation of teaching effectiveness focuses on the continuous acquisition of information and the continuous improvement of the teaching model through data and information, i.e. on the process as opposed to the results. The overall evaluation is more global in nature and requires the evaluation of teaching effectiveness after all teaching sessions and interactions have been completed. The essence of multidimensional assessment of teaching effectiveness is to ensure that each student gains in knowledge or life experience by means of scoring, so it is important to examine the teacher's level while still basing the assessment on the student's experience. Based on the above description, the steps to identify the characteristics of the elements of multidimensional teaching effectiveness evaluation are completed.

3.2 Information Integration Optimisation System Software Features

Information integration presupposes that all data and information is open and transparent in order to ensure that users in the system are independent of each other [7, 8]. The ultimate goal is that both the heterogeneous data sources and the distribution of educational information are transparent, but not directly accessible to the users of the system. The transparent distribution of educational information required by information integration consists of two aspects, namely the transparency of the location where the data is located and the transparency of the storage path [9]. This is to ensure that the system host can extract the required data at any time without the need for additional sorting. At the same time, in order to enhance the sense of use for system users, it is necessary to set up separate storage paths for local data and remote data, and to provide the corresponding device directories by the system for system users to choose from. Teaching data and teaching information can usually form the structure of a single data image after the information has been integrated. The concept of neural network is introduced in the system software to optimise the data and information integration structure of the system software with the following expression formula:

$$G = \sum \exp(\alpha - 1)^2 \times \frac{\alpha}{d} \qquad (1)$$

In Eq. (1), α indicates the number of nodes in the network adoption number, and d indicates the output weight of the data information in the system. In a multidimensional teaching effectiveness evaluation system, student examination results can directly reflect the effectiveness of teaching at a certain stage. In order to produce more accurate evaluation results, the relationship between the difficulty of the examination paper and the examination results is defined, resulting in the expression formula:

$$H = \frac{L}{\delta} \times \left(\frac{d}{r}\right) - L \tag{2}$$

In Eq. (2), L represents the average time taken by the students in the paper, δ represents the number of questions, r represents the number of knowledge points and d has the same meaning as in Eq. (1). Combining formula (1) and formula (2) yields the formula for the fitness function, as shown below:

$$y = \frac{h}{\eta} + \frac{1}{\phi} + \frac{L}{h} + \sum |\phi - \eta| \tag{3}$$

In Eq. (3), h represents the error of data integration in the system, η represents the data integration rate, ϕ represents the data weights and L has the same meaning as Eq. (3).

In the system software, information integration technology is used to make a series of data on the browsing traces of teachers, students and school administrators in the system and to set up a visualisation module of the system to facilitate system improvement and upgrading during subsequent use. The data stored in the database is first extracted and collated into the log information required by the user, and duplicate data is eliminated when carrying out multidimensional teaching effectiveness evaluation. Amongst other things, there are also many-to-one and many-to-many correspondences between teachers and students in the system database. To ensure that the data structure is standardised, the intermediate tables of the system logs are extracted and revised. In addition, the diversity of the evaluation of teaching effectiveness, lecture progress and classroom atmosphere has to be highlighted in the system to meet the needs of curriculum reform [10]. For the user, as long as the local data and the heterogeneous data sources are transparent, it is possible to ensure that the data remains consistent with the original data after the next login to the system. Based on the basic functions of the system software, the multidimensional teaching effectiveness evaluation can be divided into four modules, namely the information login module, the system management module, and the registration evaluation module. Based on this, the steps to optimise the functionality of the system software using information integration are completed.

4   System Testing

As the multidimensional teaching effectiveness evaluation system is designed to be used primarily in a teaching environment, both ease of application and operability need to be taken into account. The following is a test of the system based on practical application needs.

4.1   Test Preparation

The hardware configuration of this system test is mainly: Windows 10 Education Edition, 2.5 GHZ, i7 4720hq + GTX 965 m, memory 4G, hard disk 350G; development language: PHP, Bootstrap front-end framework, C#, python, CSS; database: Mysql database; development platform: Zend Studio.

The language environment for this test is not limited by hardware conditions and has the advantage of being highly readable and at the same time highly portable. A compiler with automatic management is embedded in the system to ensure automatic allocation of system registers. As the data structures in the programming process take up different space, the variables need to be assigned after full consideration of the application scenario. Different data ranges are set for different types of data to drive reasonable allocation by the compiler.

To ensure that different numbers of IIC bus devices can be bundled smoothly onto the same root bus, synchronous communication is used for programming. In some modules, the system is required to provide power-down protection, in which case a chip-switching function is introduced in time. Data with significant changes are extracted to the host display and displayed on the corresponding digital tube according to the needs of the system test. Also, in order to solve the compatibility problem between the plug-in and the system, an additional SDK installation package needs to be downloaded and imported into the system. In this experiment, the main board of the system hardware was modelled in a stand-alone manner, docking the interfaces and components of each part.

4.2 Analysis of Test Results

The test selected the RBF-based multidimensional teaching effectiveness evaluation system, the UML-based multidimensional teaching effectiveness evaluation system, and the SSH-based multidimensional teaching effectiveness evaluation system, and the multidimensional teaching effectiveness evaluation system designed here for comparison testing. The CPU usage of the four types of multidimensional teaching effectiveness evaluation systems were tested under different concurrent user numbers, and the experimental results are shown in Tables 1, 2, 3 and 4.

According to the experimental results in Table 1, when the number of concurrent users is 80, the highest CPU utilization rate of the multidimensional teaching effectiveness evaluation system designed in this article is 3.115%. The highest CPU utilization rate of the SSH based multidimensional teaching effectiveness evaluation system is 5.447%. The highest CPU utilization rate of the UML based multidimensional teaching effectiveness evaluation system is 5.206%. The highest CPU utilization rate of the RBF based multidimensional teaching effectiveness evaluation system is 5.048%, And the average CPU usage of the multi-dimensional teaching effectiveness evaluation system designed in this article is lower than the other three multi-dimensional teaching effectiveness evaluation systems: 1.729%, 1.635%, and 1.755% respectively.

According to the experimental results in Table 2, when the number of concurrent users is 240, the highest CPU utilization rate of the multidimensional teaching effectiveness evaluation system designed in this article is 7.152%, the highest CPU utilization rate of the SSH based multidimensional teaching effectiveness evaluation system is 14.112%, the highest CPU utilization rate of the UML based multidimensional teaching effectiveness evaluation system is 13.546%, and the highest CPU utilization rate of the RBF based multidimensional teaching effectiveness evaluation system is 13.005%, And the average CPU usage of the multi-dimensional teaching effectiveness evaluation

Table 1. CPU usage (%) for systems with 80 concurrent users

Number of experiments	Multidimensional teaching effect evaluation system based on RBF	Multidimensional teaching effect evaluation system based on UML	Multidimensional teaching effect evaluation system based on SSH	The multidimensional teaching effectiveness evaluation system designed for this occasion
1	4.510	3.188	4.516	3.002
2	3.694	4.007	3.418	2.499
3	4.225	3.654	5.216	2.106
4	5.048	4.901	4.815	3.114
5	4.602	3.788	5.447	2.848
6	4.129	4.615	3.744	2.316
7	4.317	3.815	4.006	2.009
8	5.008	5.206	3.447	3.115
9	3.679	3.991	3.599	2.557
10	3.775	4.887	5.044	2.133

Table 2. CPU usage (%) for systems with 240 concurrent users(%)

Number of experiments	Multidimensional teaching effect evaluation system based on RBF	Multidimensional teaching effect evaluation system based on UML	Multidimensional teaching effect evaluation system based on SSH	The multidimensional teaching effectiveness evaluation system designed for this occasion
1	12.165	13.545	12.649	6.461
2	11.084	12.165	13.474	7.152
3	10.916	13.119	12.545	6.169
4	9.668	12.774	13.774	5.023
5	11.215	13.004	12.912	5.889
6	12.446	12.466	13.660	5.474
7	11.548	12.877	12.748	6.324
8	12.779	13.546	12.466	5.159
9	11.464	12.779	13.899	6.337
10	13.005	13.545	14.112	5.852

system designed in this article is lower than the other three multi-dimensional teaching effectiveness evaluation systems: 5.645%, 6.998%, and 7.240%, respectively.

Table 3. CPU usage (%) for systems with 480 concurrent users(%)

Number of experiments	Multidimensional teaching effect evaluation system based on RBF	Multidimensional teaching effect evaluation system based on UML	Multidimensional teaching effect evaluation system based on SSH	The multidimensional teaching effectiveness evaluation system designed for this occasion
1	26.316	25.910	23.645	12.336
2	24.715	24.778	24.818	11.089
3	22.004	26.141	24.461	11.748
4	23.991	25.097	23.847	11.524
5	25.008	24.711	22.007	12.775
6	24.719	25.912	23.649	11.519
7	23.656	25.446	22.541	12.073
8	22.076	24.078	23.877	12.334
9	23.449	24.933	24.059	11.872
10	24.578	23.545	23.748	12.069

According to the experimental results in Table 3, when the number of concurrent users is 480, the highest CPU utilization rate of the multidimensional teaching effectiveness evaluation system designed in this article is 12.775%, the highest CPU utilization rate of the SSH based multidimensional teaching effectiveness evaluation system is 24.818%, the highest CPU utilization rate of the UML based multidimensional teaching effectiveness evaluation system is 25.912%, and the highest CPU utilization rate of the RBF based multidimensional teaching effectiveness evaluation system is 26.316%, And the average CPU usage of the multidimensional teaching effectiveness evaluation system designed in this case is lower than the other three multidimensional teaching effectiveness evaluation systems: 12.117%, 13.121%, and 11.731%, respectively.

According to the experimental results in Table 4, when the number of concurrent users was 800, the multidimensional teaching effectiveness evaluation system was lower than the other three multidimensional teaching effectiveness evaluation systems in terms of CPU usage: 20.095%, 22.691% and 19.420% respectively. According to the experimental results, when the number of concurrent users increased, the CPU usage of all four systems increased, but the results of the experiment were always lower than those of the other three systems, indicating better operability in use.

To further validate the practicality of the method proposed in this article, a multidimensional teaching effectiveness evaluation system based on RBF, a multidimensional

Table 4. CPU usage (%) for systems with 800 concurrent users(%)

Number of experiments	Multidimensional teaching effect evaluation system based on RBF	Multidimensional teaching effect evaluation system based on UML	Multidimensional teaching effect evaluation system based on SSH	The multidimensional teaching effectiveness evaluation system designed for this occasion
1	45.161	45.356	39.464	22.745
2	44.315	44.112	41.499	23.466
3	43.202	46.913	42.077	21.849
4	44.969	44.202	40.336	20.591
5	45.416	45.913	41.549	23.007
6	43.915	44.735	42.883	22.493
7	39.466	46.004	41.964	21.748
8	38.452	45.822	42.718	22.946
9	39.078	44.913	43.553	23.467
10	41.167	43.125	42.349	21.877

teaching effectiveness evaluation system based on UML, and a multidimensional teaching effectiveness evaluation system based on SSH were used as experimental indicators for comparative testing. The test results are shown below.

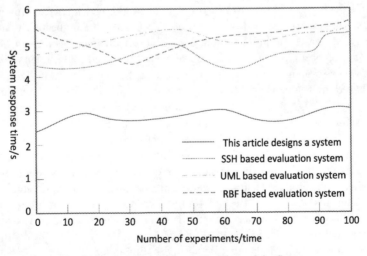

Fig. 3. Comparison of Response Time of the System

According to Fig. 3, it can be seen that the response time of the system designed in this article is around 3 s, while the response time of the comparison system is around 5 s. The response time of the system designed in this article is significantly lower than that of the comparison method, indicating that the efficiency of the system in this article is higher and more practical.

5 Conclusion

With the continuous development of information technology, corresponding teaching models and methods of evaluating teaching effectiveness are also undergoing changes. Combining the existing multimedia technology and applying the emerging information integration technology to teaching evaluation is an effective way to improve teaching. Therefore, studying a multidimensional teaching quality evaluation system suitable for the transformation of applied technology universities is of great significance. In this regard, this paper studies the design of multidimensional teaching effectiveness evaluation system based on information integration. The multidimensional teaching effectiveness evaluation system in this paper integrates a variety of influencing factors and provides reliable materials for the academic community to carry out relevant research. At the same time, it provides a feasible basis for the application of information integration technology to a certain extent, and also responds to the call for teaching reform in essence. At the technical level, the functionality of the system software has been optimized, and the efficiency of the multidimensional teaching effectiveness evaluation system has been improved. Due to limited research conditions, the accuracy of the multidimensional teaching effectiveness evaluation system designed in this study still needs further testing and improvement based on test data.

References

1. Hotaman, D.: The effect of formative assessment on the academic achievement levels of prospective teachers. J. Curriculum Teach. **9**(3), 33–39 (2020)
2. Xu, N., Fan, W.: Research on interactive augmented reality teaching system for numerical optimization teaching. Comput. Simul. **37**(11), 203–206+298 (2020)
3. Ma, L., Fan, P., Xu, S.: Exploring of teaching effect of course "vehicle chassis structure" based on the teaching mode of divided class. Int. J. Innov. Educ. Res. **8**(6), 230–234 (2020)
4. Chen, Y.: Evaluation of teaching effect of internet of things education platform based on long-term and short-term memory network. Int. J. Contin. Eng. Educ. Life Long Learn. **31**(1), 15–23 (2021)
5. Boswell, S.S., Sohr-Preston, S.L.: I checked the prof on ratemyprofessors: effect of anonymous, online student evaluations of professors on students' self-efficacy and expectations. Soc. Psychol. Educ. **23**(4), 943–961 (2020)
6. Li, Y.: Deviation of a key indicator in university teaching evaluation and its improvement——starting from "'Attach Importance to Undergraduate Teaching.'" J. Hebei Normal Univ. (Educ. Sci. Ed.) **22**(2), 4–8 (2020)
7. Liu, Y., Zhang, R.: A study on the policy of undergraduate teaching evaluation in China from the perspective of historical institutionalism. J. High. Educ. Manag. **14**(5), 115–124 (2020)
8. Ebrahimzade, A., Abedini, M.R., Ramazanzade, K., et al.: Effect of integrated teaching on students' learning. Strides Dev. Med. Educ. **18**(1), 1011 (2022)

9. Quispe, J., Quispe-Aviles, N.L., Aymachoque-Aslla, L., et al.: Effect of chess teaching on mathematical, attention and concentration abilities in school-aged children of the Peruvian Amazon. Apuntes Universitarios 11(1), 1–22 (2021)
10. Tian, Y.: Knowledge service oriented online teaching effect evaluation. Inf. Sci. 38(9), 129–136 (2020)

Evaluation Method of Higher Vocational Online Education Effect Based on Data Mining Algorithm

Mengxing Niu[1]([✉]) and Xiaoli Wang[2]

[1] Sanmenxia Polytechnic, Sanmenxia 472000, China
niumengxing1989@163.com
[2] Sanmenxia College of Social Administration, Sanmenxia 472000, China

Abstract. Conventional online education effect evaluation methods in higher vocational education mainly use the random subjective evaluation framework to obtain evaluation factors, which is vulnerable to the impact of micro personalized differences, resulting in low comprehensive evaluation indicators of education effect. Therefore, a new online education effect evaluation method in higher vocational education needs to be designed based on data mining algorithms. That is to say, the online education effect evaluation system is determined, the online education effect evaluation model of higher vocational education is constructed based on the data mining algorithm. The case analysis results show that the designed evaluation method for online education effect of higher vocational education has good evaluation effect, high comprehensive evaluation index, reliability and certain application value, and has made certain contributions to the follow-up optimization and reform of online education of higher vocational education.

Keywords: Data Mining Algorithm · Vocational School · On-line · Education · Effect Evaluation

1 Introduction

Education is entering the era of "Internet+", with more emphasis on the cultivation of skilled and application-oriented talents [1], which puts forward higher requirements for the talent education model of higher vocational colleges. The precision teaching mode is to improve the traditional teaching method [2] with the help of advanced multimedia equipment and network technology, reasonably integrate rich teaching resources, and push different teaching resources to different students through the precise positioning of different student groups [3], so as to truly realize the full sharing of teaching resources. The precise teaching mode breaks through the time and space constraints of the traditional teaching mode [4], supports teachers to carry out various teaching activities online, opens up new channels for students' learning after class, and also enriches the ways of communication and interaction between teachers and students [5], creates an independent and relaxed learning environment more suitable for contemporary college

L. Yun et al. (Eds.): ADHIP 2023, LNICST 548, pp. 51–63, 2024.
https://doi.org/10.1007/978-3-031-50546-1_4

students, and effectively stimulates students' enthusiasm and interest in learning, A "interactive" teaching mode [6] with efficient interaction between teachers and students has been established.

For a long time, in the teaching evaluation system of higher vocational education, the evaluation of students' learning effect mainly adopts summative evaluation, that is, after a certain stage of teaching activities in the curriculum, teachers adopt the "generic" evaluation scheme [7], and finally give quantitative evaluation results, ignoring the individual differences of students, so it is difficult to teach students in accordance with their aptitude. The talent training goal of higher vocational colleges is to provide highly skilled talents [8] for the society. However, many skills cannot pass several examinations or show accurate quantitative results. If we equate the examination results with the teaching effect evaluation, we can judge that the learning achievements of vocational college students by the quantitative results ignore the differences of students and the actual needs of society for each post [9]. This teaching effect evaluation model has greatly affected the talent cultivation effect of vocational colleges.

The traditional teaching effect evaluation model mainly includes the forms of on-site supervision, student evaluation, and teaching materials inspection. Teachers' teaching objectives, teaching strategies, teaching implementation, and teaching evaluation run through the whole process of teaching activities [10]. Limited by time, manpower, and other factors, it is difficult for schools to fully understand the whole process of teachers' teaching, Therefore, it is difficult to make an objective and fair evaluation of each teacher's teaching effect. This traditional teaching effect evaluation model has a certain randomness and subjectivity. It focuses on the macro "face" assessment and ignores the micro personalized differences, which can easily reduce students' enthusiasm for learning, also restrict the improvement of teaching quality, and can not achieve the talent training goal of higher vocational colleges to provide highly skilled talents for society and industry, In order to solve the existing problem of education effect evaluation and improve the quality of online education in higher vocational education, this paper designs an effective evaluation method of online education effect in higher vocational education based on data mining algorithm.

2   Design of Online Education Effect Evaluation Method for Higher Vocational Education Based on Data Mining Algorithm

2.1   Determine Online Education Effect Evaluation System

Classroom teaching evaluation is to use mathematical modeling methods to interpret data or materials on a quantitative basis, or to conduct empirical evaluation on a quantitative basis to distinguish between high and low grades. Generally speaking, the rating scale is a standardized scale for evaluating classroom teaching behavior. In this paper, numerical scale method is used to design [11]. The advantage of this scale is that it is easy to form and quantify, and it is the most frequently used scale among all evaluation scales. The evaluation scale mainly focuses on the evaluation of teachers' teaching process. Generally, the teaching process is divided into several indicators. Each level indicator is divided into thousands of level two indicators, and some even have three indicators [12].

Each level indicator is described and graded. Users of the evaluation scale for classroom teaching effect evaluation include teachers, students and administrators. According to the above basic principles, this paper has built an online education effect evaluation system, as shown in Fig. 1 below.

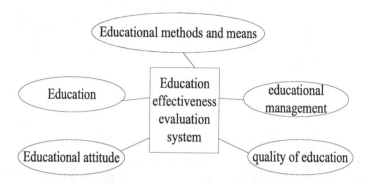

Fig. 1. Online education effect evaluation system

It can be seen from Fig. 1 that the hierarchical structure of teaching effect is to decompose complex problems into various components of elements, and divide these elements into several groups according to different attributes to form different levels. The elements at the same level, as criteria, dominate some elements at the next level, while they are also dominated by elements at the upper level. This dominance relationship from top to bottom forms a hierarchy. The top level usually has only one element, which is usually the predetermined goal or ideal result of the analysis problem. The middle level is generally the criteria and sub criteria level. The lowest level includes the decision-making scheme [13]. The governing relationship of elements between levels is not necessarily complete, that is, such elements can exist, which do not dominate all elements of the next level. Moreover, the number of layers is related to the complexity of the problem and the level of detail needed to be analyzed. There are generally no more than 9 elements in each level, because one level contains too many elements, which will make it difficult to compare and judge pairs. A good hierarchy is extremely important for solving problems. Hierarchical structure is based on the decision-makers' comprehensive and in-depth understanding of the problems they face. If they are hesitant about the division of levels and the determination of the dominant relationship between levels, it is better to re analyze the problem and clarify the relationship between the various parts of the problem to ensure that a reasonable.

The most basic form of the hierarchy model is a tree with one node connected to several nodes. In the hierarchy model, the node of the upper layer is a variable, which is determined by the variable of the lower layer. According to the actual situation of vocational colleges, this paper analyzes and decides the final evaluation indicators based on the situation of our college through experienced teachers, expert group members, leaders of teaching management, and members of the teaching supervision group. The evaluation indicators include five parts: teaching methods and means, teaching situation, teaching attitude, teaching management, and teaching quality.

1. Evaluate teaching methods and means in accordance with the school's teaching policies, reflecting the integration of theory with practice in teaching. In teaching evaluation, attention should be paid to whether teachers' teaching methods and means can be close to reality, and the combination of theory and practice should be emphasized. Flexible and diverse teaching methods can stimulate students' interest in learning and improve teaching effectiveness [14]. The interaction between teachers and students in teaching, based on the student-centered teaching principle, expands the scope of knowledge, better provides employment for students, and cultivates sustainable skilled talents for society.

2. Teaching situation evaluation is the evaluation of a teacher's teaching situation. The evaluation factors include whether the curriculum outline meets the requirements of the curriculum system, and whether the teaching objectives and content of the course are reasonable. Meanwhile, considering that students come from all over the country, whether Mandarin can be used for teaching is an important factor to ensure that students can understand and participate in the classroom. The rectification of teaching materials is also a part of the evaluation of teaching situations. Based on the actual situation, textbooks are compiled to better adapt to teaching characteristics and make the teaching content conform to the principles of students' sufficiency and applicability.

3. The main evaluation factors of teaching attitude include whether the teacher carefully prepares lessons before teaching, the preparation of teaching aids, the rigor of teaching attitude, the clarity of teaching content, and the moral cultivation and exemplary role of the teacher. Teachers should demonstrate a positive attitude during the teaching process, guide students to actively participate, stimulate their enthusiasm for learning, and set an example to become students' role models.

4. The evaluation of teaching management mainly includes classroom discipline management, strict requirements for student management, and after-school tutoring. Teachers should have good classroom management skills to ensure that students can listen quietly and abide by discipline. In addition, teachers should also provide after-school guidance, including homework guidance, practical training guidance, practical operation guidance, mathematical modelling guidance, etc., to help students consolidate their knowledge and solve problems.

5. Teaching quality evaluation is an overall evaluation based on teaching quality, mainly including students' learning situation, ability development, learning outcomes (including theory and practice), and students' preferences for courses. The teacher's teaching effect is good, and students naturally enjoy this course and are willing to work hard. This way, students' learning situation will improve, their grades will also improve, and their abilities will also be enhanced.

In summary, the evaluation of online education in higher vocational education should comprehensively consider multiple factors such as teaching methods, teaching contexts, teaching attitudes, teaching management, and teaching quality, in order to comprehensively evaluate the effectiveness and quality of online education and provide effective guidance and support for the sustainable development of higher vocational education.

2.2 Building the Evaluation Model of Online Education Effect in Higher Vocational Education Based on Data Mining Algorithm

The data mining education effect evaluation model built in this paper belongs to the gray relational model, so it needs to be processed: dimensionless processing. Due to the difference between the dimension and value of the measurement data, the data of each index cannot be compared, so dimensionless processing must be used to eliminate the dimension. In this paper, the range method is used for dimensionless processing. Calculation of correlation coefficient of comparison series and reference series. Calculate the correlation degree [15]. The traditional calculation formula of correlation degree is to calculate the average value of correlation coefficient, which does not fully reflect the importance of each index in the overall index evaluation system, resulting in inaccurate results. Therefore, this paper multiplies the correlation coefficient of each index by the weight, and adds them to obtain the final correlation degree r_k, as shown in (1) below.

$$r_k = \sum_{i=1}^{m} \zeta_k \omega_k \tag{1}$$

In formula (1), ζ_k Represents the correlation coefficient, ω_k Representing the evaluation weight, in order to further improve the evaluation accuracy of the university teaching quality evaluation model, it is necessary to build an effective evaluation system before designing the university teaching quality evaluation model. Firstly, according to the actual teaching quality of colleges and universities, the informatization evaluation dimension is selected to determine the relevant index types of the evaluation index system; Secondly, in order to ensure the operability of indicators, this paper divides the specific factors that affect the teaching quality of colleges and universities, and constructs indicator groups, as shown in Fig. 2 below.

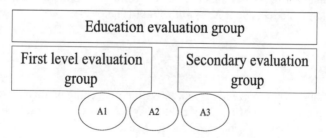

Fig. 2. Schematic Diagram of Indicator Group

It can be seen from Fig. 2 that the specific scope of teaching quality evaluation can be judged from Fig. 1, which effectively improves the accuracy of teaching quality evaluation in colleges and universities [16]

In this paper, data mining technology is used to design an algorithm for evaluating the teaching quality of colleges and universities. This algorithm can mine data rules according to the identification relationship between various data, and judge the degree of its impact on teaching quality S, as shown in (2) below.

$$S = \frac{\sqrt{D_A - (D_B + E)}}{V^2} \tag{2}$$

In formula (2), D_A Represents the value of comprehensive influencing factors, D_B Represents the value of primary influencing factors, E Represent the weight of teaching quality information, V Represents the judgment coefficient.

When applying the data mining algorithm to complete the analysis of evaluation factors, we need to pay attention to the calculation changes in different stages. In data mining and data preparation, we must compare with the real college teaching quality data to achieve data interaction and ensure the effectiveness of mining information. Different feature units may appear when using the above algorithm for decision-making. In order to ensure the optimization of decision-making [17], the model designed in this paper combines CR theory to optimize the model. The designed decision-making model can be used Charnes Cooper transformation. At this time, the expression model obtained is optimal, and the expression model can be used for decision-making to ensure the optimization of decision-making results. Correlation evaluation error factor at this time e As shown in (3) below.

$$e = \frac{1}{g}\sqrt{f} \tag{3}$$

In formula (3), g Represents the grey relational evaluation index, f It represents the actual correlation degree. This formula can effectively calculate the gray correlation value of each evaluation factor, and compare the relevant evaluation factors inside the model with the actual gray correlation value. If the comparison value is higher than 1.00, it proves that there is an error factor. Otherwise, it proves that there is no error factor, which can achieve effective evaluation. At this time, the online education effect data mining evaluation model is built K As shown in (4) below.

$$K = \frac{\Delta y}{\Delta x} - \sqrt{\frac{\Delta x}{\Delta y}} \tag{4}$$

Formula (4) Total, Δx Represents the output value of teaching quality, Δy It represents the comprehensive income of teaching quality. When applying the evaluation model to evaluate, we need to pay attention to the comprehensive input, that is, we need to bring all the factors that affect the decision-making of teaching quality evaluation into the overall evaluation, or treat the evaluation process as a decision-making whole, so as to ensure that the input of each internal weight is consistent with the decision-making amount, and improve the comprehensive evaluation level of the model. It is found in the actual use that the model is also linear programming. Therefore, the model can be further optimized according to the linear programming standard of the model [18], and all decision units can be extracted for comprehensive solution to ensure that the decision units within the decision model are consistent with the actual DEA, so as to achieve high-precision teaching quality assessment. In the actual assessment process, There may be impact on the accuracy of dynamic assessment, so it is necessary to calculate the correlation coefficient of dynamic assessment r, as shown in (5) below.

$$r = \frac{1}{n} \sum_{k=1}^{n} \xi(k) \tag{5}$$

In formula (5), n Represents the evaluation reference coefficient, $\xi(k)$ Represents the standard evaluation weight, and basic evaluation indicators can be generated by combining the above dynamic evaluation correlation coefficient g, as shown in (6) below.

$$g = \frac{\sqrt[2]{s}}{a} \tag{6}$$

In formula (6), s Represents the evaluation feedback value, a Represents the comprehensive evaluation coefficient. Decision tree is a common and important data mining method. The implementation of this algorithm is to use the top-down greedy algorithm to summarize the given data samples, extract classification rules from unordered data tuples, and recursively generate a tree structure from the top root node. Each branch node of the tree structure represents a test or selection result, Through reasonable classification of each selection result, the process continues until all attributes are traversed to finally generate a decision tree [19]. The decision tree algorithm mainly includes two processes: constructing the tree and pruning the decision tree. The former refers to taking the input training data as the function value of the established algorithm, generating each branch from the output different attribute values, and each branch continues to operate recursively to the lower level, finally forming a decision tree; For the newly established decision tree, a considerable number of branch nodes are generated because the input training sample data contains abnormal content, which is why the decision tree must be pruned. The whole decision tree process is shown in Fig. 3.

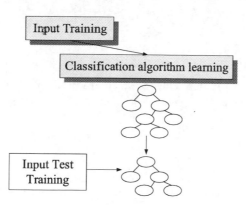

Fig. 3. Decision Tree Process

It can be seen from Fig. 3 that the non incremental data sample set in the online education process has better classification and statistics capabilities, and is more suitable for application in the field of education data mining. The ID3 algorithm is briefly discussed below. First, the expected value of the sample can be calculated n, as shown in (7) below.

$$n = \sum_{i=1}^{m} P(y_i) \log_2 P(y_i) \tag{7}$$

In formula (7), $P(y_i)$ It represents the evaluation value of the sample attribute. Teaching evaluation is an important part of the teaching process. It is based on teaching objectives, formulates scientific evaluation standards, uses all effective technical means to measure the process and results of teaching and learning activities, and gives value judgments. Teaching quality is the basis for the survival and development of colleges and universities, and is also an inevitable requirement for the internationalization of higher education. Colleges and universities have accumulated a large amount of data in their daily teaching and management, but these data have not been fully utilized. Most colleges and universities only obtain the surface information of the data through statistics or sorting. They only carry out simple numerical calculation and qualitative summary of these information, and then report these surface information to teachers as the basis for promoting professional titles, appraising excellence, performance pay, etc. The internal information implied in these data has not been deeply mined and considered. How to transform these data into information that can be further used to provide decision support for school administrators is an urgent problem in college education.

Data mining is the process of extracting hidden, unknown, but potentially useful information and knowledge from a large number of incomplete, noisy, fuzzy, random data. In the long-term teaching and management [20], with the accumulation of more and more data, people are more and more interested in mining relevant knowledge from these data. Mining association rules (knowledge) is to search for valuable relationships between data items from a given data set. This paper conducts a sample weighted average based on data mining to obtain the entropy of the evaluation data set at this time $E(D_f)$ As shown in (8).

$$E(D_f) = \sum_{s=1}^{q} I(N) \tag{8}$$

In formula (8), $I(N)$ It represents the value of the sample evaluation attribute, and the current data sample evaluation description proportion qe As shown in (9) below.

$$qe = - \sum_{i=1}^{m} p_{is} \log P(y_i) \tag{9}$$

In formula (9), p_{is} Represents the evaluation gain coefficient. By calculating the information gain of all attributes, ID3 algorithm forms the attribute with the largest gain into the test attribute in the data sample set, and then generates the branch point. This branch node is marked as an indicator attribute and classified into a given sample set.

Conduct online evaluations of teachers' teaching content, teaching level, political level, scientific research level, etc. based on the different needs of different users. Thus, students' evaluation of teachers, teachers' evaluation of teachers, and experts' evaluation of teachers have been achieved.

3 Example Analysis

3.1 Overview and Preparation

In order to verify the actual evaluation effect of the college teaching quality evaluation method designed in this paper based on data mining technology, a university is selected for example analysis, and an example analysis platform for college teaching quality evaluation is built. The CPU is 16Core, the memory is 128 GB, the disk is 3 TB, the bandwidth is 1000 Mb/s, the hard disk is 12 TB, and the operating system is CentOS 7, This paper mainly collects the teaching quality information of the film and television planning specialty of the university as experimental data. The experimental data set collected includes 300 data samples, and 20 students in the data set are randomly selected. After the case analysis environment and samples are selected, the comprehensive evaluation index of education effect can be used as the case analysis index T As shown in (10) below.

$$T = \frac{a_0 + c_a + f_a}{U} \tag{10}$$

In formula (10), a_0 Representative reviews the evaluation coefficient, c_a Represents the qualified evaluation coefficient, f_a Represents the random evaluation coefficient, U It represents the standard evaluation parameters. The standard value of the known comprehensive evaluation index of education effect is 1.00. The closer to 1.00, the better the evaluation effect of the education effect evaluation method is. Otherwise, it proves that the evaluation effect is relatively poor.

3.2 Application Effect and Discussion

In combination with the above overview and preparations, we can analyze the evaluation methods, that is, select different education effect evaluation indicators, respectively use the online education effect evaluation method based on data mining algorithm designed in this paper, and the online education effect evaluation method based on association rules (method 1), As well as the evaluation method of online education effect of higher vocational education based on random classification (method 2), formula (10) is used to calculate the comprehensive evaluation indicators of different evaluation methods. The example analysis results are shown in Table 1 below.

It can be seen from Table 1 that the comprehensive evaluation index of the online education effect evaluation method of higher vocational education designed in this paper based on data mining algorithm under different evaluation numbers is relatively high, close to the standard index of 1.00. The conventional online education effect evaluation method of higher vocational education based on association rules (method 1), As well as the evaluation method of online education effect of higher vocational education based on random classification (method 2), the comprehensive evaluation indicators under different evaluation option numbers are relatively low, and there is a large gap with the standard indicator 1.00. The above example analysis results prove that the evaluation method of online education effect of higher vocational education designed in this

Table 1. Application Effect

Evaluation option number	Standard	This paper	Method 1	Method 2
X1	1.00	0.95	0.45	0.65
X2	1.00	0.91	0.41	0.54
X3	1.00	0.92	0.52	0.42
X4	1.00	0.93	0.56	0.65
X5	1.00	0.92	0.49	0.53
X6	1.00	0.95	0.55	0.56
X7	1.00	0.96	0.62	0.69
X8	1.00	0.93	0.51	0.54
X9	1.00	0.92	0.52	0.68
X10	1.00	0.97	0.63	0.56

paper based on data mining algorithm has good evaluation effect, reliability and certain application value.

Further analyzing the performance of the method proposed in this article, the two datasets were randomly divided into three groups, and the output speed of the evaluation results was used as a comparison indicator. The comparison of the test results of the three methods is shown in Table 2:

Table 2. Evaluation Result Output Speed/ms

Data set	Group	The method of this paper	Method 1	Method 2
3DMatch data set	1	12	57	37
	2	13	59	39
	3	11	62	45
3DRegNet data set	1	14	66	55
	2	12	68	56
	3	15	65	58

Analyzing Table 2, it can be seen that the evaluation results output speed of the method in this article is the fastest, with an average of 12.83 ms, while the evaluation results output speed of Method 1 and Method 2 is relatively slow, with an average of 62.83 ms and 48.33 ms, respectively. This indicates that applying the method in this article to evaluate the effectiveness of online teaching in vocational colleges can achieve efficient evaluation results output, which is conducive to quickly achieving teaching evaluation.

4   Conclusion

With the development of the times, the country also pays more attention to the development of education. Mixed teaching, as a new teaching method, is gradually rising in China, and more and more schools have begun to use mixed learning mode. Moreover, the new learning model has also been highly praised by teachers and students. The hybrid teaching mode mainly adopts the combination of online learning and offline learning, which shows some problems difficult to understand by students in the traditional teaching mode in the form of images and three-dimensional, enhancing students' memory and expanding their imagination. For students in higher vocational colleges, using blended learning can improve their learning quality and enjoy more learning resources. Because there are many high-quality educational resources on the Internet, the new education model enables students to make full use of the resources on the Internet to improve their learning ability and help students solve learning problems. The online teaching platform can not only strengthen the interaction of learning and the cohesion of knowledge, but also play a good role in assisting and supplementing traditional teaching. This hybrid teaching mode can penetrate the characteristics of online and offline learning, constantly improve students' learning motivation, and also enhance students' learning interest. According to the current online education background of higher vocational education, this paper uses data mining algorithm to design a new evaluation method of online education effect of higher vocational education. Through case analysis, the results show that the evaluation method designed for online education effect evaluation in higher vocational education has good evaluation effect, reliability and certain application value, and has made certain contributions to improving the quality of online education in higher vocational education. This study aims to design a new method for evaluating the effectiveness of online education in higher vocational education using data mining algorithms against the current background of online education in higher vocational education. Through case analysis verification, this evaluation method has good evaluation effectiveness, reliability, and application value, making a contribution to improving the quality of online education in higher vocational education. Compared with traditional subjective evaluation frameworks, this method is based on data mining algorithms and overcomes the impact of personalized differences, providing more accurate comprehensive evaluation indicators. By determining the evaluation system, constructing an evaluation model, and generating an evaluation platform, this method has outstanding innovation in the field of online education effectiveness evaluation in higher vocational education. Personalized recommendation technology plays an important role in online education. Future research can combine data mining algorithms with personalized recommendation techniques to provide students with more personalized learning resources and paths based on their individual characteristics and learning needs, further improving the effectiveness and satisfaction of online education.

References

1. Feng, L., Liu, B.: Research and simulation of fault data mining algorithms for serial communication transmission. Comput. Simul. (005), 039 (2022)

2. Mei, Y.: Simulation of Chinese online teaching invocational colleges based on complex evolution and improved neural network. J. Intell. Fuzzy Syst. Appl. Eng. Technol. **4**, 40 (2021)

3. Liu, Y., Qin, Y.: The innovation research and practice of the hybrid teaching mode in colleges and universities based on computer technology. J. Phys. Conf. Ser. **1744**(4), 042055 (4pp) (2021)

4. Tuomikoski, A., Holopainen, A., Halvari, J., et al.: Social, health and rehabilitation sector educators' competence in evidence-based practice: a cross-sectional study. Nurs. Open **8**(6), 3222–3231 (2021)

5. Bi, H., Gu, Z., Liu, D.: Advantages of multimedia network teaching in ice and snow sports education in higher vocational colleges. J. Phys. Conf. Ser. **1852**(4), 042062 (7pp) (2021)

6. Huang, Y., Yao, J., Huang, G.: Application of intelligent information technology in the reform of hybrid teaching courses in colleges and universities. J. Phys. Conf. Ser. **1852**(2), 022065 (2021)

7. Efendi, R., Lesmana, L.S., Putra, F., et al.: Design and implementation of computer based test (CBT) in vocational education. J. Phys. Conf. Ser. **1764**(1), 012068 (12pp) (2021)

8. Silvana, T.S., Ekohariadi, Buditjahjanto, I.G.P., Rijanto, T., Munoto, Nurlaela, L.: Study of the implementation of online learning models in vocational schools. J. Phys. Conf. Ser. **1810**(1), 012066 (2021)

9. Putra, A., Insani, N., Winarno, A., et al.: The innovation of intelligent system e-consultant learning to improve student mindset of vocational education in the disruptive Era 4.0. J. Phys. Conf. Ser. **1833**(1), 012033 (6pp) (2021)

10. Yu, Y.: Based on the apartment community management mode to explore the education strategy of higher vocational colleges—take Polus International College as an example. Open J. Soc. Sci. **09**(2), 322–333 (2021)

11. Constantinou, C., Wijnen-Meijer, M.: Student evaluations of teaching and the development of a comprehensive measure of teaching effectiveness for medical schools. BMC Med. Educ. **22**(1), 113 (2022)

12. Lakeman, R., Coutts, R., Hutchinson, M., et al.: Appearance, insults, allegations, blame and threats: an analysis of anonymous non-constructive student evaluation of teaching in Australia. Assess. Eval. High. Educ. **47**(8), 1245–1258 (2022)

13. Okoye, K., Arrona-Palacios, A., Camacho-Zuñiga, C., et al.: Towards teaching analytics: a contextual model for analysis of students' evaluation of teaching through text mining and machine learning classification. Educ. Inf. Technol. **27**, 3891–3933 (2022). https://doi.org/10.1007/s10639-021-10751-5

14. Heffernan, T.: Sexism, racism, prejudice, and bias: a literature review and synthesis of research surrounding student evaluations of courses and teaching. Assess. Eval. High. Educ. **47**(1), 144–154 (2022)

15. Tan, R.Z., Markus, C., Vasikaran, S., Loh, T.P., APFCB Harmonization of Reference Intervals Working Group: Comparison of four indirect (data mining) approaches to derive within-subject biological variation. Clin. Chem. Lab. Med. **60**(4), 636–644 (2022)

16. Yates, D., Islam, M.Z.: Data mining on smartphones: an introduction and survey. ACM Comput. Surv.Surv. **55**(5), 1–38 (2022)

17. Khalaf, L.I., Aswad, S.A., Ahmed, S.R., et al.: Survey on recognition hand gesture by using data mining algorithms. In: 2022 International Congress on Human-Computer Interaction, Optimization and Robotic Applications (HORA), pp. 1–4. IEEE (2022)

18. Koukaras, P., Tjortjis, C., Rousidis, D.: Mining association rules from COVID-19 related twitter data to discover word patterns, topics and inferences. Inf. Syst. **109**, 102054 (2022)

19. Singh, H., Bashir, N.Z., Virdee, S.S.: Evaluation of the quality of undergraduate full veneer crown preparations at a UK dental teaching hospital. Eur. J. Prosthodont. Restor. Dentist. **31**(1), 31–39 (2023)
20. Sacre, H., Akel, M., Haddad, C., Zeenny, R.M., Hajj, A., Salameh, P.: The effect of research on the perceived quality of teaching: a cross-sectional study among university students in Lebanon. BMC Med. Educ. **23**(1), 31 (2023). https://doi.org/10.1186/s12909-023-03998-8

Wireless Networks for Social Information Processing, Civilian Radar Signal Processing

Processing Method of Civil Radar Echo Signal Based on Kalman Filter Algorithm

Jia Pan[✉]

Guangxi Science and Technology Normal University, Laibin 546100, China
jiapan_gxkjsfxy@163.com

Abstract. With the popularization of civil radar application, it has great development potential in the situations of earthquake disasters and engineering accidents, such as personnel search and rescue, medical detection, and urban anti-terrorism. A civil radar echo signal processing method based on Kalman filter algorithm is designed. The Kalman filter algorithm is used to suppress the noise of the acquired radar echo signal. According to the amount of information obtained for the target being explored in different stages, target detection is regarded as the second stage of civil radar echo signal processing. Based on the Faster R-CNN detection framework, the context information and multi-scale Faster R-CNN target detection method are designed to determine the presence or absence of the target based on the denoised signal. Implement reference signal reconstruction, multipath clutter suppression, target location and tracking, and obtain some basic parameters to determine the target. The test results show that the tracking distance error of this method for stationary target, inching target and moving target is small.

Keywords: Kalman Filter Algorithm · Civil Radar · Faster R-CNN · Multipath Clutter Suppression

1 Introduction

The basic working principle of the radar is to send a series of electromagnetic waves to the target to be measured through the transmitter. After the transmission, the electromagnetic waves will collide with the target to be measured, and then part of the electromagnetic waves will be absorbed by the target to be measured, and the other part will be reflected back. Radar is to detect the target to be measured according to the reflected electromagnetic wave. When the target to be measured is detected, the feature information about the moving target to be measured can be extracted from the reflected electromagnetic wave for analysis, and finally some parameters [1] of the target to be measured can be determined. If the emitted electromagnetic wave is completely absorbed by the object, the target will be invisible and cannot be detected by radar. In short, radar is to use the waveform information returned from detection to obtain target information. Based on the working characteristics of radar, radar technology has been widely used in weather, transportation, national defense and other aspects. With the diversification of radar system applications, multi-function speed radar, navigation radar and detection radar have

L. Yun et al. (Eds.): ADHIP 2023, LNICST 548, pp. 67–82, 2024.
https://doi.org/10.1007/978-3-031-50546-1_5

sprung up like mushrooms. In the civil field, radar is widely used. For example, weather detection radar can provide timely and accurate meteorological information for weather prediction, and remote sensing detection radar can provide various detailed parameters for the detection of earth resources and the imaging of targets. It can be said that the application of civil radar has penetrated into every corner of people's life.

The performance of radar signal processing is one of the important signs to measure the performance of a radar system. Its main role is to suppress interference as much as possible in complex environments and obtain more important and valuable information [2]. In the radar echo signal, the composition of the radar echo signal is more complex. Clutter, noise and jamming will seriously affect the detection performance of the system, so after obtaining the radar echo signal, the back-end signal processing, target search and other algorithms are particularly important.

A considerable amount of research has been achieved on the processing of radar echo signals. Scholars have proposed the use of wavelet threshold shrinkage to solve the problem of noise interference in the process of ground penetrating radar recording echo signals in a wide frequency band. A preprocessing method for ground penetrating radar echo signals is designed to remove radar echo noise, and the method is implemented through DSP circuits and software. Scholars have also proposed a simulation method for echo generation and signal processing of ground-based radar. Through radar system simulation technology, the echo generation and signal processing of ground-based radar are processed, and the application of the method is achieved through various aspects such as radar system analysis, design, testing, and evaluation. However, the above methods have significant tracking distance errors for stationary targets, micro moving targets, and moving targets, which cannot meet the application requirements of civilian radar echo signal processing.

In order to solve the shortcomings of the above methods, this paper takes the research project of X band Ground Based Radar (GBR) software simulation system proposed by the scientific research contract unit as the background, conducts a more in-depth study on the design and implementation of the echo generation and signal processing simulation subsystem, and proposes a civil radar echo signal processing method based on Kalman filter algorithm, In order to reduce the error in tracking distance during the processing of civilian radar echo signals. The specific research contents are as follows: the working principle and mode of the X band ground-based radar simulation system are introduced, the functions of the echo generation and signal processing subsystem in the entire simulation system are emphatically analyzed, and the overall design of the echo generation and signal processing simulation subsystem is carried out according to the system parameters. Under the application background of the X band ground-based radar simulation system, this paper proposes an implementation scheme of the echo generation and signal processing simulation subsystem, and models, simulates and tests it. The test results prove the rationality of the modeling.

2 Design of Echo Signal Processing Method for Civil Radar

2.1 Noise Suppression

Civil radar signal echo refers to the signal that the electromagnetic waves emitted by the radar system return to the radar receiver after interacting with the target object. These signal echoes carry information about the target object, including its position, velocity, size, and scattering characteristics. Compared with other signal echoes, civilian radar signal echoes have special signal characteristics and target reflection characteristics. Through distance and velocity measurement capabilities, they provide a wide range of application scenarios and play an important role in aviation, navigation, meteorology, geological exploration, and other fields. Therefore, in order to ensure the effectiveness of the radar echo signal, this paper implements noise suppression on the obtained radar echo signal based on the Kalman filter algorithm [3].

If we say random process $\{y_n\}$ Is a q The autoregressive process of order is generally expressed as $AR(q)$, it satisfies the following difference equation:

$$y(n) + c(1)y(n-1) + \ldots + c(q)y(n-q)$$
$$= \alpha(n) \tag{1}$$

In formula (1) $c(1), \ldots, c(q)$ Is a constant; $\alpha(n)$ It has a mean of zero and a variance of σ_α^2 Stationary white noise sequence. Next, we will use the AR model to estimate the parameters.

Establish the following model for the acquired radar echo signal:

$$\beta_y(t) = \beta_s(t) + \beta_N(t) + \omega_1(t) \tag{2}$$

In formula (2) $\beta_s(t)$ Represents pure signal; $\beta_N(t)$, $\omega_1(t)$ Represents a stationary additive noise signal, assuming that the three are uncorrelated; $\beta_y(t)$ Indicates a noisy signal.

Assuming that the radar echo signal is short-time stable, it can be used q Order AR model. Kalman filter is an optimal estimation method based on Bayesian inference, which can optimally estimate the state of the system under the condition of given observation data. By converting the AR model into the Equation of state and observation equation of Kalman filter, we can take advantage of the advantages of Kalman filter to obtain more accurate and stable state estimation, and can better deal with these nonlinear and non Gaussian situations, improving the adaptability and accuracy of the model.

Which will be used to represent the radar echo signal q The order AR model is transformed into the state equation and observation equation of Kalman filter.

In this method, we first estimate the power spectrum of the observed signal and the power spectrum of the noise, obtain the power spectrum of the pure signal required in the experiment through spectral subtraction, and then calculate the AR coefficients of the noise and the signal through the power spectrum respectively.

The specific parameter estimation process is shown in Fig. 1.

The measurement matrix, state transition matrix, noise covariance matrix of the measurement equation and noise covariance moment of the state equation are obtained according to the AR coefficient obtained above.

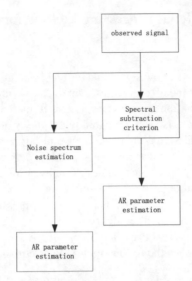

Fig. 1. Specific parameter estimation process

The parameters of Kalman filter are obtained, and then Kalman filter is applied to the signal.

The time update equation and the state update equation are obtained by continuous iteration.

The Kalman filter algorithm requires initial values x (0 | 0) and P (0 | 0). The initial estimated value of x (0 | 0) can be 0 or other values obtained from prior information, and P (0 | 0) can be the identity matrix. In this study, x (0 | 0) and P (0 | 0) are 0 and unit matrices, respectively.

2.2 Object Detection

According to the amount of information obtained from the explored target in different stages, target detection is regarded as the second stage of civil radar echo signal processing, that is, the presence or absence of targets is judged based on the denoised signal. Based on the Faster R-CNN detection framework, a context based and multi-scale Faster R-CNN target detection method is designed. On the one hand, the method uses multi-scale feature fusion, which uses deep semantic features for detection, and also fuses shallow details to enrich the information of the proposed features. On the other hand, the bidirectional GRU in RNN is used to process multi-scale features to achieve reasonable fusion of context information in SAR images, so that the candidate region generation phase and detection phase are assisted by context information. Under the joint action of multi-scale information fusion and context information, more informative target features can be obtained to improve target detection performance [4].

The proposed SAR image target detection framework is shown in Fig. 2. The detection network consists of four parts: feature extraction module, context information fusion module, candidate region generation module, and detection module. Firstly, the feature extraction module uses CNN to extract features from SAR images. At the same time, in order to enrich the feature information, the module also fuses the output features of different levels in the network, so that the final output features include deep semantic information and shallow details. Secondly, the context information fusion module composed of bidirectional GRU network is used to process the output features of the feature extraction module to achieve the fusion and utilization of the context information in the feature map. Then, the candidate region generation module processes the multi-scale context features processed by the context information fusion module, and outputs the classification and regression results of anchor boxes through the convolution operation, so as to select candidate regions. Finally, the detection module processes the corresponding features of the selected candidate region through the full connection layer to obtain the final target category and target location coordinates, and complete target detection in SAR images [5].

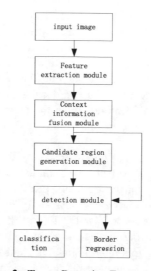

Fig. 2. Target Detection Framework

The network training in the proposed method is in the form of joint optimization. In the training process, after the input image is extracted by the feature extraction module and the context information fusion module, the loss of the candidate area generation module and the loss of the detection module are obtained by the candidate area generation module and the detection module. Add the two losses to get the total loss of the network, and update the network weight through BP algorithm and gradient descent. The specific training process is as follows:

(1) The training data set is generated and expanded. Due to the large size of the original SAR image, it is not conducive to network training. Therefore, it is necessary to cut the large image in the original SAR image dataset to obtain several SAR image sub images, as shown in the following formula:

$$A = \{a_1, a_2, \ldots, a_n\} \tag{3}$$

In formula (3) a_n It refers to the No n SAR image sub images.

And mark them at the target level, including the target category and target border. At the same time, due to the small amount of SAR image data, the sub images obtained by clipping are expanded by adding noise, filtering, rotating, and flipping. Among them, Gaussian noise and salt pepper noise with additive mean of 0 and variance of 0.01 are used for noise addition; Median filter, Gaussian filter and mean filter are adopted for filtering; Rotate to rotate at three angles, 90°, 180° and 270° respectively; The flip mode is horizontal; At the same time, rotate the horizontally flipped pictures by 90°, 180° and 270° respectively. Integrate the sub images obtained after the above processing to obtain the final training dataset;

(2) Load the pre training model. In order to accelerate the model convergence and improve the final performance, the weight parameters of the network are initialized by loading the pre training model. Because the feature extraction network part of the feature extraction module uses the structure of. VGG16, the VGG16 network parameters trained by ImageNet dataset for classification tasks are loaded onto the feature extraction network, while the feature fusion part uses random initialization to initialize the parameters; The weight parameters of context information fusion module, candidate area generation module and detection module are initialized randomly;

(3) Get multi-scale context features. The training data set is input into the network, and the image is processed by the forward calculation of the feature extraction module to obtain multi-scale features. Input the obtained multi-scale features into the context information fusion module, and get the multi-scale context features through operation;

(4) Generate the candidate region and obtain the loss of the candidate region generation module. The multi-scale context features are input into the candidate region generation module for processing, and the regression prediction values [6] of the foreground background prediction probability and position of the preset anchor frame on the feature map are obtained respectively. The loss of the candidate region generation module is calculated by using the category label, the position label of the target box and the prediction value in the training data, and several candidate regions are selected as the input of the detection module according to the prediction results. Several candidate regions selected according to the prediction results are as follows:

$$H = \{h_1, h_2, \ldots, h_m\} \tag{4}$$

In formula (4) h_m It refers to the No m Selected candidate regions.

(5) The selected candidate areas are detected to obtain their category information and location coordinates, and the loss of the detection module is calculated at the same time. Label the selected candidate area according to the image tag, and input the selected candidate area and multi-scale context features into the detection module to obtain the fine regression value of the target category prediction probability of the candidate area and the target rectangular box coordinates, so as to obtain the interested target and complete the detection. At the same time, the loss of detection module is calculated according to the target detection results and the labels of candidate regions;

(6) Update the network parameters and complete the training. Add the loss of candidate area generation module and the loss of detection module as the overall loss of the

network. BP algorithm and gradient descent method are used to optimize the loss of the network and update the parameters. After one update, repeat the process from (3) to (6) to update the network parameters iteratively. When the training times reach the preset training times, the network completes the training [7].

Use the trained model to test the test data. During the test, the original SAR image of a large scene is tested, so the test process is as follows:

(1) For the SAR test image in a large scene, the original SAR image is clipped from left to right and from top to bottom by sliding window capture, the original SAR test image is divided into several test sub images, and the corresponding position coordinates of the cut test sub images on the original large image are recorded;
(2) Input the cut test subgraphs into the target detection network after training, and obtain the target detection results of each subgraph through forward calculation of network parameters;
(3) Set the threshold of classification confidence f The test results will be further screened, that is, the test results meeting the following formula will be retained as the test results of each sub graph.

$$Z > f \tag{5}$$

In formula (5) Z It refers to classification confidence.
(4) According to the position of each sub image on the original SAR image, the detection results on the sub image are mapped back to the original image to obtain the detection results on the original large image;
(5) The overlapping duplicate boxes in the obtained large scene detection results are further filtered, that is, the overlapping target boxes in the original large scene SAR image are removed by the NMS method, and the final detection results are obtained.

The overall structure of the network and the process of training and testing are described above. Next, the feature extraction module, context information fusion module, candidate region generation module and detection module in the network structure are described and introduced in detail.

The feature extraction module is the basic part of the proposed SAR image target detection network. This module extracts SAR image features through a series of convolution and pooling operations, and fuses features at different scales. While using deep semantic features for detection, it also fuses shallow details, making SAR image feature information more abundant. The feature extraction network structure in this module is based on VGG1618.

The module includes a feature extraction network composed of five convolutions and a feature fusion part composed of four convolutions. Among them, the five convolution parts include a total of 13 convolution layers, 13 ReLU layers and 4 maximum pooling layers. The first convolution part and the second convolution part respectively include two convolution layers, each of which is followed by a ReLU layer, and at the same time, the last of each convolution part is followed by a maximum pooling layer. The third to fifth convolution sections respectively include three convolution layers, and each convolution layer is also connected with a ReLU layer. In addition, a maximum pooling layer is connected at the end of the third and fourth convolutions. In the network structure of this part, the size of the convolution kernel is 3 × 3, and the step size is 1. In order to avoid reducing the size of the feature map, the convolution layer is filled with 1 pixel. The step size of the maximum pool layer is set to 2, and the pool area size is 2 × 2 [8]. The detailed network structure parameters are shown in Table 1.

Table 1. Structure parameters of feature extraction network

Convolution part	Hierarchy name	Convolutional core/pool Sizing	Number of input channels	Number of output channels	step
Conv1	Conv1_1	9 × 9	1	64	3
	Conv1_2	9 × 9	32	64	3
	Max-pooling1	4 × 4	32	64	1
Conv2	Conv2_1	9 × 9	32	128	3
	Conv2_2	9 × 9	64	128	3
	Max-pooling2	4 × 4	64	128	1
Conv3	Conv3_1	9 × 9	64	256	3
	Conv3_2	9 × 9	128	256	3
	Conv3_3	9 × 9	128	256	3
	Max- pooling3	4 × 4	128	256	1
Conv4	Conv4_1	9 × 9	128	512	3
	Conv4_2	9 × 9	256	512	3
	Conv4_3	9 × 9	256	512	3
	Max- pooling4	4 × 4	256	512	1
Conv5	Conv5_1	9 × 9	256	512	3
	Conv5_2	9 × 9	256	512	3
	Conv5_3	9 × 9	256	512	3

In the table, only the network parameters of the feature extraction network part in the module are given, and the activation function layer is omitted. By default, each convolution layer is connected to a ReLU layer.

The context information fusion module is the core part of the proposed method. In SAR images, the environment in which targets appear is not random, and different targets often appear in specific scenes. For example, vehicle targets generally appear in roads, parking lots and other scenes; aircraft targets generally appear in airport scenes, and ship targets often appear in the ocean and near shore. It can be seen that there is a certain dependency between the target in the SAR image and the surrounding environment. This relationship is the context information of the target. In the target detection task, the reasonable use of the target context information can better distinguish the target and background in the detection process, reduce the error results, and improve the target detection performance. After extracting features from SAR images, the context information of the target is reflected in the dependency between feature pixels. Because RNN can learn the dependency between sequence data elements well, it can be used to fuse the context between feature pixels, so as to achieve the fusion of context information around the target. Therefore, this module uses the bidirectional GRU network in RNN and the corresponding convolution layer to achieve the fusion of context information in SAR image features.

The composition of context information fusion module is shown in Table 2.

Table 2. Composition of context information fusion module

S/N	structure	number
1	Two way GRU network	4
2	Convolution layer	2
3	ReLU layer	1

The module first inputs the SAR image features obtained by the feature extraction module into the bidirectional GRU network 1 and bidirectional GRU network 2, respectively, to obtain the horizontal context information feature map and the vertical context information feature map. Secondly, after splicing the two along the channel dimension, the context feature 1 can be obtained by feature fusion through 1×1 convolution operation. Then input context feature 1 into two-way GRU network 3 and two-way GRU network 4 respectively, and repeat the same operation as two-way GRU network 1 and 2. Finally, the output features of bidirectional GRU networks 3 and 4 are concatenated with the original SAR image features in the channel dimension. After 1×1 convolution and ReLU activation layer processing, the context feature 2 can be obtained, which is used as the output of the context information fusion module to complete the fusion of the context information in the SAR image features.

The candidate region generation module is an important part of the proposed method, which is used to extract the possible target regions in SAR images, and calculate the loss of the candidate region generation module required for network training.

In the candidate region generation module, in order to generate a region of interest that may be the target, after inputting multi-scale context features, the module will use a small network for sliding window processing. The sliding window size in the small network is nxn. The sliding small network will process the features in the sliding window and get the output vector by activating the function layer. The output vector will continue to be input into two parallel full connection layers, one of which is used to predict the target and background classification of the candidate area, and the other is used to predict the regression value of the target box position of the candidate area. By combining the results of classification and regression, candidate regions that may be targets are obtained. It is worth noting that since this small network processes the feature map through a sliding window, and the parameters of the small network are shared during the sliding process, this structure can actually be regarded as an nxn convolution layer. The convolution step is the sliding step of the sliding window, and the full connection layer of the subsequent processing output vector can also be regarded as a 1 × 1 convolution operation. Therefore, the candidate region generation module is essentially a full convolutional network.

The composition of candidate region generation module is shown in Table 3.

Table 3. Composition of candidate region generation module

S/N	structure	One
1	Convolution layer	Three
2	ReLU layer	

The detection module is the key part of the proposed method to realize the detection function. The module uses multi-scale context features and the proposed candidate area to obtain the multi-scale context features corresponding to the candidate area through the RoI Align operation. The candidate area features are classified and finely adjusted in the target category and position through the corresponding full connection layer.

2.3 Parameter Extraction

The acquisition of some basic parameters of the target is regarded as the final stage of radar signal processing, and one of the key issues is the reconstruction of reference signal. First, the signal starting point is determined synchronously, then the multipath effect is removed by channel estimation and equalization. Then, according to the civil radar signal standard, the pure code stream is obtained through constellation mapping and channel coding, and then the coding and modulation process at the transmitter is repeated, including channel coding, constellation mapping, OFDM modulation, and framing to obtain the reconstructed signal. Synchronization, channel estimation and equalization and channel decoding are the key steps.

Synchronization plays an important role in SDPR, and its performance directly affects the subsequent reference signal extraction. SDPR system synchronization mainly includes the following three aspects: symbol synchronization, carrier synchronization and sampling rate synchronization.

Symbol synchronization refers to finding the starting point of OFDM symbols in the sampled data, that is, determining the DFT window position during demodulation. It is the first step of subsequent carrier frequency offset estimation and sampling rate synchronization. In the SDPR system, the synchronization starting point is usually the clutter suppression starting point, corresponding to the arrival time of the direct wave, which directly affects the estimation of the target bistatic distance. Conventional adaptive filtering method is used for clutter suppression. If the estimation of symbol synchronization start point is inaccurate, that is, the reconstructed reference signal may not be causal with some multipath clutter in the monitoring signal, and multipath clutter may not be completely suppressed, affecting target detection [9].

Carrier synchronization refers to the estimation and compensation of the deviation between the local carriers between the transmitter and the receiver. For SDPR, because the transmitter is not controlled, the crystal oscillators at the transmitter and receiver cannot match completely. Therefore, the carrier frequency offset is relatively common in SDPR, and it is necessary to estimate and compensate the signal at the receiver in real time. In the OFDM modulation system, if the frequency offset is normalized to the subcarrier spacing, the frequency offset obtained is as follows:

$$\xi = \psi + \zeta \tag{6}$$

In formula (6) ψ It refers to integer octave offset; ζ It refers to decimal octave offset. Among ψ It will not cause ICI, but will cause 50% error probability of the information that is called out. and ζ The sampling point will not be at the vertex of the subcarrier, thus destroying the orthogonality of the subcarrier and introducing ICI. Therefore, the receiving end must ψ And ζ Estimate and compensate.

Sampling rate synchronization is to estimate and compensate the sampling time deviation between the transmitter and receiver. Considering that the transmitter and receiver usually use the GPS module to provide a highly stable clock source, the impact of sampling rate deviation is small and can be ignored.

Generally speaking, rough symbol synchronization is performed first to find the approximate position of the synchronization signal, and then fractional octave offset is estimated using the synchronization signal, and then integer octave offset is estimated using the compensated synchronization signal, and finally fine symbol synchronization is performed.

In order to meet the real-time requirements of the SDPR system, in addition to considering using the GPS module to provide a highly stable clock source, in signal processing, carrier frequency offset estimation and symbol fine synchronization can be done once per frame, and this processing can also eliminate the cumulative error caused by sampling frequency offset.

Channel estimation is implemented based on pilot symbols. The channel estimation method based on pilot symbols is to use known discrete pilot symbols to calculate the impulse response of subcarriers at discrete pilots, and then estimate the channel response at data subcarriers through interpolation algorithm according to the arrangement rule of discrete pilots in the frequency domain direction. Finally, the estimated subcarrier channel response is used to equalize the subcarrier data in the frequency domain.

Let the received signal be expressed as $R(k)$, the channel response at the pilot is:

$$\tilde{F}(k_p) = \frac{R(k)}{G(k)} \tag{7}$$

be based on $\tilde{F}(k_p)$ The linear interpolation method is used to obtain the k_p+1 Subcarrier response at $\tilde{F}(k_p+1)$. Finally, the channel response at the pilot is averaged in the frequency domain for each subcarrier.

The constellation is focused directly according to the signal modulation mode to obtain the reconstructed reference signal.

Multipath clutter is a type of interference signal in radar systems, which is the echo signal generated by the reflection, refraction, and scattering of signals emitted by radar when encountering obstacles, terrain, buildings, etc. during the propagation process. These echo signals propagate through different paths to the radar receiver and are superimposed with the main direct signal, forming multiple echo signals with different time delays. Multipath clutter introduces additional time delay, which expands the time-domain characteristics of the echo signal, resulting in strength measurement errors of the target and reducing the detection performance of the radar system. And due to the different path lengths of multipath propagation, the phase of the echo signal may change. This can cause phase distortion in coherent processing and reduce the target resolution ability of the radar system.

Therefore, this article is based on the frequency domain subcarrier multipath clutter cancellation method to achieve the suppression of multipath clutter. The frequency domain multi-path clutter cancellation method is based on the principle that multi-path clutter is completely correlated on the same subcarrier, while the target is suppressed by multi-path clutter due to the existence of Doppler frequency shift and its non correlation.

After acquiring the monitoring signal after carrier frequency offset compensation, the channel response of multipath clutter is estimated through the least square algorithm $E(w)$, the frequency domain monitoring signal after clutter suppression can represent:

$$D_0(w) = D(w) - C(w)E(w) \tag{8}$$

In the above formula $D(w)$ Is the monitoring signal after carrier frequency offset compensation; $C(w)$ Is the frequency domain reference signal vector [10].

Finally, by locating and tracking the target, some basic parameters to determine the target are obtained. The target is located by TDOA positioning method [11]. For the convenience of expression, the problem of target location on a two-dimensional plane is considered. Assume that the coordinate position of the target is:

$$\chi = [x, y]^v \tag{9}$$

In the above formula v Represents the coordinate threshold; x, y Represents its coordinate value.

The coordinate position of the transmitting station is:

$$\delta_x = [x_0, y_0]^v \tag{10}$$

In the above formula x_0, y_0 Represents its coordinate value.

The coordinate positions of the N receiving stations are:

$$\begin{cases} \varepsilon_{x_i} = [x_i, y_i]^v \\ i = 1, 2, \ldots, N \end{cases} \tag{11}$$

In the above formula x_i, y_i Represents its coordinate value.

Calculate the baseline distance between the transmitting station and the receiving station according to the above formula, and calculate the bistatic distance of target P measured by the receiving station [12].

After obtaining the position and motion state of the target, it can be regarded as the secondary measurement value as the input of the tracking process [13]. At this time, the common tracking methods such as Kalman filter and Bayesian theory can achieve target tracking.

3 Experimental Research

3.1 Experimental System and Platform

For the civil radar echo signal processing method designed based on Kalman filter algorithm, the method is tested through the designed experimental system and platform. The experimental equipment used includes the UWB radar system with NVA-6100 as the core and a laptop computer. The working mode of the selected transmitter is IF transmission, and the bandwidth is 0.85–9.55 GHz, meeting the ultra wideband standard. The radar data is connected to the notebook computer through the USB interface and the signal is preprocessed.

The experimental system is shown in Fig. 3.

Fig. 3. Experimental System

In order to verify the effectiveness and accuracy of this method, an experimental platform is built. The platform is mainly composed of the track system and the target are composed of two parts. The track system is 5m long and 3m wide, and the two long rails are connected through a transmission rod to achieve synchronous movement. At the same time, the track system is equipped with a track control box, which can move the track to a fixed distance or conduct regular movement, and can carry out standardized detection of the accuracy of the radar system test distance.

3.2 Comparison Method Design

The echo signal pre-processing method of ground penetrating radar and the echo generation and signal processing simulation method of ground-based radar are used as the comparison method to test together, and are represented by method 1 and method 2 respectively.

3.3 Static Target Experimental Verification

First, carry out the static target related experiments on the experimental platform. During the experiment, the radar system is placed on a table 0.85 m from the ground, the target is facing the radar system, and the target is moved on the slide rail to five different positions 1 m, 2 m, 3 m, 4 m, 5 m from the radar by controlling the power motor of the track, and the position is marked and tested by the radar system. The distance test error results obtained are shown in Table 4.

Table 4. Test Results of Distance Error of Static Target

group	Set Distance (m)	Distance error (m)		
		Design method	Method 1	Method 2
A	1.0	0.025	0.056	0.052
B	2.0	0.012	0.056	0.041
C	3.0	0.001	0.039	0.046
D	4.0	0.011	0.041	0.035
E	5.0	0.021	0.035	0.050

According to the range error test results of stationary targets in Table 4, when the target is stationary, the range error of the design method is less than that of Method 1 and Method 2, indicating that its signal processing effect is better and more suitable for civil radar applications.

3.4　Experimental Verification of Inching Target

In order to verify the tracking and extracting ability of the design method for the micro moving target, the micro moving target experiment is conducted on the experimental platform. In the experiment, first move the target to five different positions at 1 m, 2 m, 3 m, 4 m and 5 m, then control the track motor to make it move at a constant speed in the direction of the radar system at a speed of 0.5 cm/s. The distance test error results obtained are shown in Table 5.

Table 5. Test results of range error of inching target

group	Set Distance (m)	Distance error (m)		
		Design method	Method 1	Method 2
A	1.0	0.038	0.075	0.063
B	2.0	0.032	0.068	0.078
C	3.0	0.028	0.068	0.082
D	4.0	0.030	0.063	0.064
E	5.0	0.039	0.071	0.063

According to the distance error test results of micro moving targets in Table 5, when the target is a micro moving target, the distance error of the design method is still significantly smaller than that of Method 1 and Method 2, indicating that its signal processing effect is better than that of Method 1 and Method 2.

3.5 Experimental Verification of Moving Targets

In order to verify the tracking and extraction ability of the design method for moving objects, the experimental platform is used for moving object experimental testing. In the experiment, first move the target to five different positions at 1 m, 2 m, 3 m, 4 m and 5 m, and then control the track motor to make it move at a constant speed of 30 m/min towards the radar system. The distance test error results obtained are shown in Table 6.

Table 6. Distance Error Test Results of Moving Target

group	Set Distance (m)	Distance error (m)		
		Design method	Method 1	Method 2
A	1.0	0.068	0.098	0.086
B	2.0	0.063	0.092	0.094
C	3.0	0.069	0.090	0.096
D	4.0	0.064	0.096	0.092
E	5.0	0.067	0.092	0.088

According to the distance error test results of moving targets in Table 6, when the target is a moving target, the distance error of the design method is higher than the distance error between the stationary target and the inching target, but still less than the distance error of Method 1 and Method 2, indicating that its signal processing effect is better than that of Method 1 and Method 2. However, the difference between the distance error of the design method and the distance error of Method 1 and Method 2 is significantly smaller.

4 Conclusion

In order to reduce the tracking distance error in the processing of civil radar echo signal, this paper proposes a civil radar echo signal processing method based on Kalman filter algorithm. The AR model is converted into the Equation of state and observation equation form of Kalman filter. Based on the Faster R-CNN detection framework, a target detection method based on context information and multi-scale Faster R-CNN is designed, and the multi-path clutter is suppressed based on the frequency domain subcarrier multi-path clutter cancellation method to achieve the processing of civil radar echo signals. And through experimental verification, the tracking distance error of this method is small, and the signal processing effect is good. In the following research, the focus will be on the transmission process of civilian radar echo signals to improve the practical application effect of civilian radar.

References

1. Zhang, R., Huang, C., Li, Z., et al.: Maritime moving target detection technique via space-time integration of passive radar with geosynchronous illuminator. J. Sig. Process. **38**(7), 1405–1415 (2022)
2. Zhao, T., Zheng, Y., Yang, L., et al.: A translation compensation method for ISAR based on second-order WVD. J. Sig. Process. **38**(4), 870–878 (2022)
3. Buzzi, S., Grossi, E., Lops, M., Venturino, L.: Radar target detection aided by reconfigurable intelligent surfaces. IEEE Sig. Process. Lett. **28**, 1315–1319 (2021)
4. Delgado, A.V., Sanchez-Fernandez, M., Venturino, L., et al.: Super-resolution in automotive pulse radars. IEEE J. Sel. Top. Sig. Process. **15**, 913–926 (2021)
5. Huang, P., Xia, X.-G., Zhan, M., Liu, X., Liao, G., Jiang, X.: ISAR imaging of a maneuvering target based on parameter estimation of multicomponent cubic phase signals. IEEE Trans. Geosci. Remote Sens. **60**, 1–18 (2022). https://doi.org/10.1109/TGRS.2021.3091645
6. Peters, S.T., Schroeder, D.M., Haynes, M.S., et al.: Passive synthetic aperture radar imaging using radio-astronomical sources. IEEE Trans. Geosci. Remote Sens. **59**, 9144–9159 (2021)
7. Fedorov, R., Berngardt, O.: Monitoring observations of meteor echo at the EKB ISTP SB RAS radar: algorithms, validation, statistics. Solar-Terrestrial Phys. **7**(1), 47–58 (2021)
8. Ji, X.X., Sun, Y.: SAR Image target recognition based on monogenic signal and sparse representation. Wirel. Commun. Mob. Comput. **2021**(2), 1–11 (2021)
9. Chen, S., Pan, M.: Analytical model and real-time calculation of target echo signals on wideband LFM radar. IEEE Sens. J. **21**, 10726–10734 (2021)
10. Cai, J., Wang, R., Hu, C.: Maneuvering targets radar detection based on minimum entropy. J. Sig. Process. **38**(7), 1416–1423 (2022)
11. Zhang, R., Meng, C., Wang, C., et al.: Compressed sensing reconstruction of radar echo signal based on fractional Fourier transform and improved fast iterative shrinkage-thresholding algorithm. Wirel. Commun. Mob. Comput. **2**, 1–15 (2021)
12. Zhang, T., Ren, J., Li, J., Nguyen, L.H., Stoica, P.: Joint RFI mitigation and radar echo recovery for one-bit UWB radar. Sig. Process. **193**, 108409 (2022)
13. Rommel, T., Huber, S., Younis, M., et al.: Matrix pencil method for topography-adaptive digital beam-forming in synthetic aperture radar. IET Radar Sonar Navig. **10**, 15 (2021)

Frequency Offset Estimation of X-band Marine Radar Sampling Signal Based on Phase Difference

Jianming Wang[⊠]

Maritime College, Tianjin University of Technology, Tianjin 300384, China
wangjianming@email.tjut.edu.cn

Abstract. Due to the relative radial movement between the transmitter and receiver of marine radar, the frequency of radar sampling signal is prone to deviation, which reduces the quality of radar sampling signal. In order to ensure the effective transmission of radar signals, a frequency offset estimation method of marine radar sampling signals in X-band based on phase difference is proposed. The AD9225 chip is selected to acquire the marine radar signal, and the undersampling theorem is used to determine the marine radar signal sampling frequency, so as to prevent the radar signal from mixing. After digital down conversion processing, two baseband signals are obtained, and the phase information of the radar sampling signal is extracted. Based on the Midamble code, the frequency offset estimation program of marine radar sampling signal is designed. The frequency offset estimation result of the signal can be obtained by executing the established procedure, and the frequency offset estimation of the sampling signal of the X-band marine radar can be realized. Experimental data show that after the application of the proposed method, the minimum signal to noise ratio of radar sampling signal is 4 dB, the minimum mean square error of frequency offset estimation is 4%, and the minimum time of frequency offset estimation is 2 s, which fully confirms that the proposed method has better application performance.

Keywords: X-Band · Sampling Signal · Frequency Offset Estimation · Marine Radar · Phase Difference · Radar Signal Processing

1 Introduction

Navigation radar is an indispensable navigation instrument [1] on the bridge of a ship, which plays a vital role in navigation. Marine radar is a device that uses electromagnetic wave technology to detect targets, so it is not affected by the weather environment, and can detect small and weak targets in a long distance. Navigation radar can accurately detect the distance and direction of target objects, and can achieve target tracking, predict the location and time of encounter with the ship, so as to guide the ship navigation and avoid collision. In the process of marine radar communication, the relative radial movement between the transmitter and receiver will change the frequency of the received

L. Yun et al. (Eds.): ADHIP 2023, LNICST 548, pp. 83–99, 2024.
https://doi.org/10.1007/978-3-031-50546-1_6

signal. This phenomenon is called Doppler effect, and the change of this frequency is called Doppler frequency shift. In addition, the accuracy of the local oscillator and the receiver oscillator will also cause a deviation in the frequency of the received signal, which is called frequency offset, thus affecting the effective transmission of data [2]. However, many factors such as temperature changes, equipment aging, and electromagnetic interference in practical applications may affect the sampling signal of marine radar, resulting in frequency deviation. The frequency offset issue may have a negative impact on the performance of marine radar. For example, frequency offset may lead to errors in distance measurement, inaccurate speed measurement of ships, and even inaccurate target positioning. Therefore, it is very important to accurately estimate and compensate for the frequency offset of marine radar sampling signals.

At present, many scholars have done a lot of research on the estimation algorithm of frequency offset, and also have a lot of research results and a large number of algorithms that can be implemented in practice, and these practical and feasible algorithms and research results will provide important reference and guarantee for improving the performance of frequency offset estimation in the future. Among the frequency offset estimation methods, there are mainly two common methods. The first is the frequency offset estimation method based on the least squares principle. This method is mainly aimed at the frequency offset estimation problem under the medium and high SNR channel conditions. Based on the transformation of additive complex Gaussian noise, the frequency offset estimation process is completed based on the least squares principle [3]. However, this algorithm requires that the phase used for frequency offset estimation cannot change during the whole estimation process; The second is the frequency offset estimation method based on differential phase [4], which uses continuous data frames to reduce the mean square error of the frequency offset estimation value to the GCRLB bound, which is a lower bound than the CRLB of traditional estimation algorithms. Although the above two methods have their own advantages, they also have certain defects, which cannot meet the requirements of marine radar sampling signal frequency offset estimation and restrict the development and application of marine radar. In addition, there is also a blind and simple estimation technique for carrier frequency offset (CFO) of a universal filtered multi carrier (UFMC) system under specified carrier allocation proposed in reference [5]. Reference [6] proposes carrier frequency offset estimation and compensation for MBOFDM ultra wideband systems based on ICA. The Doppler frequency offset estimation for 5G-NR downlink based on deep learning proposed in reference [7]. The autocorrelation method proposed in reference [8, 9], the Maximum likelihood estimation method proposed in reference [10–12], and the Fourier transform method proposed in reference [13–16]. Therefore, a method of marine radar sampling signal frequency offset estimation based on phase difference is proposed. On the basis of obtaining the navigation radar signal, select an appropriate sampling frequency, extract the phase information of the sampled signal, and then perform frequency offset estimation by comparing the phase difference between the reference signal and the sampled signal. By accurately estimating the frequency offset of marine radar sampling signals through phase difference calculation, the performance and accuracy of the radar system can be improved.

2 Frequency Offset Estimation of Marine Radar Sampling Signal

2.1 Marine Radar Signal Acquisition

According to the characteristics of marine radar signal, AD9225 is selected as the marine radar signal acquisition equipment to illustrate the marine radar signal acquisition process. The AD9225 acquisition chip mainly has the following technical performance indicators:

(1) Resolution

AD9225 resolution refers to the fineness of chip scale description, which is usually determined by the number of bits used by AD9225 when outputting digital signals. For example, AD9225 is a 12 bit acquisition chip, and the total voltage range is described by 12 independent binary codes. The absolute voltage unit recognized by the system is 1bit [17, 18]. The resolution of AD conversion chip is defined as the ratio of the minimum input analog voltage corresponding to the binary minimum variable to the internal reference voltage of AD, and the expression is

$$2^N = \frac{\alpha}{\beta} \tag{1}$$

In formula (1), N represents the resolution bits of AD9225 chip; α indicates the least significant bit; β indicates the maximum voltage that the AD9225 chip can recognize the analog voltage, and it is also the reference voltage of the AD9225 chip.

(2) Integral Nonlinearity

The integral nonlinearity is defined as the relative deviation between the ideal input conversion level and the actual input conversion level, representing the maximum error between the values collected by AD9225 chip at all sampling points and the true values, in LSB. In order to ensure that AD9225 does not lose code, it is generally specified that the maximum and minimum values of integral nonlinearity are 0.5 LCB and -0.5LSB at 25 °C. The integration nonlinearity is caused by the nonlinearity of ADC's analog front-end, sample holder and ADC's transfer function. The integration nonlinearity is the precision index of AD9225 chip. The smaller the integration nonlinearity, the higher the precision. The calculation formula of integral nonlinearity is

$$\gamma_t = \frac{\hat{Q}_t - \tilde{Q}_t}{\varepsilon_o} \tag{2}$$

In formula (2), γ_t represents the integration nonlinearity of AD9225 chip; $\hat{Q}_t$ and $\tilde{Q}_t$ represents the collected value and the real value at the sampling point of AD9225 chip; ε_o represents the integral nonlinearity calculation error term.

(3) Differential Nonlinearity

Differential nonlinearity is defined as the relative deviation between the actual converted code width and the ideal code width, which is the maximum error of analog quantity between adjacent quantization units of AD9225 chip, and the unit is LSB. In order to ensure that AD9225 does not lose code [19], it is generally specified that the maximum and minimum values of differential nonlinearity are 0.5 LCB and −0.5 LSB at 25 °C. Due to the deviation of circuit structure and manufacturing process of AD9225 chip itself, the quantization voltage at some points in the range is greater or less than the standard quantization voltage. The resulting error is called differential nonlinearity, and the calculation formula is

$$\gamma_{t-\max} = \frac{\max(\hat{Q}_t - \tilde{Q}_t)}{\varepsilon_{o-\max}} \tag{3}$$

In formula (3), $\gamma_{t-\max}$ represents differential nonlinearity; $\max(\hat{Q}_t - \tilde{Q}_t)$ means the maximum value of $(\hat{Q}_t - \tilde{Q}_t)$; $\varepsilon_{o-\max}$ represents the maximum value of integral nonlinearity calculation error term.

(4) Conversion Rate

The time required for AD9225 chip to convert analog signal to digital signal is called conversion time, and its reciprocal is the conversion rate unit, usually SPS, that is, the number of samples per second. For example, the sampling rate of AD9225 is 25 Msps, that is, the number of samples per second can reach 2.5×10^7. The sampling rate is one of the important technical indicators of the ultra-high speed AD acquisition chip.

(5) Signal-to-Noise Ratio

The signal-to-noise ratio is the dB number of the ratio of the effective value of the signal level to the effective value of various noises (including quantization noise, thermal noise, self noise, etc.) [20]. The signal-to-noise ratio depends on the quantization bits. The larger the number of bits, the smaller the quantization noise. Due to the noise and distortion of the actual AD9225, the actual resolution of the AD9225 is affected and the bits of the AD9225 are reduced. The signal to noise ratio calculation formula is

$$\zeta = \frac{q_1}{q_2} \times 100\% \tag{4}$$

In formula (4), ζ represents the signal to noise ratio; q_1 indicates the number of effective signals in the collected signal; q_2 indicates the number of noise signals in the collected signal.

(6) Noise Power Ratio

The noise power ratio refers to the power ratio of signal to noise. The signal noise ratio is extended to measure the transmission characteristics of frequency division multiplexing communication systems. The noise power ratio is generally represented by the noise power ratio curve [21]. When the input noise level is very low, the noise in the stopband is mainly quantization noise. When the input noise increases, the noise power ratio also increases linearly. The higher the resolution of the AD9225, the smaller the quantization noise and the higher the noise power ratio.

AD9225 chip is a high-performance analog signal acquisition chip with the acquisition rate of 25 MSPS and the resolution of 12 bits. It has built-in on-chip high performance reference voltage source and sample and hold amplifier. It adopts a multi-stage differential pipeline architecture with built-in output error correction logic. It can provide 12 bit accuracy at 25 MSPS sampling rate and ensure no code loss in the whole operating temperature range. The timing of AD9225 is shown in Fig. 1.

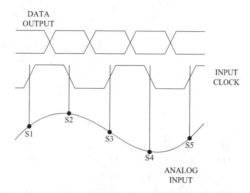

Fig. 1. Timing Diagram of AD9225

As shown in Fig. 1, the sequence diagram is one of the important information to help understand the chip. According to the AD9225 data manual, the sampling data needs to be delayed for 3 clock cycles to complete conversion and output. When the rising edge of the first clock cycle occurs, the sampling and holding circuit inside the AD chip collects the analog voltage at point S1. After voltage comparison and data coding, the analog voltage at point S1 at the rising edge of the fourth clock cycle is converted into a digital signal, and then t_{OD} the data of S1 point sampled at the rising edge of the first clock cycle is DATA1 output [22].

The AD9225 chip selected above is used to collect and acquire marine radar signals, which lays a solid foundation for the subsequent selection of marine radar signal sampling frequency.

2.2 Selection of Marine Radar Signal Sampling Frequency

Since the research target is the marine radar signal in the X-band, the undersampling theorem is used to determine the marine radar signal sampling frequency to prevent the occurrence of radar signal aliasing and facilitate the subsequent frequency offset estimation.

Under sampling theorem [23]: suppose there is a band limited to (f_l, f_h). Analog signal of $x(t)$, its center frequency is f_o, when the sampling frequency meets:

$$f_s = \frac{4f_o}{(2n+1)} \tag{5}$$

In formula (5), f_s represents the sampling frequency; n represents the total number of analog signals, meeting $f_s \geq 2(f_h - f_l)$. It should be noted that $f_o = \frac{f_l + f_h}{2}$.

The sampled radar signal spectrum will periodically extend, and $\frac{f_s}{2}$. The frequency spectrum of the signal at will be symmetrically folded. To avoid overlapping of the lower sidebands, the sampling frequency f_s. It shall meet the following requirements:

$$\frac{2f_h}{n+1} \leq f_s \leq \frac{2f_l}{n} \tag{6}$$

According to the characteristics of the periodic extension of the sampled signal spectrum, select an appropriate filter to filter out the low-frequency components of interest in this paper, thus completing the spectrum shift, that is, the first down conversion. Of course, if the above conditions are met, the premise of no spectrum aliasing after sampling is that the spectrum of the original signal is only limited to a certain frequency band and does not exist in other frequency bands, otherwise, aliasing will occur in the process of spectrum shifting after sampling. For this reason, the actual X-band signal from the radar receiver needs to be passband filtered before A/D sampling to limit the X-band signal to a certain frequency band, and then X-band sampling is carried out according to the above sampling theorem to prevent aliasing.

Based on the above factors and the parameters of the radar receiver studied in this project, according to the conditions that the undersampling theorem should meet, after calculation and comparison, it can be concluded that when $n = 1$ h, sampling rate (f_s) 80 MHz sampling will not cause spectrum aliasing.

The above process completes the determination of marine radar signal sampling frequency, and provides support for the subsequent extraction of marine radar sampling signal phase information.

2.3 Extraction of Phase Information of Marine Radar Sampling Signal

Obtaining marine radar sampling signal under 80 MHz frequency sampling operation $y(t)$, obtained after digital down conversion $I(t)$ and $Q(t)$ for two baseband signals, in order to process clutter and extract targets from radar signals, this paper needs to obtain the characteristic information of radar echoes. Analyzing the amplitude and phase information of the noncoherent radar signal can directly extract more target information from the signal, including the moving speed and target characteristics of the target and other information. In order to obtain the amplitude and phase of the echo signal, it is necessary to extract the envelope and instantaneous phase of the marine radar sampling signal. The envelope and instantaneous phase of the signal can be given by the following formula:

$$\begin{cases} A(t) = \sqrt{I^2(t) + Q^2(t)} \\ \phi(t) = arctg\frac{Q(t)}{I(t)} \end{cases} \tag{7}$$

In formula (7), $A(t)$ represents the envelope of marine radar sampling signal; $\phi(t)$ represents the instantaneous phase of marine radar sampling signal.

The extraction of radar signal amplitude and phase features discussed in this section uses CORDIC (coordinate rotation digital calculation) algorithm, which is implemented in FPGA using CORDIC IP core. Relevant parameters are set through graphical user interface settings (GUI). The functions selected in this paper are Translate function, structure selection parallel mode, pipeline mode selection maximum mode, output signal selection amplitude and phase output. The calculated value is not truncated, and the input and output are 16 bits, with cache. After the setting is completed, the top-level module - marine radar sampling signal feature extraction module is generated, and its pin definitions are shown in Table 1.

Table 1. Definition of pins of marine radar sampling signal feature extraction module

Pin name	Meaning description
I_in	$I(t)$ Baseband signal input
Q_in	$Q(t)$ Baseband signal input
clk	Baseband signal detection module working clock
phase_out	Calculated phase output, 16 bit signed number output
Z_out	Calculated instantaneous modulus, 16 bit unsigned digital output
rdy	The output data is valid, and the lower module is notified to receive the data

According to the principle of CORDIC algorithm [5, 24], the data input to CORDIC core needs to be converted into data format first. Similarly, the amplitude and phase data output after CORDIC core calculation must also be converted into corresponding data format before entering other subsequent modules for processing.

Through the above process, the phase information of marine radar sampling signal is extracted, which provides a basis for the realization of frequency offset estimation of marine radar sampling signal.

3 Frequency Offset Estimation of Marine Radar Sampling Signal

The phase information of marine radar sampling signal extracted from the above $\phi(t)$ based on the Midamble code, the frequency offset estimation [25] program of marine radar sampling signals is designed. The signal frequency offset estimation results can be obtained by executing the developed program, which provides some help and support for the acquisition and processing of marine radar sampling signals. Since the user data in the data domain of the transmission information will carry the information of frequency offset as the Midamble code sequence, this paper proposes a block frequency offset estimation algorithm based on the Midamble code in the data domain to improve the accuracy of frequency offset estimation. This algorithm iteratively compensates and estimates each sub block of the data field block by block in the same way until the data compensation of the last sub block is completed. Since the frequency offset values generated by adjacent sub blocks are relatively similar, the accuracy of frequency offset estimation can be improved by reducing the estimation error block by block.

Assume that the navigation radar system has achieved ideal timing synchronization, and here the Midamble code of a subframe received by the receiver after passing through the transmission channel is expressed by formula (8):

$$r_m = \left[\sum_{q=1}^{Q} \left(\chi_{mq}(t) \otimes \delta_{mq}(t) \right) \right] \cdot e^{j2\pi f_D t} + \varepsilon_m(t) \tag{8}$$

In formula (8), r_m represents the Midamble code of a subframe of marine radar sampling signal; $\chi_{mq}(t)$ indicates that the basic Midamble code is rotated in the default way to get the q Midamble code of code track; $\delta_{mq}(t)$ refers to the No q The channel impulse response experienced by the code channel signal, and the window length of the channel impulse response is 16 chips; f_D represents the frequency offset generated by the marine radar sampling signal; $\varepsilon_m(t)$ represents additive complex Gaussian noise [26, 27] with zero mean.

For the signal of a single path, assume that the optimal sampling is achieved in formula (5), that is:

$$g(t) = g_o(t) \otimes g_o(t) = \begin{cases} 0 & t = \kappa T_C \\ 1 & t = 0 \end{cases} \tag{9}$$

In formula (9), $g(t)$ represents the best sampling result of marine radar signal; $g_o(t)$ represents the waveform of shaping filter RRC; κ Represents a random integer; T_C represents the signal period of marine radar.

In order to obtain the signal used for frequency offset estimation, it is necessary to $g(t)$ to signal r_m. The impact of will be eliminated $g(t)$ and r_m divide, then you can get;

$$Z_m = \frac{r_m}{g(t)} = e^{j(2\pi f_D T_C + \theta_m)} + \varepsilon_m(t) \tag{10}$$

In formula (10), Z_m represents a signal used for frequency offset estimation; θ_m represents the initial phase of the channel of the Midamble code.

Based on the output result of formula (10), calculate the corresponding average phase difference value. The expression is

$$\Delta\phi_m = \frac{1}{L_m - 1} \sum_{k=0}^{L_m - 2} \arctan\left(\frac{imag\left(Z_m(k+1)Z_m(k)\right)}{real\left(Z_m(k+1)Z_m(k)\right)} \right) \tag{11}$$

In formula (11), $\Delta\phi_m$ means average phase difference value of Z_m; $L_m - 1$ means the maximum value of k; $imag(Z_m(k+1)Z_m(k))$ represents the predicted marine radar sampling signal; $real(Z_m(k+1)Z_m(k))$ represents the actual marine radar sampling signal.

According to the calculation result of formula (11), the initial frequency offset estimation result of marine radar sampling signal is

$$f_D = \frac{1}{2\pi T_C} * \Delta\phi_m \tag{12}$$

Data field G Equally divided into m Block, and each sub block is $G_1, G_2, \cdots, G_m$, the estimated value of the initial frequency offset f_D Data compensated to the first sub block

G_1 obtain G_1', because the phase difference between two adjacent symbols in the data domain is $\Delta\phi_{G_1}$ Therefore, the average phase difference between adjacent symbols in the first sub block is f_{D1}. Compensate this frequency offset estimation value to the data of the second sub block G_2 So as to get G_2', and then use the average phase difference method to estimate the frequency offset. Similarly, the average phase difference of the second sub block is $\Delta\phi_{G_2}$, then it can also be obtained that the frequency offset estimation value of the second sub block is f_{D2} [28, 29], and then perform compensation and estimation for each sub block in turn until the m Subblock. G_m is estimated to end when the data compensation of is completed.

Accumulate and average the average phase difference of each sub block mentioned above, and record it as the final frequency offset estimation result of marine radar sampling signal. The expression is

$$\hat{f}_D = \frac{f_{D1} + f_{D2} + \cdots + f_{Dm}}{\tau^o \times \upsilon^a \times m} \tag{13}$$

In formula (13), $\hat{f}_D$ represents the final result of frequency offset estimation of marine radar sampling signal; τ^o and υ^a represents the auxiliary calculation parameter, which determines the accuracy of frequency offset estimation. The value range is [0, 1].

The above process completes the frequency offset estimation of marine radar sampling signals, which provides convenience for the subsequent processing and application of radar signals.

4   Experiment and Result Analysis

The dataset used in the experiment is the MSTAR dataset. MSTAR (Moving and Stationary Target Acquisition and Recognition) is a dataset widely used in radar target detection and recognition research. It contains various types of radar echo data, covering various targets and background conditions. The target types in the dataset include vehicles, buildings, tanks, etc. The MSTAR dataset is commonly used for research on radar target generation, detection, classification, recognition, and other related tasks. Based on this part of the data, the specific preparation and experimental results are shown below.

4.1   Experiment Preparation Stage

The preparation stage is the basis and premise for the smooth experiment. In order to ensure the smooth progress of the experiment, determine the auxiliary calculation parameters τ^o and υ^a. The best value of, set up a variety of experimental conditions, do a good job of experimental preparation. The credibility of the experimental conclusions obtained from a single experimental condition is low. Therefore, 10 experimental conditions are set in this study to maximize the credibility of the experimental conclusions, as shown in Table 2.

As shown in Table 2, the radar sampling signal volume and signal to noise ratio values are inconsistent among the 10 experimental conditions set, indicating that the background environment corresponding to each experimental condition is quite different, which meets the application performance test requirements of the proposed method. It

Table 2. Setting table of test conditions

Test conditions	Radar sampling signal volume/MB	Signal to noise ratio/dB
1	1252	45
2	1024	23
3	985	35
4	1563	30
5	1478	28
6	1952	29
7	1800	32
8	1475	31
9	1632	34
10	1452	37

should be noted that the prepared experimental data has a high signal-to-noise ratio and needs to be processed, which is also the key to verify the application effect of the proposed method.

Auxiliary calculation parameters τ^o and υ^a is closely related to the frequency offset estimation accuracy of the proposed method, so it is necessary to calculate the auxiliary parameters before the experiment τ^o and υ^a determine the best value.

Obtain auxiliary calculation parameters through test τ^o and υ^a. The relationship between and frequency offset estimation accuracy is shown in Fig. 2.

As shown in Fig. 2, when auxiliary calculation parameters τ^o When the value is 0.4, the accuracy of frequency offset estimation reaches 90% of the maximum value. Therefore, determine the auxiliary calculation parameters τ^o the optimal value is 0.4; When auxiliary calculation parameters υ^a. When the value is 0.5, the accuracy of frequency offset estimation reaches the maximum of 96%. Therefore, determine the auxiliary calculation parameters υ^a. The best value is 0.5.

4.2 Analysis of Experimental Results

In order to comprehensively measure the application performance of the frequency offset estimation method of the X-band marine radar sampling signal, the signal to noise ratio of the radar sampling signal, the mean square error of frequency offset estimation and the time of frequency offset estimation are selected as evaluation indicators. Set the frequency offset estimation method based on the least squares principle in reference [3] and the frequency offset estimation method based on differential phase in reference [4] as comparison methods 1 and 2, and design the frequency offset estimation comparison experiment of the X-band marine radar sampling signal. The specific analysis process of the experimental results is as follows:

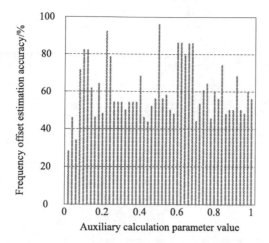

(1) τ^o Relation with frequency offset estimation accuracy

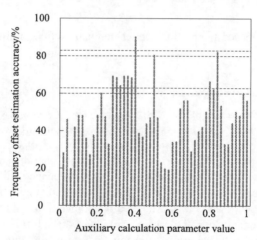

(2) υ^a Relation with frequency offset estimation accuracy

Fig. 2. Schematic diagram of the relationship between auxiliary calculation parameters and frequency offset estimation accuracy

4.2.1 Signal to Noise Ratio Analysis of Radar Sampling Signal

The SNR values of radar sampling signals obtained through experiments are shown in Table 3.

As shown in the data in Table 3, after the proposed method, comparison method 1 and comparison method 2 are applied, the signal-to-noise ratio is significantly reduced compared with the prepared experimental data. Among them, the signal-to-noise ratio of the radar sampling signal obtained by the proposed method is generally lower than that of the comparison method 1 and 2, and the minimum value reaches 4 dB, which

Table 3. Signal to noise ratio of radar sampling signal/dB

Test conditions	Propose method	Comparison method 1	Comparison method 2
1	8	15	16
2	6	13	12
3	5	10	10
4	4	11	14
5	10	14	17
6	8	17	19
7	8	10	12
8	7	10	11
9	7	11	12
10	8	14	15

indicates that the proposed method has better denoising effect on the radar sampling signal.

4.2.2 Mean Square Error Analysis of Frequency Offset Estimation

Frequency offset is Frequency modulated wave Frequency swing. Generally speaking, the maximum frequency offset affects the frequency spectrum bandwidth of the FM wave. But it does not mean that the larger the maximum frequency offset is, the wider the spectrum bandwidth will be Modulation index Problems. Generally speaking, the larger the modulation index, the wider the bandwidth of the frequency shift wave spectrum. The maximum frequency offset is a decisive factor of the modulation index, so it affects the spectrum bandwidth of the FM wave. As a phenomenon in frequency modulation, frequency offset is conducive to the transmission of audio information. Under normal frequency offset, the transmission of audio information is guaranteed and is required by audio information in the specified value. The mean square error of frequency offset estimation is easily affected by the X-band frequency. The X-band is mainly divided into the downlink band and the uplink band. The downlink band is 7.25–7.75 GHz, and the uplink band is 7.9–8.4 GHz.

The mean square error of frequency offset estimation in the downlink and uplink bands is obtained through experiments, as shown in Fig. 3.

As shown in the data in Fig. 3, after the proposed method is applied, under the background of downlink and uplink frequency bands, the mean square error of frequency offset estimation obtained is generally lower than that of comparison methods 1 and 2, with the minimum values of 4% and 2% respectively, indicating that the proposed method has better accuracy of frequency offset estimation of radar sampling signals.

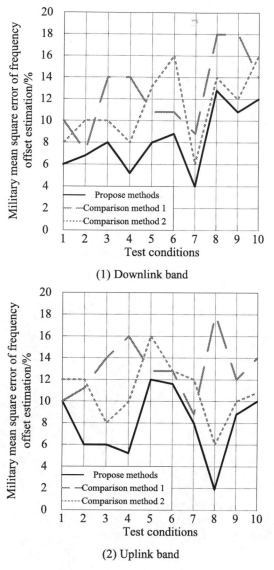

Fig. 3. Schematic diagram of mean square error of frequency offset estimation

4.2.3 Frequency Offset Estimation Time Analysis

Consistent with the mean square error of frequency offset estimation, the frequency offset estimation time will also be affected by the X-band frequency. The X band is divided into the downlink band and the uplink band. The downlink band is 7.25–7.75 GHz, and the uplink band is 7.9–8.4 GHz.

The frequency offset estimation time obtained through experiments is shown in Fig. 4.

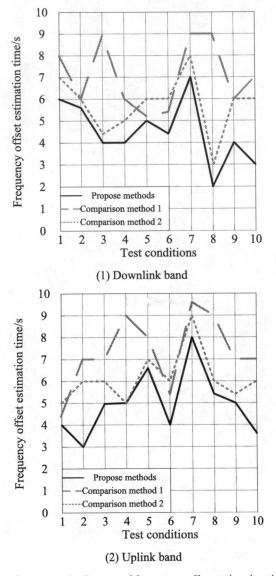

(1) Downlink band

(2) Uplink band

Fig. 4. Schematic diagram of frequency offset estimation time

As shown in the data in Fig. 4, after the proposed method is applied, under the background of downlink and uplink frequency bands, the estimated time of frequency offset obtained is generally lower than that of comparison methods 1 and 2, with the minimum values of 2 s and 3 s respectively, indicating that the proposed method is more efficient in estimating the frequency offset of radar sampling signals.

Set the signal sampling rate to 10 kHz, the signal frequency range to 2 GHz–4 GHz, the frequency offset size to 100 Hz, and the noise level to 20 dB. Introducing a frequency offset of 100 Hz, three methods were used to test the cumulative phase offset at each sampling point, and the results are shown in Fig. 5.

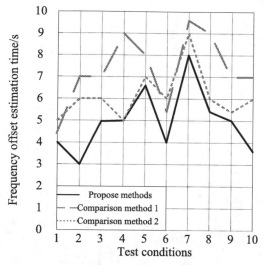

Fig. 5. Schematic diagram of cumulative phase offset

As shown in Fig. 5, the cumulative phase shift of the proposed method is less than 10°, which is lower compared to the two comparison methods. This verifies that the proposed method for estimating the frequency offset of X-band marine radar sampling signals based on phase difference has good estimation performance.

5 Conclusion

With the rise of radar equipment and the expansion of its application range, radar signal processing requirements are getting higher and higher, especially the accuracy of frequency offset estimation. Therefore, the research on frequency offset estimation method of X-band marine radar sampling signal based on phase difference is proposed. The proposed method can reduce the signal to noise ratio of radar sampling signal, reduce the mean square error of frequency offset estimation, and shorten the time of frequency offset estimation, which provides a reference for the frequency offset estimation of marine radar sampling signal.

References

1. Pehlivan, M., Yegin, K.. X-band low-probability intercept marine radar antenna design with improved bandwidth and high isolation. IEEE Trans. Antennas Propag. **69**(12), 8949–8954 (2021)
2. Yang, F., Ma, M.: Data transmission delay compensation algorithm for wireless multi-hop communication network. Comput. Simul. **39**(4), 146–149, 253 (2022)

3. Ali, A., Magarini, M., Pirzada, N., et al.: Direction of arrival and least square error technique used in massive MIMO for channel estimation. Int. J. Math. Comput. Sci. **16**(2), 647–657 (2021)

4. Tajbakhsh, K., Ebrahimi, S., Dashtdar, M.: Low-coherence quantitative differential phase-contrast microscopy using Talbot interferometry. Appl. Opt. **61**(2), 398–402 (2022)

5. Lin, T.T., Hwang, F.H.: Design of a blind estimation technique of carrier frequency offset for a universal-filtered multi-carrier system over Rayleigh fading. IEEE Wirel. Commun. Lett. **11**(5), 1027–1031 (2022)

6. Rajanandhini, C., Babu, S.: ICA based estimation and compensation of carrier frequency offset for MBOFDM based UWB systems. Microprocess. Microsyst. **82**(9), 103824 (2021)

7. Yang, L.H., Zenghao, W., Jie, Z., et al.: Deep learning based Doppler frequency offset estimation for 5G-NR downlink in HSR scenario. High Technol. Lett. **28**(2), 115–121 (2022)

8. Liang, Z., Li, Y., Niu, J., et al.: Fast radar detection method based on two-dimensional trilinear autocorrelation function for maneuvering target with jerk motion. J. Appl. Remote Sens. **15**(2), 026508-1–026508-15 (2021)

9. Ikeda, T., Tsuji, T., Konishi, C., et al.: Spatial autocorrelation method for reliable measurements of two-station dispersion curves in heterogeneous ambient noise wavefields. Geophys. J. Int. **226**(2), 1130–1137 (2021)

10. Peng, J., Rajeevan, H., Kubatko, L., et al.: A fast likelihood approach for estimation of large phylogenies from continuous trait data. Mol. Phylogenet. Evol. **161**(4), 107142 (2021)

11. Candan, Ç., Çelebi, U.: Frequency estimation of a single real-valued sinusoid: an invariant function approach. Sig. Process. **185**, 108098 (2021)

12. Zhang, J.F., Wang, F.Y., Ye, S.U., et al.: Research on power grid primary frequency control ability parallel computing based on multi-source data. Acta Automatica Sinica **48**(6), 1493–1503 (2022)

13. Li, R., Xuan, J., Shi, T.: Frequency estimation based on symmetric discrete Fourier transform. Mech. Syst. Sig. Process. **160**, 107911 (2021)

14. Liang, Y., Zhang, Z., Li, H., et al.: A robust and accurate discrete Fourier transform-based phasor estimation method for frequency domain fault location of power transmission lines. IET Gener. Transm. Distrib. **16**(10), 1990–2002 (2022)

15. Chen, G., et al.: Real-time reception of NHS-OFDM signal with SPA-enhanced channel estimation for intensity-modulated direct-detection systems. Chin. Opt. Lett. **20**(5), 050601 (2022)

16. Chui, C.K., Jiang, Q., Li, L., Jian, Lu.: Analysis of an adaptive short-time Fourier transform-based multicomponent signal separation method derived from linear chirp local approximation. J. Comput. Appl. Math. **396**, 113607 (2021)

17. Tse, J.R., Shen, L., Shen, J., et al.: Nyquist sampling theorem and Bosniak classification, version 2019: effect of thin axial sections on categorization and agreement. Eur. Radiol. **32**(12), 8256–8265 (2022)

18. Qi, B., Zhang, H., Zhang, X.: Time-frequency DOA estimation of chirp signals based on multi-subarray. Digit. Sig. Process. **113**, 103031 (2021)

19. Qi, Hu., Keating, S., Innanen, K.A., Chen, H.: Direct updating of rock-physics properties using elastic full-waveform inversion. Geophysics **86**(3), MR117–MR132 (2021)

20. Changela, A., Zaveri, M., Verma, D.: Mixed-radix, virtually scaling-free CORDIC algorithm based rotator for DSP applications. Integration **78**(May), 70–83 (2021)

21. Kiefer, M., et al.: IMK/IAA MIPAS temperature retrieval version 8: nominal measurements. Atmos. Meas. Techn. **14**(6), 4111–4138 (2021)

22. Hadei, S.A., Abolfazl Hosseini, S., Miri, M.: Performance analysis of the low-complexity adaptive channel estimation algorithms over non-stationary multipath Rayleigh fading channels under carrier frequency offsets. IET Commun. **16**(9), 988–1004 (2022)

23. Wang, D., Zhang, S., Liu, J., Yuan, L., Gao, G.: Modified frequency-domain channel-estimation based on the dual-dependent pilots for polarization-division-multiplexed coherent optical orthogonal frequency division multiplexing systems with offset-quadrature-amplitude modulation. Opt. Eng. **61**(06), 066109-1–066109-14 (2022)

24. Nachaoui, M., Laghrib, A., Afraites, L., et al.: A non-convex non-smooth bi-level parameter learning for impulse and Gaussian noise mixture removing. Commun. Pure Appl. Anal. **21**(4), 1249–1291 (2022)

25. Jeong, S., Farhang, A., Perovic, N.S., Flanagan, M.F.: Low-complexity joint CFO and channel estimation for RIS-aided OFDM systems. IEEE Wirel. Commun. Lett. **11**(1), 203–207 (2022)

26. Wang, C., Zhang, X., Li, J.: FDA-MIMO radar for DOD, DOA, and range estimation: SA-MCFO framework and RDMD algorithm. Sig. Process. **188**, 108209 (2021)

27. Shi, Q., Liu, M., Hang, L.: A novel distribution system state estimator based on robust cubature particle filter used for non-gaussian noise and bad data scenarios. IET Gener. Transm. Distrib. **16**(7), 1385–1399 (2022)

28. Wang, Y.Y., Yang, S.J.: Estimation of carrier frequency offset and channel state information of generalize frequency division multiplexing systems by using a Zadoff-Chu sequence. J. Franklin Inst. **359**(1), 637–652 (2022)

29. Yaşar, C.F.: Algebraic estimator of Parkinson's tremor frequency from biased and noisy sinusoidal signals. Trans. Inst. Measure. Control **43**(3), 679–686 (2021)

Terrain Echo Signal Enhancement Technology of Marine Radar Based on Generalized Filtering

Jianming Wang[✉]

Maritime College, TianJin University of Technology, Tianjin 300384, China
wangjianming@email.tjut.edu.cn

Abstract. In order to solve the problem that the terrain echo signal of marine radar is affected by noise during transmission, which leads to poor enhancement effect, a terrain echo signal enhancement technology of marine radar based on generalized filtering is proposed. Design the graphic processing pipeline of programmable GPU, and draw the terrain echo image of marine radar. The generalized weighted median filter and Wiener filter are used to process high and low frequency signals to avoid some useful signals being filtered out. According to the high and low frequency signal processing results of the echo, the polynomial fitting sliding window is used to obtain the least square error fitting results to smooth the radar echo data. The echo signal enhancement structure is constructed, and the echo signal gain is processed to achieve the purpose of pseudo signal attenuation. Call the OpenGL read pixel function, and complete the echo signal enhancement processing according to the linear mapping relationship between the echo map and the coordinates of the DEM when processing in the GPU segment. From the experimental results, it can be seen that the echo gain effect of this technology is actually consistent, and there is only a maximum error of 1 dB between the echo signal strength and the actual data.

Keywords: Generalized Filtering · Marine Radar · Terrain Echo · Signal Enhancement

1 Introduction

Navigation radar is a commonly used equipment in ship navigation, which can detect the surrounding environment by transmitting and receiving radar waves, and obtain information about terrain, targets, and obstacles. Navigation safety is one of the most important issues in navigation. For ships, accurately identifying and analyzing terrain echo signals can help crew members predict and avoid potential collisions and dangerous situations, and improve navigation safety. However, in practical applications, the terrain echo signal of navigation radar may be affected by various factors, such as terrain complexity, changes in sea conditions, meteorological conditions, etc., resulting in a decrease in the quality of the echo signal and causing difficulties for ship navigation. In order to solve this problem, research on enhancing the terrain echo signal of maritime radar has received widespread attention.

© ICST Institute for Computer Sciences, Social Informatics and Telecommunications Engineering 2024
Published by Springer Nature Switzerland AG 2024. All Rights Reserved
L. Yun et al. (Eds.): ADHIP 2023, LNICST 548, pp. 100–115, 2024.
https://doi.org/10.1007/978-3-031-50546-1_7

In reference [1], an echo signal enhancement algorithm based on fuzzy logic is proposed. This algorithm constructs a membership function by extracting characteristic parameters, and sets a threshold according to this function to enhance the radar abnormal echo signal; Reference [2] proposed echo signal enhancement technology based on regional multi frame joint processing. Under the condition of strong scattering interference sources, the echo of LFMCW radar presents the phenomenon of echo signal spectrum spread outside the range. It brings serious technical challenges to radar weak target detection. Based on the correlation analysis of various components of the echo data of multiple adjacent radar beams in the center area of the detected target, through weighted compensation, the multi frame data from the scanning area near the target is used for multi frame joint processing to enhance the echo signal. Although the above two methods can enhance the radar echo signal, they are vulnerable to the influence of false signals in the signal, resulting in poor signal enhancement effect. In addition to the above methods, the following methods can also be used: (1) low-pass filters [3], bandpass filters [4], and notch filters to remove or weaken noise and interference in the echo signal. (2) Processing the echo signal in the time domain, such as smoothing, removing outliers or Outlier [5], can reduce the fluctuation and noise in the signal and make the signal clearer and more stable. (3) Processing the echo signal in the frequency domain, such as Fourier transform [6, 7], spectrum analysis, etc., can extract the frequency characteristics of the signal, further analyze and understand the signal. These methods can be selected and combined according to specific applications and needs to achieve enhancement and optimization of echo signals.

Generalized filtering can be adapted to various signal types and frequency ranges. Some signals may have complex frequency characteristics, which need to be comprehensively considered in time domain and frequency domain. Generalized filtering can capture these characteristics more accurately. For this reason, the terrain echo signal enhancement technology of marine radar based on generalized filtering is proposed. On the basis of drawing the terrain echo image of the navigation radar, the generalized filtering method is used to process the high and low frequency signals of the echo signal; Using the least squares error fitting method for data processing, and applying the terrain echo signal enhancement technology of marine radar, effective enhancement and optimization of the terrain echo signal of marine radar are achieved, improving navigation safety and ocean survey accuracy.

2 Plotting Terrain Echo Image of Marine Radar

The radar transmits an electromagnetic beam through the antenna to detect the direction and distance of the target and display it on the radar screen. The navigation radar antenna is a directional antenna. The transverse beam width of the transmitted electromagnetic wave [8] is about $2°$, and the longitudinal beam width is about $20°–30°$. Objects within the beam range reflect (backscatter) electromagnetic waves back to the radar antenna. The radar measures the target distance by measuring the round-trip time of the electromagnetic wave. The antenna pointing is the target orientation. When the antenna is in a certain direction, all echoes within the beam range are quantified as a sequence of echo intensity values arranged by time (distance), called scanning lines. The antenna

keeps rotating, and the antenna direction changes in a circular order to complete 360°
circumferential scanning. Before drawing the terrain echo image of marine radar, it is
necessary to design the graphic processing pipeline of programmable GPU, as shown in
Fig. 1.

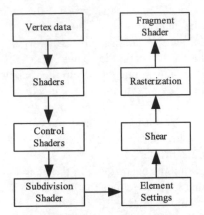

Fig. 1. Graphic processing pipeline of programmable GPU

The information such as vertex attributes and camera settings of the scene graph
is sent to the vertex shader. In the vertex shader, the world coordinates, normal vector,
model view transformation matrix and projection matrix of the vertex are read. The
radar detection distance is the distance from the object to the antenna. If the viewpoint
is set at the radar antenna, the radar detection distance is the distance from the vertex
to the origin under the view coordinate system. Then the radar detection range can be
calculated by formula (1):

$$l = \sqrt{x^2 + y^2 + z^2} \tag{1}$$

In formula (1), (x, y, z) Represents the apparent coordinate system. The echo intensity is related to the radar detection range and normal of the vertex, and is transmitted
to the subsequent process in the form of output variables. At the end of the vertex
shader, the vertex coordinates under the view coordinate system should be projected and
transformed, and the vertex coordinates after the projection transformation should be
assigned to the built-in variables of the vertex shader to ensure that the scene is processed
according to perspective projection. Next, the GPU [9] will perform primitive assembly,
cutting and rasterizing operations. Vertices are assembled into primitives and discretized
into several pieces. In the slice element shader, the radar detection distance needs to be
assigned to the built-in variable of the slice element shader. The reserved slice element
after the depth test becomes a pixel, and the radar detection distance written to the built-
in variable will eventually be written to the depth cache. In the slice shader, it is also
necessary to convert the echo intensity [10] into an RGB color that can represent the
echo intensity and write it into the built-in variable of the slice shader. Similarly, the
color value of the reserved slice after depth testing will eventually be written into the
color cache. The marine radar terrain echo image drawn from this is shown in Fig. 2.

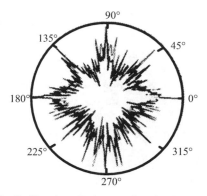

Fig. 2. Terrain echo image of navigation radar

The pixel color value is written to the scanline, which generates a scanline. Several scan lines can be generated at one time, and radar images [11] can be generated by drawing these scan lines onto the screen according to their orientations.

3 Echo High and Low Frequency Signal Processing Based on Generalized Filtering

3.1 High Frequency Signal Processing with Generalized Weighted Median Filter

In practical applications, the number of terrain echo signals from navigation radar is often large, and calculating indicators one by one through each window requires a huge amount of computation, which cannot meet the real-time requirements of signal processing. The noise points and signal details have not been effectively distinguished, and the signal quality still significantly decreases after filtering. Based on the shortcomings of weighted median filtering [12, 13], an improved adaptive weighted median filtering method is proposed to improve the noise detection and classification, noise weight determination, and noise filtering of the filtering algorithm. The improvement ideas are as follows:

The detection and classification of image noise is the premise of noise filtering [14]. The detection of noise can be divided into two steps: coarse detection and fine detection. For coarse detection, first count the number of pixels with maximum or minimum pixel value in the window. If a pixel's gray value is not extreme, it is a non noise point. Otherwise, it is a suspected noise point to be marked. The marking formula is:

$$F[I] = \begin{cases} 1, & I = 255/0 \\ 0, & other \end{cases} \tag{2}$$

In formula (2): I It is the gray value of a pixel of the window point. Then fine detection is carried out. After rough detection, the image suspected noise points contain noise points and image edge details. It is necessary to further detect this. First, count the number of suspected noise points. If the gray value of a pixel of the window point is smaller than the set window size, the suspected noise points in the window can be filtered as noise points; Otherwise, increase the window size and continue to judge until the requirements are met.

After noise detection and classification, image noise is basically detected and marked. If median filtering is directly applied to noise points, "overfiltering" will occur. To overcome this phenomenon, remove the number of noise points in the window, multiply the gray values of the other non noise points by the weight values, and then sort the size on this basis to get a new set of sequences and take the value, that is, the output value of the window's central pixel can be expressed as formula (3):

$$I = Med\left\{\frac{1}{a^2}g_0(k-m), \cdots, \frac{1}{a^2}g_m(k-m+m')\right\} \tag{3}$$

In formula (3), $g_0(k-m), \cdots, g_m(k-m+m')$ Assign values to each pixel in the window; m' Indicates the number of noise points; a Indicates the size of the filtering window; k Indicates window sorting.

Combine the formula to output the filtering results, continue to migrate the filtering template, and complete the overall image filtering.

3.2 Generalized Adaptive Wiener Filtering Low Frequency Signal Processing

Wiener filter [15] can filter image noise by assuming that the noise image signal is the sum of image signal and noise signal, and the second-order statistical characteristics of both are known. The filter parameters can be obtained according to the relevant error criteria, which can be expressed as formula (4):

$$E = D(b, h) - \xi \cdot F(b, h) \tag{4}$$

In formula (4), $D(b, h)$ Represents the b that 's ok h Image signal of the column; $F(b, h)$ Represents noise signal; ξ Is the Wiener filter, which can be expressed as Formula (5):

$$\xi = \frac{F^2(b, h) - \sigma^2}{F^2(b, h)} \tag{5}$$

In formula (5), σ^2 Represents the noise pixel variance value. The adaptive Wiener filter is used to filter the low-frequency sub image in the wavelet domain. Although the low-frequency image contains most of the image information, it is basically not affected by noise. The noise pollution degree of low-frequency sub image in wavelet domain is relatively light, but it can not be ignored. An improved adaptive Wiener filter potential for this part of image will retain a lot of useful information.

3.3 Fitting Based on Least Squares Error

According to the high and low frequency signal processing results of the echo, the formula (6) can be obtained by adding a sliding window to the data at all range points of the radar echo at a certain time:

$$C_a^t[l] = o - a, \cdots, o + a \tag{6}$$

In formula (6), l Represents distance; o Represents the center of the sliding window. Combined with the generalized filtering processing results, the sliding window is polynomial fitted according to formula (7):

$$C[l] = \sum_{z=0}^{o} \eta_z l^z \tag{7}$$

In formula (7), η Represents the polynomial coefficient; z Indicates the degree of the polynomial. Based on this, the result of least square error fitting is Formula (8):

$$\vartheta_z = \sum_{z=0}^{o+a} \left(C[l] - C_a^t[l] \right)^2 \tag{8}$$

According to the basic principle of calculus ϑ_z Then, the original data is characterized by Eq. (7), and all fitting points of the original echo data are obtained by moving the sliding window from front to back. Because the clutter and noise data deviating from the polynomial curve will be discarded in the fitting process, this method can smooth the radar echo data in the range.

4 Terrain Echo Signal Enhancement Technology of Marine Radar

4.1 Echo Signal Enhancement Structure Design

The traditional radar simulator image generation mostly uses the scanning line intersection algorithm to calculate all echo polygons within the range and draw them to video memory. The scanning effect of the radar simulator is achieved through OpenGL template cache. A small angle sector area is drawn with the ship's position as the center and the range as the radius as the template. Only the echoes overlapped with the sector template can be seen in video memory. The center angle position of the sector template changes at a certain speed to form a dynamic scanning picture. This mechanism will increase the redundant operation of the computer CPU to a certain extent, which is specifically shown in the following two aspects:

(1) A small scanning angle is smaller than the whole scene, but when judging echo occlusion, it needs to calculate the distance, angle and other imaging geometric information of all polygons in the scene, which results in a large amount of redundant calculation.
(2) When using DEM to enhance echo image, it is necessary to overcome the influence of latitude gradual growth rate. As long as the image is divided into multiple regions in latitude, the energy accumulation of adjacent regions will be incomplete during echo calculation. It is necessary to overlap a small part of adjacent regional echoes, which will also generate redundant calculation.

As the resolution increases and the radar range increases, the amount of echo data that needs to be simulated increases sharply. The echo signal enhancement structure built based on this is shown in Fig. 3.

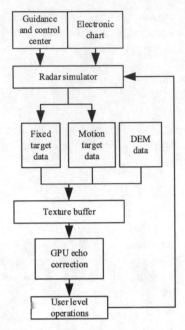

Fig. 3. Echo signal enhancement structure

Read the electronic chart data of the target area, store the contour lines as a linked list, store the longitude and latitude boundary, the ship's information, and radar parameters given by the guidance and control center to the radar simulator, and transfer the longitude and latitude boundary, the ship's information, and radar parameters to the GPU constant register; The DEM data in this area is stored in the GPU texture buffer to facilitate subsequent processing.

4.2 High and Low Frequency Echo Signal Gain

The terrain echo signal of marine radar is formed by nonlinear coupling of two high and low frequency echo signals. If periodic signals, noise sources and nonlinear bistable systems match each other, stochastic resonance can be generated. The three-dimensional diagram of nonlinear coupling potential function of the echo signal is shown in Fig. 4.

As the high and low frequency echo signals are continuously added to the weak periodic signals, the echo signals will change periodically and lose balance. Under the action of the driving force, the potential function will also change, and the potential barrier at the central height will also change differently. However, the energy of Brownian particles is too small due to the weak energy of periodic signals in the process of motion. With the continuous addition of noise, the energy of Brownian particles increases, making them quickly cross the barrier to reach another potential well within the range of a single potential well. At this time, the stochastic resonance phenomenon occurs when the movement period is consistent with the weak signal period.Because the nonlinear coupling potential function of the echo signal is affected by the nonlinear coupling

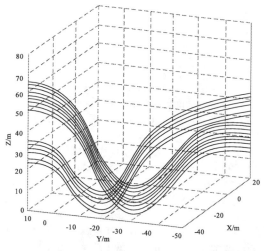

Fig. 4. 3D Diagram of Nonlinear Coupling Potential Function of Echo Signal

coefficient, the periodic signal and the noise motion direction should be in the same direction under the action of the driving force, which can generate appropriate nonlinear coupling coefficient.

When the nonlinear coupling coefficient is determined, the signal is processed with gain, as shown in Formula (9):

$$SF(A, B) = \int_{-\infty}^{+\infty} x(A)x(B)H(D)\exp(-2\pi t)dt \tag{9}$$

In formula (9), $x(A)$, $x(B)$ Represent the time domain of high-frequency and low-frequency signals respectively; $H(D)$ Indicates the width of the emitted electromagnetic wave. If a certain time point in the time domain of a signal to be analyzed t Characteristics research, through t After signal gain processing at time point, the purpose of pseudo signal attenuation can be achieved.

4.3 Echo Signal Enhancement Processing

Call OpenGL's read pixel function to read the color value of the scene from the computer's graphics card memory and save it to the texture buffer; Read the echo drawn at the current time, that is, capture the width and height of the screen, data format and type, and store them in the GPU texture buffer. The ship's position is updated in real time by the data transmitted from the guidance and control center. The rectangular coordinate system is the tangent plane of the point. Within the radar range, the radar echo coincides with a small area of the DEM. The geographical coordinates of the DEM echo image in the texture buffer are projected and clipped to a rectangle on the screen coordinates, and the coordinate mapping of the two images is calculated in the horizontal and vertical directions. When the echo image and the DEM are processed in the GPU segment, the coordinate linear mapping relationship is as follows (10):

$$\begin{cases} x_a = \alpha_1 x_e + \beta_1 \\ y_a = \alpha_2 y_e + \beta_2 \end{cases} \tag{10}$$

In formula (10), (x_e, y_e) Represents radar echo coordinates; (x_a, y_a) Represents radar DEM coordinates; α_1, α_2 Represent horizontal and vertical coordinate conversion coefficient respectively; β_1, β_2 Represents the horizontal and vertical coordinate conversion random number.

Due to the influence of latitude gradual growth rate, the geographical distance represented by each pixel on the vertical axis of DEM data increases with the increase of latitude. The echo image to be corrected is divided into rectangular areas with equal pixels on the vertical axis, and the influence of latitude gradual growth rate is eliminated according to formula (10); The greater the number of rectangular area partitions, the more accurate the latitude of DEM data correction. But to ensure the calculation speed, α_1, α_2, β_1, β_2 It can be expressed by the following formula:

$$\begin{cases} \alpha_1 = \dfrac{x_h - x_0}{x_h' - x_0'} \\ \alpha_2 = \dfrac{y_h - y_0}{y_h' - y_0'} \end{cases} \tag{11}$$

$$\begin{cases} \beta_1 = x_0 - x_0' \alpha_1 \\ \beta_2 = y_0 - y_0' \alpha_2 \end{cases} \tag{12}$$

In the above formula, (x_h, y_h), (x_h', y_h') Represent the coordinates of the lower right corner of the echo map range and the lower right corner of the echo map corresponding to the DEM respectively; (x_0, y_0), (x_0', y_0') Represent the ship's DEM coordinates and echo map coordinates respectively. Criteria for echo image enhancement with DEM data: pixels without DEM echo data can generate echo only when the radar scans the marked artificial buildings; There are echo pixels in both DEM and radar, and the gray value of DEM is multiplied by the echo value for correction: DEM has echo data but radar has no echo, and DEM pixels are used to modify the contour boundary of radar echo; After image enhancement, the radar echo image is modified from the original irregular quadrilateral binary image to the radar echo image filled with gray value. The larger the gray value, the higher the elevation, indicating the better the echo enhancement effect.

5   Experiment and Analysis

5.1   Design of Signal Capture Structure Experimental Platform

The experimental signal capture is carried out on the LabVIEW platform, and its structure is shown in Fig. 5.

The structure shown in Fig. 5 is mainly used to receive signal center frequency, level and storage data duration. Through this platform, the application effectiveness of terrain echo signal enhancement technology of navigation radar based on generalized filtering is verified. Analyze the received echo signal and demodulate the radar electromagnetic wave.

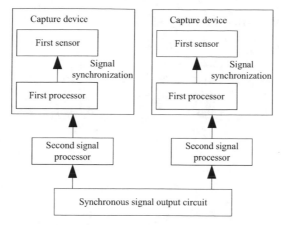

Fig. 5. Signal Capture Structure Experiment Platform

5.2 Experimental Process Simulation

Two experimental situations are set. One is that when no signal is received or the signal is blocked, the platform will first check the pending table after executing the OS code. The pending table can be used to analyze whether a signal is received. If not, return to the user status and execute the next code.

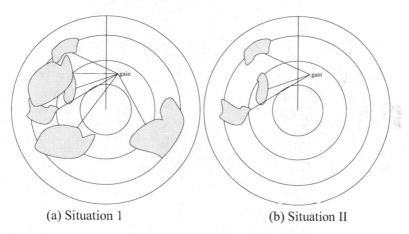

(a) Situation 1   (b) Situation II

Fig. 6. Echo gain in two cases

Case 2: When a signal is received and the signal is not blocked, the second position in the pending table will change after the platform executes the OS code. Assume that the received signal is No. 2. After executing the OS code, check the pending table. If the received signal is found, check the block table. If signal 2 is blocked, it will return to user mode as above; If it is not blocked, the corresponding processing function will be executed.

5.3 Analysis of Experimental Data

The data used in the experiment was sourced from the RADARSAT Geophysical Processor System (RGPS) dataset. This data includes a wide range of terrain and ocean application scenarios, such as landform measurement, glacier monitoring, and ocean storm observation. By using radar echo signals, surface elevation, surface coverage type, water body boundaries, and other geographic information can be obtained.

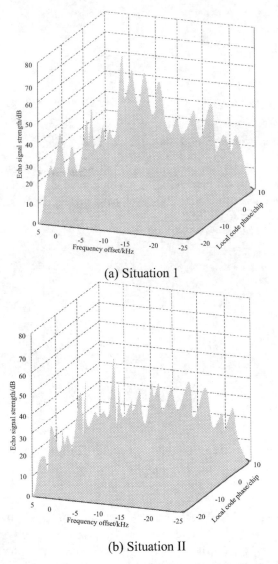

(a) Situation 1

(b) Situation II

Fig. 7. Terrain echo signal enhancement results of navigation radar

In order to verify the reliability of the terrain echo signal enhancement technology of navigation radar based on generalized filtering, the echo gain in two cases is analyzed, as shown in Fig. 6.

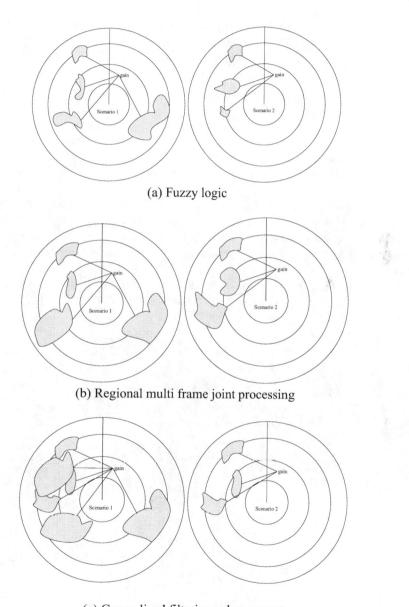

(a) Fuzzy logic

(b) Regional multi frame joint processing

(c) Generalized filtering enhancement

Fig. 8. Echo Gain Comparison Results of Three Methods

It can be seen from Fig. 6 that the echo gain in case 1 is larger than that in case 2, which means that the weaker the echo signal is, the greater the echo enhancement will be.

According to the echo gain in Fig. 6, the terrain echo signal enhancement results of navigation radar under ideal conditions are obtained, as shown in Fig. 7.

It can be seen from Fig. 7 that the maximum echo signal strength in case 1 is 55 dB, and the maximum echo signal strength in case 2 is 42 dB.

5.4 Experimental Results and Analysis

Taking the echo gain in Fig. 6 as a reference, the echo signal enhancement algorithm based on fuzzy logic, the echo signal enhancement technology based on regional multi frame joint processing and the echo signal enhancement technology based on generalized filtering are used to compare and analyze the echo gain. The comparison results are shown in Fig. 8.

It can be seen from Fig. 8 that the echo signal enhancement algorithm based on fuzzy logic and the echo signal enhancement technology based on regional multi frame joint processing are inconsistent with the actual echo gain effect, while the echo signal enhancement technology based on generalized filtering is consistent with the actual echo gain effect.

Using the echo gain in Fig. 7 as a reference, these three methods are used to compare and analyze the echo signal enhancement effect, as shown in Fig. 9.

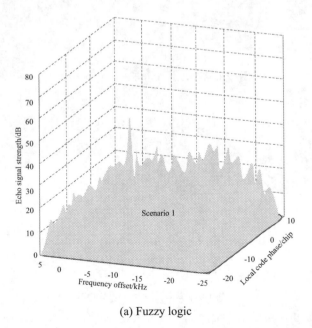

(a) Fuzzy logic

Fig. 9. Signal radar acquisition comparison results of three methods

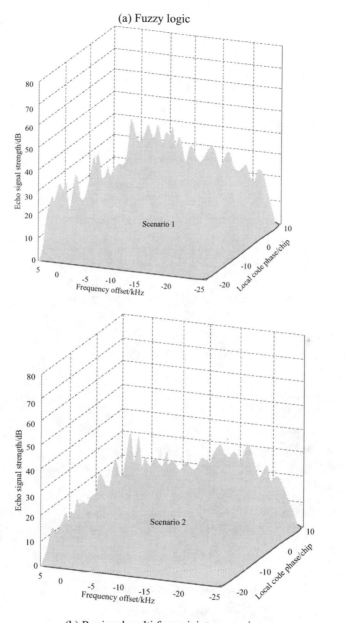

(a) Fuzzy logic

(b) Regional multi frame joint processing

Fig. 9. (*continued*)

(b) Regional multi frame joint processing

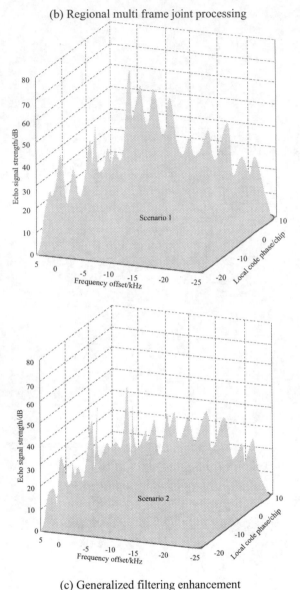

(c) Generalized filtering enhancement

Fig. 9. (*continued*)

It can be seen from Fig. 9 (a) that the maximum echo signal strength in case 1 is 35 dB, and the maximum echo signal strength in case 2 is 23 dB; It can be seen from Fig. 9 (b) that the maximum echo signal strength in case 1 is 38 dB, and the maximum echo signal strength in case 2 is 39 dB; It can be seen from Fig. 9 (c) that the maximum echo signal strength in case 1 is 55 dB, and the maximum echo signal strength in case 2 is 41 dB. Only this method has a maximum error of 1 dB with the actual data.

6   Conclusion

At present, there is a problem of poor echo signal enhancement effect in the research of radar echo signal enhancement processing methods. In order to solve this problem, a marine radar terrain echo signal enhancement technology based on generalized filtering is proposed. This technology uses generalized filtering and multinomial least square fitting to smooth clutter and noise in the echo, At the same time, the weak details of the target echo are effectively retained. On this basis, combined with the background removal, low-pass filtering and other processing in the time dimension, the radar echo signal can be enhanced.

References

1. Hao, W., Lejian, Z., Haihe, L., et al.: Radial interference echo identification algorithm based on fuzzy logic for weather radar. Acta Meteor. Sin. Meteor. Sin. **78**(01), 116–127 (2020)
2. Tao, Z., Tao, Y., Guanglei, Z.: The weak target enhancement based on the local multi-frame joint procession in the presence of strong scattering sources for LFMCW radar. Chin. J. Radio Sci. **36**(01), 142–149 (2021)
3. Long, T., Wang, S., Wen Cao, Pu., Ren, M.H., Fernandez, C.: Collaborative state estimation of lithium-ion battery based on multi-time scale low-pass filter forgetting factor recursive least squares - double extended Kalman filtering algorithm. Int. J. Circuit Theory Appl. **50**(6), 2108–2127 (2022)
4. Belen, A., Belen, M.A., Palandken, M., et al.: Design and realization of broadband active inductor based band pass filter. Chin. J. Electron. **32**(2), 1–5 (2023)
5. Chen, Q.: Stepped frequency multiresolution digital signal processing. Sci. Prog.Prog. **2021**, 1–13 (2021)
6. Prasad, D.S., Chanamallu, S.R., Prasad, K.S.: EEG signal enhancement and spectrum estimation using fourier transform magnitude response derivative functions. J. Mobile Multimedia **18**(2), 231–249 (2022)
7. Chen, J., Cao, Y., Wang, C., Li, B.: Sparse Fourier transform and amplitude–frequency characteristics analysis of vortex street signal. Meas. Control **54**(5–6), 908–915 (2021)
8. Jie, X., Jiao, C., Zhigao, L.: Simulation design of an ultra-thin wideband frequency selective absorber. Comput. Simul. **39**(3), 22–25 (2022)
9. Isotton, G., Janna, C., Bernaschi, M.: A GPU-accelerated adaptive FSAI preconditioner for massively parallel simulations. Int. J. High Perform. Comput. Appl.Comput. Appl. **36**(2), 153–166 (2022)
10. Vasenina, E., Kataoka, R., Hammert, W.B., et al.: Examination of changes in echo intensity following resistance exercise among various regions of interest. Clin. Physiol. Funct. Imaging. Physiol. Funct. Imaging **42**(1), 23–28 (2022)
11. Arieta, L.R., Giuliani, H.K., Gerstner, G.R., et al.: The influence of hydration status and segmental water content on echo intensity in older men: 334. Med. Sci. Sports Exerc.Exerc. **53**(8S), 105 (2021)
12. Hao, G.C., Feng, S.Q., Wang, W., et al.: High quality time-frequency analysis via normalized generalized Warblet-WVD. Acta Automatica Sin. **48**(10), 2526–2536 (2022)
13. Shao, C., Kaur, P., Kumar, R.: An improved adaptive weighted mean filtering approach for metallographic image processing** J. Intell. Syst.Intell. Syst. **30**(1), 470–478 (2021)
14. Zhang, Z., Yi, R., He, S., et al.: SPR signal amplification based on dynamic field enhancement at the sensor surface. IEEE Sens. J. **21**(7), 9523–9529 (2021)
15. Gharamohammadi, A., Behnia, F., Shokouhmand, A., et al.: Robust wiener filter-based time gating method for detection of shallowly buried objects. IET Signal Proc. **15**(1), 28–39 (2021)

Design and Improvement of Airborne Ocean Radar Fault Detection Algorithm

Liang Pang[✉]

Public Basic Course Department, Wuhan Institute of Design and Sciences, Wuhan 430025, China
pangliang0611@163.com

Abstract. Fault detection can ensure the safe operation of airborne ocean radar. In order to improve the fault detection performance of airborne ocean radar, the design and improvement of the fault detection algorithm for airborne ocean radar is proposed. Based on the current and voltage values of stable operation, the fault area is determined. The fault information is decomposed by wavelet transform, the fault information is reconstructed, and the fault location is determined. Through the preprocessing of the fault data, the feature matching degree of the fault data is defined, and the features of the fault data are extracted by using the information state function of the fault data. Calculate the average trajectory of the observation vector of the operating state, and combine the operating trajectories of the fault data variables at different times to detect the faults of airborne ocean radar. The experimental results show that the algorithm in this paper has certain effectiveness in detecting the faults of airborne ocean radar, and has better performance in terms of missed detection rate, false detection rate and signal-to-noise ratio of fault signal acquisition.

Keywords: Fault Detection · Airborne Ocean Radar · Fault Location · Feature Extraction · Area Determination

1 Introduction

The airborne marine radar uses a laser with a high perspective in the water as the pulse emission source, which has the advantages of mature technology, large transmission power, small size, etc., and can detect the sound speed and temperature of the marine boundary layer in a large area [1, 2]. The airborne marine radar will inevitably fail in the process of operation. Once an airborne ocean radar fails, it will output incorrect conservation monitoring information, seriously affecting the operation of the ocean exploration system, and even causing significant economic losses. Therefore, the fault detection of airborne Marine radar equipment has become a key research topic.

Wang Liangcheng et al. [3] considered that the existing methods could not complete the conversion of non-linear drive information, resulting in the low fault detection rate of motor drive system, and a method of fault detection of motor drive system under PLC technology is proposed. The frequency band components of the drive motor are

L. Yun et al. (Eds.): ADHIP 2023, LNICST 548, pp. 116–131, 2024.
https://doi.org/10.1007/978-3-031-50546-1_8

obtained by wavelet packet analysis, and the fault characteristic parameters are extracted by analyzing the changes of each frequency band component. According to the fault characteristics, under the PLC technology, the non-linear information of the motor drive fault characteristic parameters is converted into linear information by combining the kernel method and the principal component analysis method. Finally, the fuzzy kernel clustering method is used to cluster the linearized information to complete the fault detection of the motor drive system. The experimental results show that this method can realize the motor fault detection. However, the signal to noise ratio of fault signal acquisition is low. Bao Haibo et al. [4] proposed a static voltage stability fault screening and ranking method considering the uncertainty of power generation in order to effectively analyze the impact of power system equipment faults and uncertainty of power generation on system static voltage stability. A deterministic voltage stability critical point model under N-1 fault scenario is established, and the severe fault set is screened according to the obtained load margin after fault.Considering the random distribution of wind power and photovoltaic power generation, a voltage stability probability evaluation model under fault scenarios is established, and the stochastic response surface method is used to obtain the cumulative probability distribution of load margins in fault scenarios. The two types of post-failure systems are sorted for voltage stability and classified according to the accumulated probability of the load margin. The calculation results of the IEEE 118 node standard system show that the proposed fault screening and sorting method is effective, and it can screen and sort the probabilistic static voltage stability of the system under each fault scenario, that is, faults is detected, but the method has a high missed detection rate. Reference [5] proposed a fault diagnosis method based on immune algorithm (Immune Algorithm, IA) for the problem of low diagnostic efficiency of traditional fault diagnosis methods for hydro-generator units. The collected fault sample data of the hydro-generator unit is preprocessed to form the fault type code, and the relevant parameters of the algorithm are set. On this basis, the immune algorithm program developed by MATLAB is used to predict the fault. This method has the advantages of simple operation and easy capacity expansion, etc., but its false detection rate is high. Reference [6] in response to the problem of low accuracy and long recognition time caused by the lack of preprocessing of fault data in LiDAR hardware, a research method for pattern recognition of LiDAR hardware fault data is proposed. Wavelet transform is used to analyze noise data, obtain data features, and shorten fault recognition time; Perform fusion processing on high-dimensional feature data, extract association rule feature quantities of fault data for fault diagnosis and recognition. However, in the process of obtaining fault data features, there is a problem of weak continuity of fault signals. Reference [7] aiming at the problems of intermittent faults and comprehensive performance degradation of airborne radar, an early warning technology considering degradation and intermittent faults is proposed. By analyzing the principle of radar fault detection, a test Index set is set up. Apply the NDGM model to clarify the fault warning decision-making process, analyze the growth of small wave energy spectrum entropy in the decision-making process, detect the operating status of radar components, and identify fault information, however, in the process of detecting the operating status of the radar, the integrity of the status data cannot be guaranteed, resulting in low accuracy of fault detection.

Based on the above research background, a fault detection method for airborne marine radar is designed in this paper to ensure the safe operation of airborne marine radar.

2   Airborne Marine Radar Fault Detection

The overall structure of airborne marine radar fault detection is shown in Fig. 1.

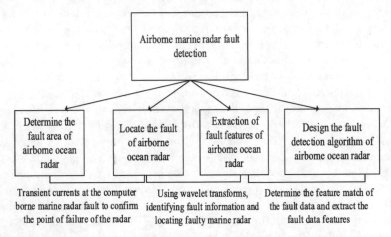

Fig. 1. Airborne marine radar fault detection architecture

2.1   Determine the Fault Area of Airborne Ocean Radar

When the airborne ocean radar fails, the current and voltage values around the fault point will change, so the transient current at the fault of the airborne ocean radar can be expressed as:

$$I_z + I_f = -I_f \frac{Z_1}{Z_1 + Z_2} \tag{1}$$

where, I_z represents the transient current at the fault location of the airborne ocean radar, I_f represents the current reflected wave at the fault of the airborne ocean radar, Z_1 represents the reflected wave impedance of the airborne ocean radar fault, and Z_2 represents the incident wave of the airborne ocean radar fault. Impedance.

The intra class aggregation degree of fault data of airborne ocean radar is calculated by using the transient current at the fault of airborne ocean radar. The formula is:

$$L_W^{(k)} = \sum_{i=1}^{M} L_i^{(k)} \tag{2}$$

where, M represents the number of categories of airborne ocean radar fault data, and $L_i^{(k)}$ is the aggregation degree of the k-dimensional airborne ocean radar fault data attribute belonging to the i category, which can be calculated by the following formula:

$$L_i^{(k)} = \sum_{x \in G_i} \left(X^{(k)} - Q_i^{(k)} \right)^2 \tag{3}$$

where, $X^{(k)}$ represents the attribute vector of the k-dimensional airborne ocean radar fault data, G_i represents the sample of the airborne ocean radar fault data in category i, and $Q_i^{(k)}$ represents the attribute vector of the k-dimensional airborne ocean radar fault data in category i.

When the airborne ocean radar is running stably, based on the current and voltage values during the operation, according to the degree of intra-class aggregation of the airborne ocean radar fault data, the fault area of the airborne ocean radar is determined. The principle is shown in Fig. 2.

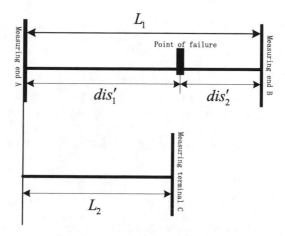

Fig. 2. Principle of determining the fault area of airborne ocean radar

If the fault location of airborne ocean radar is defined as Ω, the two fault points located by airborne ocean radar are L_1 and L_1 respectively. Install the transient current detection device on the l_1 terminal of the airborne ocean radar, and set A and B as two measurement terminals; install the voltage detection device on the ζ_1 of the airborne ocean radar, dis_1' and dis_2' represent the fault point Ω to the measurement The straight-line distance between ends A and B. Assuming that the transient current I is constant, then:

$$\begin{cases} I(T_1 - T_0) - dis_1' \\ I(T_2 - T_0) = dis_2' \\ I(t_3 - t_0) = L_2 + dis_1' \\ dis_1 + dis_2 = L_1 \end{cases} \tag{4}$$

It is obtained that the fault point Ω of airborne ocean radar is in the area of the fault line according to the above process. The calculation result is:

$$dis_1 = \frac{L_1}{2} + \frac{L_2(T_1 - T_2)}{2(T_3 - T_1)} \tag{5}$$

The fault area of airborne ocean radar is determined according to the above calculation process.

2.2 Locate the Fault of Airborne Ocean Radar

Wavelet transform is used to convert the fault information of airborne ocean radar in order to locate the fault of airborne ocean radar complete the effective identification of the fault information, and finally realize the accurate location of the fault of airborne ocean radar.

The wavelet function is used to collect the fault data of the airborne ocean radar, and the state function of the fault information of the airborne ocean radar is obtained, which is expressed as:

$$X_i = \frac{1}{p_j(k)} f_i(x_i, u_i) \tag{6}$$

where, $p_j(k)$ represents the posterior probability estimation of fault information of airborne ocean radar, $x_i \in R^n$ represents the feature state vector of fault information of airborne ocean radar, and $u_i \in R^m$ represents the feature of fault information of airborne ocean radar.

Assuming that $\phi(t)$ is a wavelet discrete function, the transformation function $\phi_{f,g}(t)$ obtained is:

$$\phi_{f,g}(t) = \frac{1}{\sqrt{f}} \phi\left(\frac{t - g}{f}\right) \tag{7}$$

where, f represents the wavelet transform factor, g represents the movement factor, $\phi_{f,g}(t)$ is obtained by the wavelet transform $\phi(t)$, and the wavelet transform function is obtained by performing the wavelet transform on f and g.

Let $s(t)$ be the fault signal of airborne ocean radar, according to formula (7), use $\phi(t)$ to perform discrete wavelet transform on $s(t)$, and obtain the coefficient of wavelet transform as:

$$T_s(f, g) = \frac{1}{\sqrt{f}} \int s(t)\phi\left(\frac{t - g}{f}\right) dt \tag{8}$$

where, f, g and t represent constants in the process of wavelet transform. Therefore, formula (8) is a continuous wavelet transform signal, which can explain that the fault signal of airborne ocean radar is continuous.

Wavelet function is used to reconstruct the fault information of airborne ocean radar [8], any fault location $e(t)$ of airborne ocean radar can be expressed as:

$$e(t) = e_l(t) + T_s\left(2^l, 2^l t\right) \lambda_{l,v} \tag{9}$$

where, $e_l(t)$ represents the signal near the fault point of airborne ocean radar, $T_s(2^l, 2^l t)$ represents the discrete wavelet signal of the fault signal of airborne ocean radar under different fault conditions, and is expressed as:

$$T_s\left(2^l, 2^l t\right) = \frac{1}{\sqrt{2^l}} T_y(f, g) \tag{10}$$

where, l is the fault weight of airborne ocean radar under different fault conditions.

Let $\phi(t)$ be the wavelet basis function, then $\lambda_{l,v}$ is the wavelet coefficient of the fault location point of the adjacent airborne ocean radar, which can be decomposed into:

$$c_{l,v} = \langle e(t), \phi_{l,v}(t) \rangle \tag{11}$$

The fault information of airborne ocean radar under different fault conditions is reconstructed according to the wavelet coefficient of formula (11), and the accurate location of airborne ocean radar fault is obtained as follows:

$$e(t) = \sum_l \sum_v \beta_l \lambda_{l,v} \phi_{l,v}(t) \tag{12}$$

where, β_l represents the location information of the fault point of the airborne ocean radar.

The location of the fault of the airborne ocean radar is obtained according to the above process.

2.3 Extraction of Fault Features of Airborne Ocean Radar

It is necessary to preprocess the fault data, convert all the fault data into the mapping interval of [0, 1] in order to ensure the integrity of the fault data of airborne ocean radar, so as to eliminate the noise in the fault data of airborne ocean radar and reduce the interference of noise to the mining of fault data of airborne ocean radar. The following methods are used to preprocess the fault data of airborne ocean radar, which is expressed as:

$$z^* = \frac{z - \min(z)}{\max(z) - \min(z)} \tag{13}$$

where, z^* represents the preprocessed airborne ocean radar fault data, $\max(z)$ represents the maximum value existing in the airborne ocean radar fault data, and $\min(z)$ represents the minimum value existing in the airborne ocean radar fault data.

After preprocessing the fault data of airborne ocean radar, the regression equation of the fault data of airborne ocean radar is constructed, and the expression is as follows:

$$\left(m^*, n^*\right) = \zeta_1 \times \frac{a_1 m_1 + a_2 + m_2^* + a_p + m_p^*}{\lambda_1} \tag{14}$$

where, ζ_1 represents the composition of fault data, (m^*, n^*) represents the regression coefficient of fault data, a_1, a_2 and a_p represent the fault points with frequent fault data, and m_2^* and m_p^* respectively represent the correlation between frequent fault points.

Specific methods is used to approximate and linearly fit the fault data of airborne ocean radar. These methods do not affect its basic form, nor do they require saving the characteristics of fault data, and can also compress the characteristics of fault data, which will eventually cause faults. There are differences in the length of the data, so the fault data must be processed with the same length.

The fault data M and B N of airborne ocean radar are set as:

$$\begin{cases} M = \{(g_1, a_1), \cdots, (g_L, a_L)\} \\ N = \{(g_1, b_1), \cdots, (g_L, b_L)\} \end{cases} \tag{15}$$

Assuming that the lengths of fault data M and N of airborne ocean radar are equal, the fault data sequences SM and SN are matched, and the expressions are as follows:

$$\begin{cases} SM = \{(g_{x1}, a_{x1}), \cdots, (g_{xm}, a_{xm})\} \\ SN = \{(g_{y1}, b_{y1}), \cdots, (g_{yn}, b_{yn})\} \end{cases} \tag{16}$$

where, the corresponding lengths of the fault data sequences SM and SN of the airborne ocean radar are m and n, respectively. The characteristic length of the fault data is quite different. Therefore, it cannot be linearly simulated by using the Euclidean distance and other metric formulas. Therefore, it is necessary to use the feature extraction algorithm to process the fault data of the airborne ocean radar in equal lengths, and obtain the equal length processing. The fault data of airborne ocean radar, namely:

$$\begin{cases} SM^* = \{(g_{l1}, a_{l1}), \cdots, (g_{ll}, a_{ll})\} \\ SN^* = \{(g_{l1}, b_{l1}), \cdots, (g_{ll}, b_{ll})\} \end{cases} \tag{17}$$

After equal length processing of fault data, the weight mapping matrix of fault data of airborne ocean radar is obtained, which is expressed as:

$$U(h) = \frac{\partial_p \times \Lambda^* \times \varepsilon(s)}{\Pi_i + f(\varsigma)} \times \ell(u) \tag{18}$$

where, ∂_p A represents the feature quantity of the airborne ocean radar in the normal operation state, Λ^* represents the fault data set of the airborne ocean radar, $\varepsilon(s)$ represents the total number of fault data sets, $f(\varsigma)$ represents the characteristic attribute of the fault data, and $\ell(u)$ represents the airborne ocean radar. A collection of feature items for fault data.

Combined with the principle of feature extraction, the feature matching degree of fault data of airborne ocean radar is defined, namely:

$$\varsigma = \frac{1}{A \cdot \phi_p(q)} \sqrt{\sum_{l=1}^{A} \left| 1 - \frac{u_l}{k_l} \right|^2} \cdot \varphi_p(q) \tag{19}$$

where, $\varphi_p(q)$ represents the correlation between attribute q and fault data, $\phi_p(q)$ represents the non correlation between attribute q and fault data, A represents the number of airborne ocean radar, u_l represents the fault data set, and k_l represents the reliability of airborne ocean radar.

It is necessary to first use the time-frequency analysis method to extract the characteristics of the fault data of the airborne ocean radar in order to extract the characteristics of the fault data of the airborne ocean radar [9, 10], and the information function of the fault data of the airborne ocean radar is obtained as follows:

$$Z(x_i) = f(x_i, z_i)k_g(i) \tag{20}$$

where, $k_g(i)$ represents the characteristic value of the fault data of airborne ocean radar, x_i represents the characteristic state vector of the fault data of airborne ocean radar, and z_i represents the characteristic number of the fault data of airborne ocean radar.

The fault data of airborne ocean radar defines as $S(x_i)$, the feature preprocessing $S(x_i)$, and the transformed fault data as $\gamma(x_i)$, the following formula is used to extract the characteristics of the fault data of airborne ocean radar, which is expressed as:

$$T_r = \frac{S(x_i) + \gamma(x_i)}{W_g \times f_z} \times \lambda_d \tag{21}$$

where, W_g represents the feature quantity of fault data of airborne ocean radar, f_z represents the generation cycle of fault data features of airborne ocean radar, and λ_d represents the feature extraction location of fault data of airborne ocean radar.

The feature matching degree of the fault data is defined through the preprocessing of the airborne ocean radar fault data, and the characteristics of the airborne ocean radar fault data are extracted by using the information state function of the fault data.

2.4   Design the Fault Detection Algorithm of Airborne Ocean Radar

After the fault feature extraction of airborne ocean radar is completed, the fault detection method is designed, which is the last step of fault detection. In the process of detecting faults of airborne ocean radar, the observation vector of the operating state of airborne ocean radar is defined, and the expression is:

$$\tau = \frac{\xi u_f}{\chi_i * \Gamma_p} \cdot \alpha_e \tag{22}$$

where, ξ represents the operating state threshold of airborne ocean radar, u_f represents the number of pivots in the fault data of airborne ocean radar, χ_i represents the covariance matrix between the pivots, and Γ_p represents the failure of airborne ocean radar under normal conditions. Data, α_e represents the historical data of airborne ocean radar.

The observation vector of the operating state of airborne ocean radar is used as the key variable of fault detection. By analyzing the historical operation of airborne ocean radar, the average trajectory of the observation vector of the operating state is calculated, namely:

$$Z_r = \frac{Y_e + Y_p}{C_p} \cdot \sqrt{X_t + Y_i} \tag{23}$$

where, Y_e represents a single element in the trajectory of the key variable, Y_p represents the change trend information of the key variable during the operation time of the airborne ocean radar, C_p represents the dynamic characteristics of the key variable, and X_t

represents the variable value of the fault detection sample of the airborne ocean radar., Y_i represents the trajectory vector of the key variable.

In view of the high degree of electronization of airborne ocean radar, the requirements for the operation trajectory of fault variables are very strict. It is necessary to give the operation trajectory of fault data variables of airborne ocean radar at different times, namely:

$$E_t = \frac{x_p \cdot G_k}{\gamma_u} + \sigma_p \tag{24}$$

where, x_p represents the weight of the airborne ocean radar fault data in the trajectory operation space, G_k represents the trajectory judgment of the airborne ocean radar fault data, γ_u represents the sampling interval of the airborne ocean radar fault data, and σ_p represents the fault data of the same period. Similarity.

Assuming that $\{a_i\}$ and $\{b_i\}$ represent the fault signals during the operation of the airborne ocean radar, the fundamental frequency vibration of the operating signal of the airborne ocean radar can be obtained by using the following formula:

$$X_i(t) = \frac{f(x) \times [\{a_i\} \cdot \{b_i\}]}{S_j(t) \times x_{ij}} J_i(t) \tag{25}$$

where, $S_j(t)$ represents the fault signal of the j airborne ocean radar, x_{ij} represents the fault signal of $S_j(t)$ at the i measuring point, and $J_i(t)$ represents the detection factor.

When the airborne ocean radar has communication failure, it is difficult to ensure its normal operation, so it is necessary to extract the key variable information of the airborne ocean radar failure, namely:

$$R_p = \frac{f_p \cdot \phi_o}{f_u + \sigma_l} \tag{26}$$

where, f_p represents the type set of fault samples of airborne ocean radar, ϕ_o represents the score matrix of fault data of airborne ocean radar in the principal component space, f_u represents the trajectory trend diagram of detection variables of fault data of airborne ocean radar, and σ_l represents the trajectory vector of fault detection data samples of airborne ocean radar.

The airborne ocean radar faults is detected according to the key variable information of airborne ocean radar faults, namely:

$$J_r = \frac{n \mp \Phi_e}{C_p} \times \frac{Z_r}{R_p} \times E_t \tag{27}$$

where, n represents the number of key variables, and Φ_e represents the process data of the key variable information of airborne ocean radar failures.

To sum up, according to the running trajectories of the airborne ocean radar fault data variables at different times, the airborne ocean radar faults are detected.

3   Experimental Analysis

3.1   Experimental Environment

An experimental platform for fault detection is built in the matlab environment in order to verify the feasibility of the method in this paper in detecting faults of airborne ocean radar. The specific configuration of this experimental platform is shown in Table 1.

Table 1. Experimental platform configuration

Serial number	Name	Parameter
1	Operating system	64-bit Windows 7
2	Memory size	8 GB
3	Main frequency	3.6 GHz
4	Processor	Intel(R) Core(TM) i8–3770
5	Matlab version	2019 A

The experimental platform built in this paper is composed of 10 PCs and the operating environment of airborne ocean radar. The experimental data is collected when the airborne ocean radar is in a fault state. The fault of airborne ocean radar is detected using the methods in this paper and the traditional methods.

3.2   Collecting Fault Data Samples of Airborne Ocean Radar

The sampling block length is 1000 (experimental space), the sampling period for airborne ocean radar fault data is 0.05 s (experimental time), the number of iterations of fault data feature extraction is 50, and the number of fault data samples is 1200. In the process of equipment failure, first, the failure data samples of airborne ocean radar are collected according to the characteristics of the failure data, as shown in Fig. 3.

3.3   Detecting Faults of Airborne Ocean Radar

The sample data in Fig. 3 is taked as the test object, combined with the idea of feature fusion, the correlation of airborne ocean radar fault data is integrated, and under different fusion coefficients, the method in this paper is used to detect airborne ocean radar faults, and detection of airborne ocean radar faults. The output is shown in Fig. 4.

The fault data of airborne ocean radar can be detected by the method in this paper according to the results in Fig. 4. When the fusion coefficient is 0.2, the early detection output of fault data has changed. The reason is the difference in the matching degree of fault features in the early stage of fault detection. As a result, it began to stabilize afterwards; when the fusion coefficient was 0.4, the detection output of the fault data was relatively stable, which verifies the effectiveness of the method in this paper.

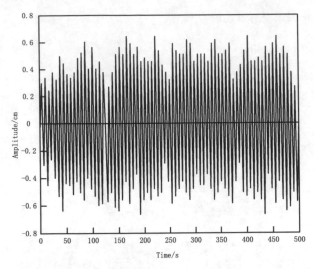

Fig. 3. Sampling of fault data of airborne ocean radar

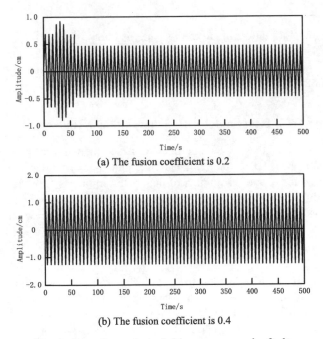

(a) The fusion coefficient is 0.2

(b) The fusion coefficient is 0.4

Fig. 4. Detection output of airborne ocean radar faults

3.4 Performance Test

The fault detection experiment of airborne ocean radar is carried out in two stages. First, the detection effect of airborne ocean radar faults is measured by using the indicators of missed detection rate and false detection rate. The missed detection rate refers to the proportion of fault signals not found in the fault detection of airborne ocean radar in the total number of faults. The false detection rate refers to the ratio of detected faults to the total number of airborne ocean radar faults; The signal-to-noise ratio index is used to measure the detection quality of fault signals of airborne ocean radar in the second stage of the experiment. The calculation formula is:

$$SNR = 10 \log \frac{\sum_{i=1}^{N} |a(i)|^2}{\sum_{i=1}^{N} |a(i) - \overline{a}(i)|^2} \tag{28}$$

where, $a(i)$ represents the fault occurrence signal function of airborne ocean radar, $\overline{a}(i)$ represents the signal strength of airborne ocean radar failure, and N represents the number of fault data.

The fault detection method based on PLC technology (literature 3) obtains the frequency band components of the motor by wavelet packet analysis, and extracts the fault characteristic parameters by analysing the changes of each frequency band component. According to the fault characteristics, linear fault information is obtained by using PLC technology and the fault detection is completed by using fuzzy kernel clustering method; the fault detection method based on generation uncertainty (literature 4) uses the random response surface method to obtain the cumulative probability distribution of the load margin under fault conditions, filter and sort out the static voltage stability and complete the fault detection. In order to verify the advantages of the method in this paper, the above two traditional methods based on PLC technology (Ref. 3) and the fault detection method based on generation uncertainty (Ref. 4) were used as comparison methods and tested against the method in this paper. The data samples in Fig. 3 were used to test the fault miss rate of the three methods for airborne marine radar. The results are displayed in Fig. 5.

It can be seen that the fault detection method based on PLC technology has a high missed detection rate when detecting the faults of airborne ocean radar from the results in Fig. 5. When the amount of fault data exceeds 60, the missed detection rate is as high as 57%. As the amount of data increases, the maximum missed detection rate reaches 70%; when the fault detection method based on the uncertainty of power generation is used, the fault missed detection rate is low, between 0 and 40%, which is significantly lower than that based on PLC technology. For the fault detection method, when the method in this paper is adopted, when the amount of fault data is 50, the fault missed detection rate reaches the maximum value of 10%. The missed detection rate begins to decrease, indicating that the judgment of the fault area can be accurate with the increase of the fault data volume. Locate the location of the fault, reducing the missed detection rate.

The test results of the fault detection rate of airborne ocean radar are shown in Fig. 6.

It can be seen from the results in Fig. 6 that the test results of the fault detection method based on PLC technology and the fault detection method based on the uncertainty of power generation in terms of the fault error detection rate of airborne ocean radar are relatively high, ranging from 0 to 75% and 0 to 40% respectively, while the error detection rate of the method in this paper in detecting the fault of airborne ocean radar is between

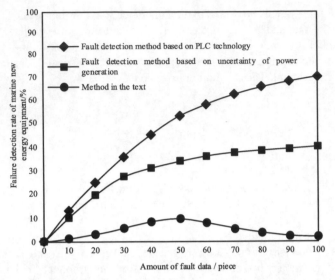

Fig. 5. The test results of the failure and missed detection rate of airborne ocean radar

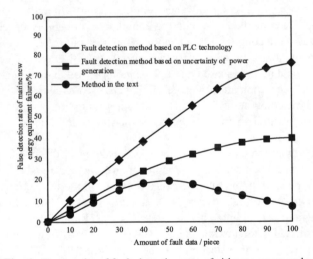

Fig. 6. Test results of fault detection rate of airborne ocean radar

0 and 20%. When the amount of fault data exceeds 50, the fault error detection rate starts to gradually decrease, When the amount of fault data is 100, the fault false detection rate is only 8%. It is stated that the method in the text can avoid the phenomenon of false detection of airborne ocean radar faults.

The results of the marine radar fault message recognition rate comparison test are shown in Fig. 7.

From the results in Fig. 7, it can be seen that the above two traditional methods based on PLC technology (Ref. 3) and the fault detection method based on generation

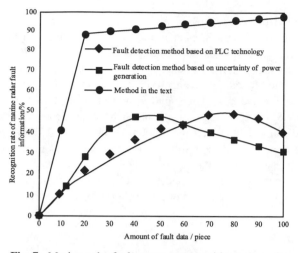

Fig. 7. Marine radar fault message recognition test results

uncertainty (Ref. 4) both The fault information recognition rate of both traditional methods is below 50%, and the information recognition effect is poor and cannot effectively locate the faulty equipment. When the number of fault data exceeds 30, the recognition rate of this method is above 90%, and when the number of data is 100, the recognition rate of this method is as high as 98%. The main reason is that this paper's method uses wavelet transform to transform the fault information, which can accurately locate the faulty marine radar, therefore, the fault information recognition rate is high.

The signal-to-noise ratio test results of the three methods in the fault signal collection of airborne ocean radar are shown in Fig. 8.

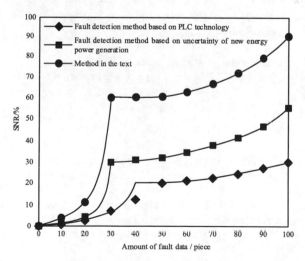

Fig. 8. SNR test results

It can be seen from the results in Fig. 8 that in the signal-to-noise ratio test of collecting fault signals of airborne ocean radar, when the amount of fault data exceeds 40, the fault detection method based on PLC technology reaches 20% signal-to-noise ratio. When the amount of fault data exceeds 30, the signal-to-noise ratio obtained by the fault detection method based on the uncertainty of power generation and the method in the paper reaches 30% and 60%. The signal-to-noise ratio of the method in this paper is still rising rapidly when collecting fault signals of airborne ocean radar with the increase of the amount of fault data. When the amount of fault data exceeds 100, the signal-to-noise ratio of the collected fault signals of airborne ocean radar is as high as 90%. This method utilizes the operating trajectories of fault data variables at different times to detect airborne marine radar faults and extract key variable information for detection. It can effectively ensure the detection quality of airborne marine radar fault signals.

4   Conclusion

A fault detection algorithm is proposed and improved for ocean radar faults. This algorithm innovatively combines feature extraction algorithms to preprocess fault data, define the feature matching degree of fault data, calculate the average trajectory of the working state observation vector, and combine the operation trajectory of fault data variables at different times to detect faults in marine radar, achieving efficient fault detection. The example analysis results show that the algorithm has good performance in fault detection of ocean radar. However, there are still many shortcomings in this study, and it is hoped that engineering case studies can be conducted in future research. Under on-site working conditions, the impact of noise factors on the operation of marine radar can be eliminated, and the working status of marine radar can be obtained. This helps marine survey personnel to timely detect radar equipment faults and accurately locate and diagnose problems. Based on this, potential fault problems can be identified in a timely manner, improving the safety of marine traffic, and ensuring the efficiency and accuracy of marine sub resource survey.

Acknowledgement.   2020 Teaching Research Project of Wuhan Institute of Design and Sciences: Case based "linear algebra" Hybrid Teaching Research and Practice (Project No.: 2020JY101).

References

1. Akta, A., Kriek, Y.: A novel optimal energy management strategy for offshore wind/marine current/battery/ultracapacitor hybrid renewable energy system. Energy **199**(15), 117425–117438 (2020)
2. Ren, Y., Huang, J., Hu, L., et al.: Denoising algorithm of hydro-generator set based on fourier decomposition and permutation entropy. Water Power **46**(10), 96–99+116 (2020)
3. Wang, L., Wang, Y., Zhang, Y.: Fault detection of new energy vehicle motor drive system based on PLC technology. Mach. Des. Manuf. 06, 199–202+207 (2022)
4. Haibo, B.A.O., Xiaoxuan, G.U.O.: Fault screening and ranking method of static voltage stability considering uncertainty of renewable energy power generation. Electr. Power Autom. Equipment **39**(7), 57–63 (2019)

5. Wumaier,·T., Mu, H., Xia, Q.: Vibrationfault diagnosis of water turbine-generator set based on artificial immune algorithm. Yangtze River **52**(5), 209–211+222 (2021)
6. Jia, Q., Guo, J., Sheng, B.: Research on pattern recognition of LiDAR hardware fault data. Laser J. **43**(4), 195–199 (2022)
7. Chen, J., Xu, H., Wu, T.: Fault early warning technology of airborne radar based on wavelet energy spectrum entropy and HMM. Fire Control Command Control **47**(10):31–35+40
8. Bian, H.U., Picheng, T.A.N., Yuan, Y.E., et al.: Design of noise signal acquisition system for bulb tubular hydro-generating unit based on acoustic characteristics. Chin. Meas. Test **47**(03), 139–143 (2021)
9. Silva, A., Zarzo, A., González, J.M.M., Munoz-Guijosa, J.M.: Early fault detection of single-point rub in gas turbines with accelerometers on the casing based on continuous wavelet transform. J. Sound Vib. **487**(10), 460–480 (2020)
10. Zhang, C., Liu, Y.: Multi-faults diagnosis of rolling bearings via adaptive customization of flexible analytical wavelet bases. Chin. J. Aeronaut. **33**(2), 407–417 (2020)
11. Zhao, J., Hou, H., Gao, Y., et al.: Single-phase ground fault location method for distribution network based on traveling wave time-frequency characteristics. Electr. Power Syst. Res. **186**(Sep.), 106401.1–106401.9 (2020)
12. Zhang, H., Xie, Y., Yi T., et al.: Fault detection for high-voltage circuit breakers based on time-frequency analysis of switching transient e-fields. IEEE Trans. Instrum. Meas. **69**(4 Pt.2), 1620–1631 (2020)

An Automatic Control Algorithm for Sampling and Timing of Civil Radar Signal Based on DSP

Juan Li[1](✉) and Lingling Cui[2]

[1] College of Intelligent Equipment and Automotive Engineering,
Wuxi Vocational Institute of Commerce, Wuxi 214153, China
343384882@qq.com
[2] Department of Physical and Health Education, Wuxi Vocational Institute of Commerce,
Wuxi 214153, China

Abstract. Aiming at the problem of fixed period sampling control of radar signal and event triggered variable period sampling control, there are few research results at present, so a DSP based automatic control algorithm for civil radar signal sampling timing is designed. Using orthogonal intermittent sampling to implement sampling modulation on radar signals, combined with improved UNet3+network and sequence data recognition method, the modulation signal recognition is completed. Based on the recognition results, a DSP processor is designed with TTA technology as the core, integrating phase-locked loop synchronization and timing sampling technology to achieve automatic control of the signal synchronization sampling process. The test results show that the time control error of the algorithm is low, the control performance is good, and the control stability is higher than 95%. With the extension of the sampling period, the control stability does not show a downward trend.

Keywords: DSP Processor · Orthogonal Intermittent Sampling · Civil Radar · Signal Sampling Control

1 Introduction

After World War I, due to the potential threat of the Nazi German Air Force, the first batch of radars developed in the 1930s. During the Second World War, pulse delay system has been used to create moving target detection (MTD) radar. With the increase of oscillator stability, it has developed into full pulse Doppler radar. In addition to detecting moving targets, it can also measure target speed. In the period after the Second World War, the theory of radar system made a lot of progress, but the hardware limited the method of pulse Doppler. With the development of very large scale integrated circuits (VLSI) in the 1970s, the application of digital technology and computers is more extensive and in-depth, and the rise of pulse Doppler radar, the synthetic aperture technology can obtain high-resolution radar images similar to visible light imaging under the weather conditions with extremely low visibility. Since the Second World War, radar technology has become

L. Yun et al. (Eds.): ADHIP 2023, LNICST 548, pp. 132–147, 2024.
https://doi.org/10.1007/978-3-031-50546-1_9

increasingly popular and its performance has become increasingly perfect. Therefore, radar has been widely used in military, such as early warning, search, surveillance radar, fire control radar, and guidance radar; Civil use, such as weather radar, control radar, remote sensing equipment, etc.

In the field of civil radar application, radar is widely used in a variety of occasions, such as unmanned monitoring for motion parameter detection and inching parameter detection of indoor human targets to achieve smart home care; It is used for intelligent driving, and can assist in reverse parking, intelligent obstacle avoidance, detection and positioning of pedestrians and vehicles [1]. In the application of civil radar, signal sampling control has always been a key research problem.

In recent decades, due to the rapid development of digital technology and computer technology, more and more digital components have been used in the process of control systems, and analog controllers have also been replaced by computer digital controllers. In order to better adapt to the changes brought about by computers and digital components, scholars have proposed the sampling control theory. With the development of sampling control theory, fixed period sampling control method and variable period sampling control method have attracted extensive attention of many scholars.

The fixed period sampling control method is to sample at a fixed time interval, and use the components with signal holding function to keep the last sampling time information during the non sampling period, so as to control the system using the sampling information. Compared with the traditional continuous controller, the fixed period sampling control has the characteristics of saving transmission resources, saving computing resources, and strong applicability. At the same time, it is also easy to apply in the actual computer digital control process. In practical applications, sampling control has also been widely used, such as the actual paper industry, high-precision electric simulation turntable system, networked traction control system, etc. [2]. The fixed period sampling control method has become a hot research direction in the control field. Compared with the fixed period sampling control method, the variable period sampling control method collects the data of the system by changing the sampling period, and uses the sampling data to design the relevant variable period sampling controller control system. The more common method is the event trigger control method. Its main idea is to set the event trigger mechanism. When the event trigger mechanism is satisfied, it needs to sample the relevant information of the system. If the trigger mechanism is not satisfied, it uses a signal holding component to keep the sampling information unchanged until the next trigger time. Therefore, only when the conditional trigger mechanism is met, the system information needs to be sampled. Based on the above analysis, due to the rapid development of digital technology and computer technology, the fixed period sampling control method and variable period sampling control method are two key and hot research topics in the current control field, and some representative research results have been achieved. However, the achievements of fixed period sampling control and event triggered variable period sampling control in the field of fuzzy adaptive control are relatively few, and no systematic theoretical achievements have been formed.

For this reason, the automatic control problem of fixed period sampling of civil radar signal is studied, and some achievements have been made in the research of this problem. Some scholars have proposed a FPGA based method of rotor vibration signal full cycle equal phase sampling control. This method uses the Keyphasor Frequency Multiplier signal as the trigger signal of A/D conversion. The Keyphasor Frequency Multiplier circuit is implemented by the internal logic device of FPGA chip, and the periodic linear interpolation prediction module is added to the circuit to improve the accuracy of Keyphasor Frequency Multiplier. The research results show that the key phase frequency multiplication circuit designed by this method can achieve the predetermined function, and has the advantages of high integration, wide frequency multiplication range, etc. It can be widely used in the full cycle equal phase sampling control of rotor vibration signals. Some scholars proposed that the accuracy of data acquisition system should include the accuracy of sampling signal amplitude and the accuracy of sampling control timing. A design scheme of trigger timing sampling control in data acquisition system was given, which realized a comprehensive program control of multi trigger mode and solved the problem of precise synchronous trigger timing sampling control. Although both methods mentioned above can achieve automatic control of sampling timing, there is a problem of significant time control error. Therefore, a DSP based automatic control algorithm for sampling and timing of civil radar signals is designed to improve the accuracy of automatic control for sampling and timing of civil radar signals. Using orthogonal intermittent sampling to implement sampling modulation on the radar signal, the orthogonal intermittent sampling modulated radar signal is obtained. Improve the UNet3+network to balance recognition accuracy and training time, solve the problem of low recognition accuracy in complex electromagnetic environments, and achieve recognition of sampled modulation signals. Based on the results, a phase-locked loop synchronous sampling control circuit is designed using DSP as the core processor, combined with phase-locked loop synchronization and timing sampling technology, to achieve automatic control of the signal synchronous sampling process.

2   Design of Automatic Control Algorithm for Sampling and Timing of Civil Radar Signal

2.1   Signal Sampling and Modulation

The radar signal is sampled and modulated by quadrature intermittent sampling, and the quadrature intermittent sampling modulated radar signal is obtained.

Civil radar signal is composed of M The frequency of each sub pulse can be expressed as:

$$\begin{cases} l_m = l_0 + m\Delta l \\ m = 0, 1, 2, \ldots, M - 1 \end{cases} \tag{1}$$

In formula (1) m Indicates the sub pulse serial number; l_m Indicates the frequency of each sub pulse; l_0 Represents the starting frequency; Δl Indicates the frequency interval between adjacent sub pulses [3].

According to the above assumptions, the radar signal can be expressed as:

$$y(r) = \sum_{m=0}^{M-1} rect\left(\frac{r}{\alpha}\right) \exp(k2\pi l_m r) \tag{2}$$

In formula (2)r Represents a fast time variable; α Indicates the pulse width; k Represents the interval parameter.

among r Meet the following formula:

$$-\frac{R}{2} \leq r \leq \frac{R}{2} \tag{3}$$

In formula (3), R is a real number.

The schematic diagram of civil radar signal is shown in Fig. 1.

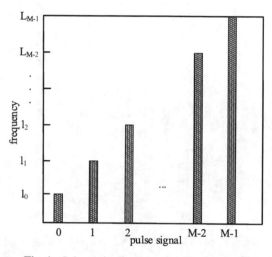

Fig. 1. Schematic diagram of civil radar signal

To ensure the orthogonality between subpulses, according to the characteristics of sin c (□) function, the frequency interval between subpulses of radar signal Δl The following formula should be met:

$$\Delta l = \frac{b}{\alpha} \tag{4}$$

In formula (4)b It is a positive integer.

In particular, it can make $\Delta l = \frac{1}{\alpha}$.

The equivalent bandwidth of civil radar signal can be expressed as:

$$C = M \, \Delta l \tag{5}$$

Therefore, it uses multiple narrow band sub pulses to achieve equivalent large bandwidth and can be used as radar imaging signal [4]. For civil radar, FFT and other signal processing methods based on Fourier transform are often used for imaging processing, and the corresponding range resolution is:

$$\Delta U = \frac{d}{2C} \tag{6}$$

In formula (6) d Is the transmission rate of electromagnetic wave.
The non blurring distance is:

$$\Delta u = \left[-\frac{d}{4\Delta l}, \frac{d}{4\Delta l} \right] \tag{7}$$

The model of quadrature coded intermittent sampling modulating civil radar signal is shown in Fig. 2.

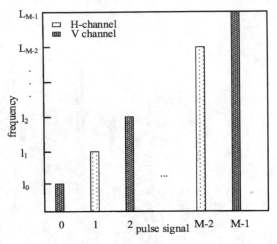

Fig. 2. Schematic diagram of quadrature coded intermittent sampling modulated radar signal

For H channel and V channel, randomly select one of them from the complete civil radar signal $\frac{M}{2}$ It is composed of mutually orthogonal sub pulses. The sub signal frequencies of the two orthogonal polarization channels can be expressed as:

$$l_{Hm} = l_0 + F_H(m)\Delta l \tag{8}$$

$$l_{Vm} = l_0 + F_V(m)\Delta l \tag{9}$$

In formula (4)F_H, F_V Is the orthogonal subset of the integer set [0: M □ 1], where the number of elements of each is $\frac{M}{2}$ [5].

Quadrature coded intermittent sampling modulated civil radar signal can be expressed as:

$$y_H(r) = \sum_{m=0}^{M/2-1} rect\left(\frac{r}{\alpha}\right) \exp(k2\pi l_{Hm}r) \tag{10}$$

$$y_V(r) = \sum_{m=0}^{M/2-1} rect\left(\frac{r}{\alpha}\right) \exp(k2\pi l_{Vm}r) \tag{11}$$

Due to the orthogonality between subpulses of civil radar signals, the same polarization component and cross polarization component of the quadrature coded intermittent sampling modulation signal can be transformed by Fourier transform of the received echo signal and l_{Hm}, l_{Vm} Index is detached.

2.2 Identification of Sampled Modulation Signal

The recognition method based on sequence data is used to realize the recognition of sampled modulation signals. Compared with the method of processing time-frequency images, sequence data generally does not need various complex transformations, and the amount of calculation is relatively small, which can largely retain the information of the original data, and can be suitable for processing large amounts of data. UNet3+network is introduced and improved to give consideration to both recognition accuracy and training time, so as to solve the problem of low recognition accuracy in complex electromagnetic environment.

Recognition of radar signals based on deep learning, the better the features extracted, the better the recognition ability of neural networks [6]. UNet3+network integrates shallow features and deep features, and better describes signal features by complementing each other to avoid the loss of key information, so that the network can judge through more features and has better recognition ability. Although the UNet3+network has a strong ability to extract features, it also has some limitations: the unimproved UNet3+network has five levels, and every additional network level requires more feature fusion, resulting in an overly complex model structure, requiring more operations, and reducing the training speed of the network.

Aiming at the problems of the above UNet3+network, an improved UNet3+network model [7] is proposed on the basis of the original network. There are two main improvements: first, in order to improve the training speed of the network, the hierarchy of the network is deleted, as shown in the following formula:

$$f = 5 - 2 = 3 \tag{12}$$

On the premise of avoiding the decline of network recognition accuracy, reduce redundant fusion and reduce the amount of computation.

The second is the introduction of attention mechanism. Although UNet3+network has the ability to extract detailed data features, not all the extracted features are helpful to judge the recognition signal. Some features are invalid features. The attention mechanism will invalidate the redundant features, so that the network can focus on key features, highlight the impact of key features, and improve network performance.

The improved UNet3+network structure includes convolutional pooling layer module, feature fusion module, attention layer and other layers. In the convolution pooling layer module, there are three one-dimensional convolution pooling layers with the same parameters. The feature is extracted through 32 convolution cores with a length of 5. The three convolution pooling layers are followed by regularization layers with coefficients of 0.2, 0.1, and 0.1 to prevent over fitting. Then through convolution layers 4, 5 and 6, 8 convolution kernels with length of 5 are used for convolution operations. In the feature fusion module is the feature fusion of full scale skip connection, where convolution layers 7 and 8 are convolutional operations through 8 convolution kernels with length of 7, and convolution pooling layer 9 is convolutional operations through 8 convolution kernels with length of 7. All the maximum pooled layers in the model have a size of 2. Finally, the regularized layer is followed by the tiling layer and the full connection layer to output the final result [8] (Tables 1 and 2)

Table 1. Module Structure of Convolution Pooling Layer

S/N	Hierarchy	Input	Output
1	Convolution pooling layer 1	(None, 1024, 1)	(None, 512, 32)
2	Regularization layer 1	(None, 512, 32)	(None, 512, 32)
3	Convolution pooling layer 2	(None, 512, 32)	(None, 256, 32)
4	Regularization layer 2	(None, 256, 32)	(None, 256, 32)
5	Convolution pooling layer 3	(None, 256, 32)	(None, 128, 32)
6	Regularization layer 3	Input: (None, 128, 32)	Output: (None, 128, 32)

Table 2. Structure of feature fusion module

S/N	Hierarchy	Input	Output
1	Feature fusion layer 1	Input 1: (None, 512,8) Input 2: (None, 256, 8) Input 3: (None, 128,8)	(None, 896, 8)
2	Feature fusion layer 2	Input 1: (None, 512, 8) Input 2: (None, 128, 8) Input 3: (None, 896, 8)	(None, 1536, 8)

The input of attention layer is (None, 1536, 8), and the output is (None, 1536, 8). The structure of other levels is shown in Table 3.

The training process of the model is as follows:

Step 1 Data input

Input radiation source data and corresponding labels into the network model.

Step 2 Sample pretreatment

Min Max normalization is performed on the sample set, and the sample value obtained will be limited to [0, 1].

Step 3 Convert Label

The labels of the eight types of signals are converted into unique hot codes, which are represented as binary vectors.

Step 4 Build a training dataset

In order to strengthen the generalization ability of the model, random seeds were used to disrupt eight types of signal samples, and the samples were divided into training sets, verification sets, and test sets at a ratio of 0.47:0.23:0.30;

Table 3. Structure of other levels

S/N	Hierarchy	Input	Output
1	Convolution layer 4	(None, 512, 32)	(None, 512, 8)
2	Convolution layer 5	(None, 256, 32)	(None, 256, 8)
3	Convolution layer 6	(None, 128, 32)	(None, 128, 8)
4	Convolution layer 7	(None, 896, 8)	(None, 896, 8)
5	BN floor 1	(None, 896, 8)	(None, 896, 8)
6	Convolution layer 8	(None, 896, 8)	(None, 896, 8)
7	Convolution pooling layer 9	(None, 1536, 8)	(None, 1536, 8)
8	BN floor 2	(None, 1536, 8)	(None, 1536, 8)
9	Regularization layer 4	(None, 1536, 8)	(None, 1536, 8)
10	Tile	(None, 1536, 8)	(None, 12288)
11	Full connection layer	(None, 12288)	(None, 8)

Step 5 Set the early stop mechanism

In network training, when the loss of a round of verification set is greater than the loss m of the previous round of verification set, and all losses after four more rounds of training are greater than m, the network training is ended;

Step 6 Compile the network

Adam is used as the network optimizer, and category is used_Crossentropy function calculates the loss;

Step 7 Add a learning rate dynamic adjustment mechanism

The learning rate is initially set to 0.0001. During network training, when the loss of verification set increases, the learning rate of this round is multiplied by 0.5 as the learning rate of the next round of training. The minimum learning rate is not less than 0.0000125;

Step 8 Classified network training

Set the number of samples selected for each batch of training to 512 and the maximum number of training rounds to 50 for training.

2.3 Sampling Timing Automatic Control

According to the result recognition, a PLL synchronous sampling control circuit is designed with DSP as the core processor, combined with PLL synchronization and timing sampling technology, to realize the automatic control of signal synchronous sampling process.

Based on TTA technology, a DSP processor is designed, which is composed of Internet and processing unit.

The Internet provides a channel for each unit in the processor to exchange data. It includes two basic modules: Socket and bus. In addition to providing data exchange function, the bus is also used to transmit control signals, such as source and target register

IDs, function unit latch signals, etc. Socket provides the connection between function unit and register file and bus. Each socket can be connected to one or more buses and one or more registers [9] of a function unit. The input socket transfers the data from the bus to the function unit and register file, and the result socket places the data from the function unit and register file on the bus. The input socket is further divided into a trigger socket and an operand socket. The only difference between them is that the trigger socket transmits data at the same time, and the operation code is also passed into the functional unit to specify the type of operation. The source address field and destination address field are transferred from the control bus to all connection points. A three input socket and two output socket interconnection network is designed, as shown in Fig. 3.

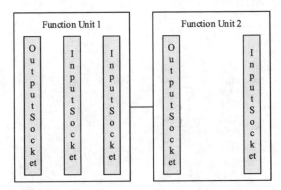

Fig. 3. Interconnection network structure design

The programming method of TTA is to specify some required data values to flow in TN. Therefore, this architecture has only one operation mode: moving from one FU to another. In general, FU is connected to TN through three different types of registers: Operand, Trigger and Result. All transfers are from register to register.

The functions of the three registers are shown in Table 4.

The function unit can be pipelined internally. In the first stage, the operand register (O) and trigger register (T) temporarily store data from the input socket. Because the data of the operand socket can be temporarily stored through the operand register, the data can be written to the operand socket in advance. When the data is transferred to the trigger socket, the operation begins. In the first stage, a register contains an operation code, which controls which operation the functional unit performs. The operation code is encoded in the target ID used to indicate that the functional unit triggers the socket. Each pipeline segment of the functional unit contains a combination logic and pipeline register. All pipeline segments of the functional unit are synchronized with the instruction stream specified by the compiler. Every time an instruction flows out, the pipeline of the functional unit will advance one step [10].

In TTA structure, FU and data transmission network are completely separate, and the pipeline of transmission network and FU is also completely independent. Optimizing FU structure means maximizing the pipeline of FU. There will be a unique instruction prefetching unit, a unique Guard unit (responsible for instruction coding and conditional execution), and a unique Valid unit (used to obtain the value in the result register of the

Table 4. Functions of three registers

S/N	Register	Function
1	Operand	The operation mode is the same as GPR except that the corresponding FU of this operand register is different after being triggered
2	Trigger	The transmission written to the trigger register can cause the operation of the corresponding FU. The transmitted value is used as the operand of this operation. If more operands are needed, they can be extracted from other operand registers. In order to perform different operations on the same FU, different address contents of trigger register can be accessed
3	Result	Most FUs are pipelined, so operations can be started once per clock cycle. When the result of one operation leaves the pipeline, it will be written to the result register of FU. Starting from this result register, it can be written to other FUs in the next cycle through TN

FU unit) in the FU. The pipelining is realized in the following way: Hybrid pipeline: It is implemented in the TTA system structure in the way of Hybrid pipeline. In this pipeline mode, lower level pipelines can perform operations in advance if they do not affect the final result, which is equivalent to the multi-function in the pipeline. Hybrid pipeline allows maximum program scheduling freedom. The only limitation is the fixed length of pipeline and the fixed speed of each stage. Data can be read in or read out at any time. Of course, some causality must be considered. For example, when no data is read into the operation register, it is wrong to read the data in the result register.

The use of Guard in the TTA structure makes conditional execution possible. The execution of each MOVE instruction is conditional. Thus, each MOVE instruction includes the Guard selection bit. The MOVE instruction encoding method is related to the setting of the guard unit parameters SINGLEGUARD, PARMOVES, GUARDSPECSIZE. The function of SINGLEGUARD indicates that if SINGLEGUARD is defined, an instruction corresponds to a guard. If SINGLEGUARD is not defined, a MOVE includes a guard. As shown in Fig. 4 and Fig. 5.

In the TTA architecture, the functional unit is the main body that performs data operations. The function unit obtains data from the input socket, and places the result on the result socket after the operation is completed.

The PLL synchronous sampling control circuit consists of five parts: DSP processor, second-order active low-pass filter circuit, voltage zero crossing comparison circuit, PLL and frequency division circuit.

The structure of the circuit is shown in Fig. 6.

The cut-off frequency of the second-order active low-pass filter circuit is Wuqi, which is responsible for filtering out the second and above second harmonics to obtain pure fundamental signal. The voltage zero crossing comparison circuit is responsible for converting the fundamental sinusoidal signal obtained by the second-order low-pass filter into a square wave signal and outputting it to the back-end phase-locked loop and frequency division circuit for fundamental frequency multiplication. The obtained frequency multiplication signal is the control signal of the sampling period [11].

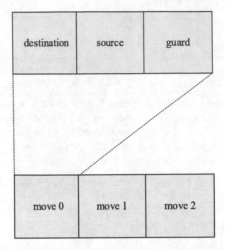

Fig. 4. Undefined instruction format of SINGLEGUARD

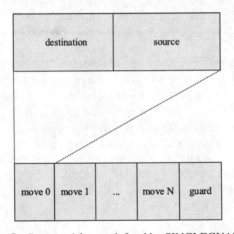

Fig. 5. Command format defined by SINGLEGUARD

The PLL frequency multiplication output frequency is 12800 Hz, and the control of system sampling is staggered, that is, 128 out of 256 samples are radar signals, and the other 128 are analog ground signals collected for zero calibration. The main chips used in the synchronous sampling control circuit of PLL include CD4046 and CD4020.

3   Case Test

3.1   Experimental Process

Relevant parameters of civil radar signals are shown in Table 5.

The design algorithm is used to automatically control the sampling and timing of the experimental civil radar signal. Firstly, the quadrature intermittent sampling is used to

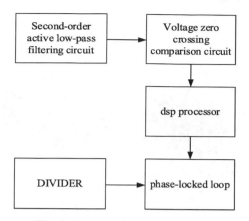

Fig. 6. Structure design of the circuit

Table 5. Relevant Parameters of Experimental Civil Radar Signal

S/N	Parameter	Parameter value	Company
1	Signal carrier frequency	30	GHz
2	Signal pulse width	45	μs
3	Signal bandwidth	8	MHz
4	Target speed	22	m/s
5	Pulse repetition period	550	μs
6	Signal IF	80	MHz
7	Sampling rate	40	MHz
8	Target slant distance	22	km

sample and modulate the radar signal of the experimental civil radar, and the quadrature intermittent sampling modulated radar signal is obtained, as shown in Fig. 7.

Then the recognition method based on sequence data is used to realize the recognition of sampled modulation signals. In recognition, the improved UNet3+network is trained first, and the training results are shown in Fig. 8 and Fig. 9.

It can be seen from the above figure that the improved UNet3+network model converges faster.

At last, the automatic control of signal synchronous sampling process is realized by PLL synchronous sampling control circuit. The time control error and control stability of the design algorithm are tested respectively. In the testing, two traditional methods, the FPGA based rotor vibration signal full cycle equal phase sampling control method and the trigger timing sampling control method in the data acquisition system, will be used as comparative testing methods, and represented by Method 1 and Method 2 respectively.

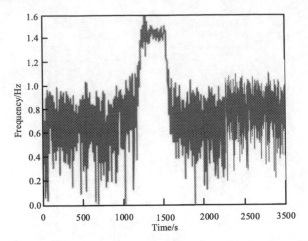

Fig. 7. Quadrature Intermittent Sampling Modulated Radar Signal Obtained

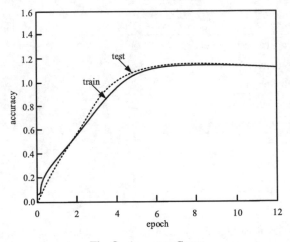

Fig. 8. Accuracy Curve

3.2 Test Results of Time Control Error

The time control error test results of test design algorithm and method 1 and method 2 under different sampling periods are shown in Table 6.

The test results in the above table show that the time control errors of the three methods are rising with the extension of the sampling period. The time control error of the design algorithm is below 2.54 ms, which is generally lower than the time control error of Method 1 and Method 2, which shows that the automatic control performance of the design algorithm is stronger.

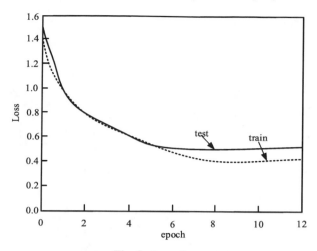

Fig. 9. Loss Curve

Table 6. Test Results of Time Control Error

Sampling period (s)	Time control error (ms)		
	Design algorithm	Method 1	Method 2
50	1.36	5.63	8.36
60	1.58	5.98	8.49
70	1.69	6.24	8.64
80	1.84	6.48	8.82
90	1.90	6.75	8.91
100	2.08	6.89	9.05
110	2.28	7.05	9.46
120	2.32	7.21	9.82
130	2.39	7.49	9.90
140	2.41	7.59	10.08
150	2.54	7.85	10.20

3.3 Control Stability Test Results

Then test the control stability of the design algorithm and method 1 and method 2 in different sampling periods. The test results are shown in Table 7.

According to the test results in the table above, the overall control stability of the design algorithm is far higher than that of Method 1 and Method 2. At the same time, with the extension of the sampling period, the control stability of Method 1 and Method 2 shows a downward trend, while the control stability of the design algorithm does not show a downward trend, which has been higher than 95%.

Table 7. Control Stability Test Results

Sampling period (s)	Control stability (%)		
	Design algorithm	Method 1	Method 2
50	95.634	86.326	82.963
60	95.874	86.302	82.908
70	95.125	86.268	82.900
80	95.365	86.228	82.586
90	95.025	86.086	82.521
100	95.985	86.005	82.498
110	95.652	85.058	82.432
120	95.458	83.219	82.405
130	95.365	83.208	81.352
140	95.158	83.196	81.304
150	95.365	83.187	81.210

4 Conclusion

An automatic control algorithm for sampling and timing of civil radar signals based on DSP has been designed, and some scientific achievements have been made. The experimental results show that the time control error of this algorithm is below 2.54 ms, and the control stability is above 95%. However, according to many other hot issues in this field, the author believes that there are still many aspects to be further studied, such as how to save resources and ensure the control effect of the periodic sampling controller, and the need for further application of the design algorithm.

References

1. Gao, M., Liang, Q., Nan, J., et al.: Blind multiband signal predistortion model based on compressed sampling structure. J. Signal Process. **38**(2), 319–328 (2022)
2. Elvira, V., Lsantamaria, I.: Multiple importance sampling for symbol error rate estimation of maximum-likelihood detectors in MIMO channels. IEEE Trans. Sig. Process. **69**, 1200–1212 (2021)
3. Solodky, G., Feder, M.: Optimal sampling of a bandlimited noisy multiple output channel. IEEE Trans. Sig. Process. **69**, 2026–2041 (2021)
4. Xie, M., Zhao, C., Hu, D.: Performance evaluation index of radar signal sorting based on pulse train. J. Signal Process. **38**(11), 2350–2358 (2022)
5. Liang, D., Chen, Y.: Frequency-quadrupled radar composite signal generation and multi-dimensional target detection enabled by microwave photonics. Acta Electron. Sin. Electron. Sin. **50**(04), 796–803 (2022)
6. Liu, S., Gao, P., Li, Y., et al.: Multi-modal fusion network with complementarity and importance for emotion recognition. Inf. Sci. **619**, 679–694 (2023)

7. Liu, Z., Liu, Y., Bao, X., et al.: The optimization design of measurement matrix based on KL-KSVD. Comput. Integr. Manuf. Syst.. Integr. Manuf. Syst. **40**(1), 388–393 (2023)
8. Heidenreich, P., Zoubir, A., Bilik, I., et al.: Editorial: introduction to the issue on recent advances in automotive radar signal processing. IEEE J. Selected Topics Sig. Process. **15**(4), 861–864 (2021)
9. Babu, E.: Denoising of MST RADAR signal using CWT and overlapping group shrinkage. Turkish J. Comput. Math. Educ. (TURCOMAT) **12**(5), 1950–1954 (2021)
10. Chang, H.L., Park, J.: Radar signal detection applying deep learning to the correlation between adjacent windows in time domain. J. Korean Institute Commun. Inform. Sci. **46**(7), 1153–1155 (2021)
11. Devi, M.R.: Low probability of intercept (LPI) radar signal identification techniques. Biosci. Biotechnol. Res. Commun. **14**(5), 365–373 (2021)

Design of Control System for Constant Speed Variable Pitch Loaded Multi Axis Unmanned Aerial Vehicle Based on Lidar Technology

Xin Zhang$^{(\boxtimes)}$ and Mingfei Qu

College of Aeronautical Engineering, Beijing Polytechnic, Beijing 100176, China
zhangxin197802@163.com

Abstract. In order to ensure the stability of drone flight and improve control performance, a constant speed, variable pitch, and heavy-duty multi-axis drone control system design based on LiDAR technology is proposed. In the hardware design of the system, the power supply circuit for drone control is designed based on the principle of lithium battery charging and discharging circuit and the working principle of SX1308; Based on the mathematical model of drone position control, a control law for the horizontal position outer loop and the horizontal position inner loop was designed. Based on the closed-loop structure diagram of the drone position controller, a drone position controller was designed to ensure that the drone can accurately hover, return, and waypoint flight. In the software design of the system, LiDAR technology is used to extract drone trajectory features, and combined with drone control algorithm design, drone control is achieved. The system testing results show that the system in the article can achieve the expected design goals and improve the control accuracy to over 90%.

Keywords: laser radar technology · constant speed and variable pitch · multi axis UAV · control system

1 Introduction

By improving the load-bearing capacity of drones, larger scale cargo transportation and equipment carrying can be achieved, improving transportation efficiency and economic benefits. At the same time, by optimizing flight control algorithms and sensor integration, the stable flight of unmanned aerial vehicles in complex environments can be guaranteed, and the risk of accidents can be reduced. In addition, the design of the control system also involves enhancing autonomous navigation and task execution capabilities, enabling drones to autonomously plan paths, execute tasks, and make corresponding adjustments when encountering obstacles or faults [1]. This will reduce the burden of manual intervention, improve work efficiency and operational safety. From the above analysis, it can be seen that the research on the control system design of constant speed, variable pitch, and heavy-duty multi-axis unmanned aerial vehicles provides strong support for the widespread application of unmanned aerial vehicles in logistics, agriculture, rescue, and other fields, promoting sustainable development of society and economy.

© ICST Institute for Computer Sciences, Social Informatics and Telecommunications Engineering 2024
Published by Springer Nature Switzerland AG 2024. All Rights Reserved
L. Yun et al. (Eds.): ADHIP 2023, LNICST 548, pp. 148–161, 2024.
https://doi.org/10.1007/978-3-031-50546-1_10

In the above context, relevant scholars have proposed some system design methods. In order to improve the accuracy of agricultural unmanned aerial vehicle navigation, Wei Hongfei et al. [2] used sliding mode control method to establish a control model for navigation line tracking and master-slave system speed coordination, and verified the control effect of the controller through virtual simulation. The simulation results show that the use of sliding mode control method can effectively reduce navigation errors, allowing unmanned agricultural machinery to travel along the navigation line during operation with small centroid deviation. When sliding mode control method is used alone, there may be certain fluctuations in error, resulting in system instability. The introduction of PID feedback regulation control algorithm for further verification of the system indicates that the PID algorithm can effectively eliminate the fluctuation of system error, improve the stability of the system, and thereby improve the quality of agricultural machinery operation. Wu Yongcheng et al. [3] used Newton's law of motion and the law of rigid body rotation to establish the six degrees of freedom motion model of UAV, and obtained model parameters by means of experimental measurement and numerical simulation. The problem of control redundancy is solved by designing control Partition coefficient. The attitude controller based on linear active disturbance rejection control is designed, and its parameters are adjusted online by fuzzy control, which improves the response speed and robustness of the control system. Simulink simulation model is built, and the longitudinal flight control simulation test of UAV verifies the rationality of the designed control Partition coefficient and the effectiveness of the attitude controller. Yang K et al. [4] proposed a unmanned aerial vehicle control system based on Pixhawk flight controller, which consists of two subsystems: flight control system and spraying system. Using PX4 based firmware to implement Pixhawk flight control. The communication protocol is analyzed in detail, and the communication control between flight controller and spray controller is realized through serial port. The software part designs the remote control decoding task of the flight control system and the spray task of the spray system, realizes the communication between the flight control system and the spray system, and successfully completes the spray function. Shi Jia et al. put forward an improved design method for the control system of auto disturbance rejection quad rotor UAV. According to the actual parameters of the built quad rotor UAV, a numerical simulation model of the attitude control system of the quad rotor UAV was constructed. By comparing with the double closed-loop PID controller, it was proved that the designed Active disturbance rejection control system has strong anti-interference ability and high control efficiency under the premise of fast response and no overshoot.

Although the above methods can effectively control the flight of unmanned aerial vehicles, the control accuracy is relatively low. Therefore, this article applies LiDAR technology to the design of a constant speed, variable pitch, and heavy-duty multi-axis unmanned aerial vehicle control system. Lidar technology can provide high-precision environmental perception capabilities. By measuring the distance of the surrounding environment and the position of obstacles, real-time information about the terrain, buildings, and obstacles around drones can be obtained. This precise environmental perception helps drones avoid collisions, improve flight safety, and improve control accuracy.

2 Hardware Design of Control System for Constant Speed and Variable Pitch Loaded Multi Axis UAV

2.1 Design of Power Supply Circuit for UAV Control

The constant speed variable pitch loaded multi axis UAV is powered by batteries, and the power supply circuit is shown in Fig. 1.

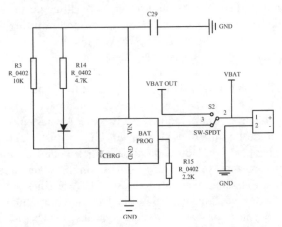

Fig. 1. Schematic diagram of lithium battery charging and discharging circuit

The figure shows the charging circuit of lithium battery. The chip is LTC4054, and pin 1 is the charge state pin. When the circuit is charged, the pin will be pulled down, and pin 1 is connected to the diode negative pole. When the lithium battery is discharged, the diode is lit, indicating that the battery is currently charging. Pin 3 is the charging current output pin, providing current and voltage. Pin 3 is externally connected to the lithium battery. Pin 5 is the standard pin of charging current. When connection point S2 contacts connection point S3, the lithium battery starts charging. When connection point S2 contacts connection point S1, the lithium battery supplies power to the system through VBAT-OUT.

In Fig. 1 the output voltage of Pin VBAT is 3.7 V, and the required voltage for system operation is 5 V. The chip SX1308 needs to increase the 3.7 V voltage to 5 V. SX1308 is a fixed frequency, SOT23-6 packaged current mode boost converter [5]. The operating frequency of up to 1.2 MHz allows smaller specifications of the external inductance and capacitance. The built-in soft start function reduces the starting impulse current. The working principle of SX1308 is shown in Fig. 2.

In Fig. 2, the output voltage is adjusted by the feedback voltage divider R_1 and R_2 to set the output voltage as:

$$VOUT = VREF \times (1 + \frac{R_1}{R_2}) \tag{1}$$

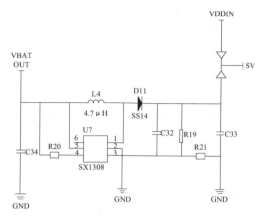

Fig. 2. SX1308 Working Principle Diagram

The chip SX1308 has 6 pins. SW is a switch node, which needs to be connected with inductance. GND is a grounding pin, FB is an output feedback pin, and the input feedback voltage is 0.6 V. EN is an energy pin. When the drive EN is greater than 1.5 V, it will activate, and when it is less than 0.4 V, it will stop. The EN pin cannot be suspended. In is an input power pin. The power supply voltage of the system chip is 3.3 V, and the 5 V voltage is reduced to 3.3 V through the voltage stabilizing chip XC6306.The voltage is detected by the voltage output pin VBAT_TST is connected to pin 10 of STM32, which has the function of ADC [6], and can be used to monitor the voltage of the system.

When using SX1308 chip, pay attention to the following in PCB layout:

- The input and output capacitors must be close to the GND pin of the IC to reduce the current loop area;
- AC current will flow through VIN, SW and VOUT cabling, so ensure that these cabling is short and wide;
- There is alternating voltage on the copper sheet at the SW pin. To prevent EMI, it needs to be controlled in a small area;
- The FB pin is a high impedance node. The FB wiring should be short enough to avoid noise causing output voltage fluctuations. The feedback resistor should be placed as close as possible to the IC. At the same time, R2 and GND should be placed as close as possible to the GND pin of the IC. The wiring from VOUT to R1 should be far away from inductance and switch nodes.

2.2 Design of UAV Position Controller

The position controller of UAV is designed to ensure that the UAV can perform position control functions such as hovering flight, homing flight and waypoint flight accurately, and it is also one of the necessary conditions for UAV to complete aerial photography tasks [7]. In the process of controller design, the difficulty of nonlinear controller design is often greater than that of linear controller design, so in order to achieve the design of position controller more easily, it is necessary to properly simplify and linearize the nonlinear mathematical model of UAV position control. According to the UAV position control mathematical model, it can be expressed as:

$$\begin{cases} \ddot{p}_x^e = -\frac{F}{m}(\sin\theta\cos\varphi\cos\psi + \sin\varphi\sin\psi) + f_x \\ \ddot{p}_y^e = -\frac{F}{m}(\sin\theta\cos\varphi\sin\psi - \sin\varphi\cos\psi) + f_y \\ \ddot{p}_z^e = -\frac{F}{m}\cos\theta\cos\varphi + g + f_z \end{cases} \tag{2}$$

Among them, φ, θ and ψ represent the three channels of the UAV attitude, f_x, f_y and f_z represent the components of the three channels, g represents the Gravitational acceleration, m represents the UAV flight distance, and F represents the resistance to the UAV flight.

From the above equations, it can be seen that the mathematical model of UAV flight control system is not a linear control system. In the process of position controller design, the introduction of nonlinear system should be avoided as far as possible, and it should be simplified. The position controller of UAV should be designed based on the simplified controlled object. If the UAV does not tilt at a large angle during flight, and the power generated by the blades on the boom is equal to the gravity of the UAV, the above nonlinear equation can be simplified as:

$$\begin{cases} \ddot{p}_x^e = -\frac{F}{m}(\theta\cos\psi + \varphi\sin\psi) + f_x \\ \ddot{p}_y^e = -\frac{F}{m}(\theta\sin\psi - \varphi\cos\psi) + f_y \\ \ddot{p}_z^e = -\frac{F}{m} + g + f_z \end{cases} \tag{3}$$

Finally, the plane rotation matrix constructed by heading angle ψ can be decoupled into a linear model of three channels, namely, two horizontal channel position controllers and one altitude channel position controller.

For the above linear position model, the position of UAV horizontal channel and altitude channel can be obtained, and the kinematic model without disturbance can be obtained, which is expressed as:

$$\dot{p} = \begin{pmatrix} \frac{dp_x}{dt} \\ \frac{dp_y}{dt} \\ \frac{dp_z}{dt} \end{pmatrix} = \begin{pmatrix} v_x \\ v_y \\ v_z \end{pmatrix} \tag{4}$$

where, p_x, p_y and p_z are the three position channels in the horizontal and altitude directions respectively, and v_x, v_y and v_z are the three speed channels in the horizontal and altitude directions respectively. Formula (4) is a Kinematics model without any external interference, so the controller can control the UAV well, so the control law of the outer position loop can be designed as:

$$\vec{v}_{exp} = K_p \vec{e}_p \tag{5}$$

In the above equation, $\vec{v}_{exp} = \begin{bmatrix} v_{x\_exp} & v_{y\_exp} & v_{z\_exp} \end{bmatrix}^T$ is the expected speed value, $\vec{e}_p = \begin{bmatrix} e_{px} & e_{py} & e_{pz} \end{bmatrix}^T$ is the outer loop error of the position controller, and K_p is the adjustable parameter value of the proportional controller.

For the mathematical model of the controlled object in the position inner loop of UAV, it can be seen from Formula (3) that it contains unknown external disturbances, so the control law of the horizontal position inner loop can be designed as:

$$\begin{cases} \vec{e}_v = \vec{v}_{exp} - \vec{v} \\ \vec{u} = K_{pv}\vec{e}_v + K_{iv}\int_0^t \vec{e}_v dt \end{cases} \tag{6}$$

Among them, K_{pv} represents the adjustable parameters of the proportional controller, K_{iv} represents the adjustable parameters of the integral controller.

According to the above calculation, the closed loop structure block diagram of UAV position controller is designed, as shown in Fig. 3.

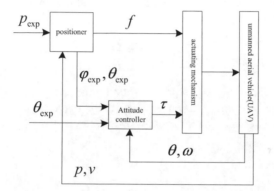

Fig. 3. Closed loop structure block diagram of UAV position controller

In conclusion, according to the internal and external loop control rate of horizontal position, the UAV position controller is designed.

3 Software Design of Control System for Constant Speed and Variable Pitch Loaded Multi Axis UAV

3.1 Extraction of UAV Track Characteristics

Using LiDAR technology to read drone trajectory image data from the CPU into the GPU, assuming (x, y) represents the pixel coordinates of the drone trajectory image, $I(x, y)$ represents the original drone trajectory image, δ represents the scale coordinates of the drone trajectory image, and the Gaussian kernel calculation formula for the drone trajectory image is:

$$G(x, y, \delta) = \frac{1}{2\pi\delta^2}e^{-(x^2y^2)/2\delta^2} \tag{7}$$

According to the above UAV trajectory image Gaussian kernel, the scale space of UAV trajectory image after filtering by Gaussian filter [8] is obtained, and the calculation formula is:

$$L(x, y, \delta) = G(x, y, \delta) * I(x, y) \tag{8}$$

On the basis of the above scale space, the Gaussian difference pyramid model of UAV trajectory image is constructed according to the UAV trajectory detected by the laser radar. The calculation formula is:

$$D(x, y, \delta) = L(x, y, \rho\delta) - L(x, y, \delta) \tag{9}$$

Among them, ρ represents the scale coefficient of UAV trajectory image.

According to the Difference of Gaussians pyramid model $D(x, y, \delta)$ of the UAV trajectory image, remove the points and boundary points with low contrast, use the laser radar detection principle to determine the specific location of the UAV trajectory image feature points, use the Hessian matrix to calculate the Principal curvature of the UAV trajectory image, and judge whether the pixel (x, y) in the model is a boundary point. The judgment formula is:

$$H = \begin{bmatrix} D_{xx} & D_{xy} \\ D_{xy} & D_{yy} \end{bmatrix} \tag{10}$$

Among them, D_{xx}, D_{xy}, D_{yy} represents the partial derivative at the characteristic point of UAV trajectory image. Then we can get the trajectory and determinant value [9–11] of the Hessian matrix $tr(H)$ and $Det(H)$ representative, the calculation formula is:

$$tr(H) = D_{xx} + D_{xy} \tag{11}$$

$$Det(H) = D_{xx} \cdot D_{xy} - \left(D_{xy}\right)^2 \tag{12}$$

By setting a threshold ψ, if the following conditions are met, it is determined that the pixel is not a boundary point and is retained. On the contrary, it is determined that it is a boundary point and removed.

$$\frac{tr(H)^2}{Det(H)} < \frac{(\psi + 1)^2}{\psi} \tag{13}$$

On the basis of determining the feature points of the drone trajectory image, LiDAR technology is used to calculate the corresponding direction of the feature points, calculate the gradient amplitude and gradient directions $m(x, y)$ and $\mu(x, y)$ at feature point (x, y) of the drone trajectory image, and generate the SIFT feature description vector of the drone trajectory image. The calculation formula is:

$$m(x, y) = \sqrt{\frac{(L(x + 1, y) - L(x - 1, y))^2}{+(L(x, y + 1) - L(x, y - 1))^2}} \tag{14}$$

$$\mu(x, y) = \arctan \frac{L(x + 1, y) - L(x - 1, y)}{L(x, y + 1) - L(x, y - 1)} \tag{15}$$

Read the UAV trajectory image data after Gaussian filtering from the CPU into the gpu, and construct the Gaussian difference pyramid model of UAV trajectory image using laser radar technology. The Hessian matrix calculates the constructed UAV trajectory image Gaussian difference pyramid model in parallel in the gpu program, calculates the principal curvature of UAV trajectory image, and judges whether the pixels in the model are boundary points. Determine feature points and feature point gradient information, accumulate the obtained image feature point gradient information into gpu program, generate UAV trajectory image SIFT feature vector, and output it from CPU.

3.2 Design of UAV Control Algorithm

Assuming that D_{sg} represents any reference point for the drone's flight, and an arc with a radius of R is drawn towards the reference point, L_{OI} represents the distance from the drone's current position to the reference point, η_{sed} represents the angle between the line and airspeed V between the drone's current position and the reference point, and a_{afgh} represents the lateral acceleration that needs to be generated, then the distance from the drone's current flight position to the expected trajectory can be calculated using Formula (16):

$$Q_{aue} = \frac{L_{OI} \cdot V \times \eta_{sed}}{a_{afgh} \times D_{sg}} \times \frac{\{R\} + d_{dhk}}{cd_{zv}} \tag{16}$$

Among them, d_{dhk} represents the weight of the UAV, cd_{zv} represents the mass of UAV.

The acceleration command signal of the drone during flight is represented by l_{zvj}, and the lateral acceleration that needs to be generated is given using Formula (17):

$$a_{ag} = \frac{l_{zvj} + z_d}{k_{zbm}} \times p_{xn} \tag{17}$$

Among them, z_d represents a viscous damped vibration system, k_{zbm} represents a group of attitude angles of UAV in flight, p_{xn} represents pitch angle.

λ_{zhk} represents the dynamic range of the drone's pitch angle, while h_{fhj} represents the real-time output of the three axis angular rate in the body coordinate system, satisfying the condition of $h_{fhj} = \left[h_{fhk}, h_{fhu}, h_{fhp} \right]$. Among them, h_{fhk}, h_{fhu}, and h_{fhp} represent the angular velocities of each axis, then the expected roll angle command of the drone can be obtained using Formula (18):

$$E_{SG} = \frac{h_{fhj} + \left[h_{fhk}, h_{fhu}, h_{fhp} \right]}{\lambda_{zhk} + r_{sg} \times a_{ag}} - s_{sgj} \tag{18}$$

Among them, a_{ag} represents the angle vector of geographical coordinate system, r_{sg} represents the angular velocity information at the current time, s_{sgj} represents the sampling time.

Assuming that ω_{sde} represents a single information fusion cycle, l_{zbmk} represents the estimation of the drone's flight attitude at time k, and l_{djh} represents the filter convergence criterion, the guidance law for drone flight attitude control is designed using Formula (19):

$$s_{sg} = \frac{l_{djh} \cdot l_{zbmk}}{\omega_{sde}} \times \frac{\{Q_{awe} + a_{ag}\}}{E_{SG}} \tag{19}$$

Assuming that k_{sgj} represents the actual error of the filter, ∂_{hp} represents the discrete state space of the UAV's lateral flight, and l_{dhk} represents the statistical characteristics of each component of the UAV's lateral flight angle, then use Formula (20) to adjust the Natural frequency of the UAV guidance law:

$$W_{wet} = \frac{l_{dhk} \times \partial_{hp}}{k_{sgj}} \times s_{sg} \tag{20}$$

Assuming that τ_{dhk} represents the real-time collection of data for a special flight state dominated by lateral flight motion, Formula (21) is used to control the lateral flight attitude of the unmanned aerial vehicle:

$$f_{sh} = \frac{\tau_{dhk} \times W_{wet}}{E_{SG}''} \pm s_{sg} \tag{21}$$

To sum up, first calculate the distance from the current flight position of the UAV to the desired trajectory, give the required lateral acceleration, get the desired roll angle command of the UAV, design the attitude control guidance law [12–14] for the UAV's lateral flight, adjust the natural frequency of the unmanned mechanism guidance law, and complete the control of the UAV's flight attitude.

4   System Test Analysis

4.1   Experimental Platform

The experimental platform used in this paper is improved on the flight control system independently developed by the laboratory team.

The flight control system independently developed by the laboratory team takes GNSS and IMU module as the core. The whole flight control system includes the main control part (AP), IMU, geomagnetism (MAG), GNSS, data recording module (FDR), voltage converter (HUB) and signal light (LED). The details of each module are shown in Table 1.

For the improvement of the basic system, this paper mainly adds laser radar, visual camera and optical flow module. The laser radar used is RPLIDAR A1, which is common in 2D laser radar. The visual sensor used is Xiaomi binocular camera, with a resolution of 752 * 480 and a focal length of 2.1 mm. The output frame rate can be adjusted according to the actual use needs. In this application, the frame rate is set to 30 frames per second. The airborne computer adopts GIGABYTE of Gigabyte Technology, and its processor is Intel Core i7, with a size of 46.8 × one hundred and twelve point six × 119.4 mm.

Connect the laser radar and binocular camera to the airborne computer via USB, calculate the original data of the sensor through the computer to estimate the current moving distance, and send the calculation results to the AP through the serial assistant.

The hardware connection of the flight control system used in this paper is shown in Fig. 4.

Table 1. Basic Experimental Platform Module List

Modular	Core components	major function
AP	STM32F407	Complete the calculation of fusion and control algorithms, and convert the calculation results into PWM values of the motor
IMU	ICM20602 + MA5611	Obtain real-time angular velocity and acceleration information of UAV
MAG	QMC5883	Measuring the current magnetic field information of UAV for heading calculation
GNSS	NE06-M8N	Get the position and speed information of UAV in outdoor environment
FDR	HM-11	Data recording module of UAV, integrated with Bluetooth module to set some parameters of UAV
HUB	TPS54560	The voltage conversion module makes the UAV compatible with various types of batteries
LED	RGB	Set multiple flashing modes to reflect the health status of UAV

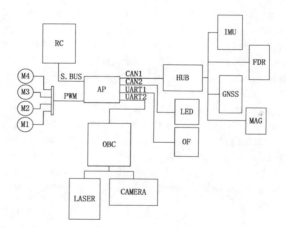

Fig. 4. Hardware Connection Diagram of UAV System

In addition, in order to obtain the real real-time position and speed information of UAV, this paper chooses to use a high-precision commercial RTK system, whose positioning error is within 0.05 m and speed error is within 0.1 m/s. The RTK system is placed on the UAV body, and the accuracy of the system in this paper is evaluated based on the comparison between the measured information and the results of the system in this paper.

The actual UAV used in this experiment is shown in Fig. 5.

Fig. 5. Physical picture of the experimental platform

4.2 Indoor Flight Test

For the indoor flight experiment, this paper selects an empty room as the flight experiment site. During the experiment, in order to ensure the normal operation of the laser radar and the visual camera, some cartons are arranged in the room as obstacles in advance.

The experimental process is: the UAV is operated by the pilot to fly slowly. The speed control mode is adopted during the flight. The speed target value of the given flight is operated by the remote controller, and the fused speed information is used as the real-time feedback value for speed control.

According to the above experimental scheme, the experimental results are as follows. In order to facilitate comparison, the position information measured by the positioning system has been transferred to the NED coordinate system in the flight control system.

Figure 6 shows the tracking of speed control during flight. The solid line in the figure is the target speed obtained by operating the remote controller, and the dotted line is the real-time speed of the merged UAV.

It can be seen from the figure that the response time of the speed control is about 2 s, which is faster than the speed control under the general GNSS environment and conforms to the characteristics of the designed system. The attitude angle comparison in the flight controller during flight is shown in Fig. 7.

Since the update frequency of attitude angle data is high, usually 400 Hz, the tracking delay of attitude angle tracking is lower than that of speed control tracking, and the angle tracking error is within ±3°.

According to the above results, the indoor flight error indicators are obtained, as shown in Table 2.

According to the analysis of the data in Table 2, the maximum error of attitude angle is 0.05 rad, and the existing scheme can achieve better attitude tracking effect. The maximum value of the speed error is 0.3, which is due to the delay of the designed speed control step response. When the speed tracking is stable, the error is kept within 0.1 m/s.

From the analysis of indoor flight results, it can be seen that the speed controller error of the system in this paper has been kept within 0.1 m/s, reaching the expected design goal.

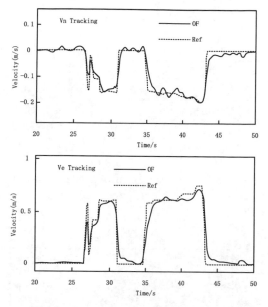

Fig. 6. Comparison of Indoor Flight Speed

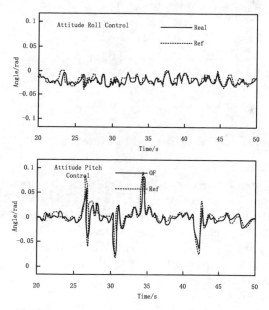

Fig. 7. Comparison of indoor flight attitude angle

4.3 Comparative Analysis

In order to avoid the singularity of the experimental results, the control system based on sliding mode controller, the control system based on fuzzy LADRC and the control

Table 2. Indoor Flight Error Indexes

Reference indicators	Maximum	average value	standard deviation
Vn Err (m/s)	0.07	0.0072	0.014
Ve Err (m/s)	0.31	0.023	0.049
Roll Err (rad)	0.029	0.002	0.004
Pitch Err (rad)	0.05	0.003	0.006

system based on Pixhawk - ScienceDirect are introduced for comparison, and the control accuracy of the constant speed and variable pitch loaded multi axis UAV is tested. The results are shown in Fig. 8.

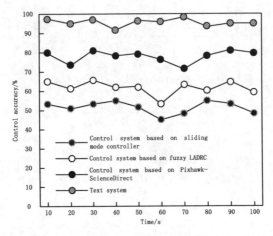

Fig. 8. Control Accuracy of Constant Speed Variable Pitch Loaded Multi axis UAV

It can be seen from the results in Fig. 8 that when the control system based on sliding mode controller and the control system based on fuzzy LADRC are adopted, the control accuracy of UAV is below 80%; When the control system based on Pixhawk - ScienceDirect is adopted, although the control accuracy of UAV is improved, it is still below 85%. When using the system in this paper, the control accuracy of UAV is more than 90%, which can ensure the stability of UAV flight.

5 Conclusion

This article proposes the design and research of a constant speed, variable pitch, and heavy-duty multi-axis unmanned aerial vehicle control system based on LiDAR technology. The results show that the system has high control accuracy, which can improve the control accuracy to over 90% and ensure stable flight of the unmanned aerial vehicle. Although this study has achieved certain results, there are still many shortcomings. In the future, the focus will be on designing more efficient and stable control algorithms to achieve precise hovering, smooth takeoff and landing, and precise flight of unmanned aerial vehicles.

Acknowledgement. School level project of Beijing Polytechnic, Project Name: Research on Control System of Constant Speed and Variable Pitch Loaded Multi-axis UAV (2022X010-KXD).

References

1. Feng, S., Li, Q.: Automatic navigation control system of agricultural UAV based on computer controller. J. Agricult. Mechaniz. Res. **44**(8), 42–46 (2022)
2. Wei, H.: Research on automatic navigation control system of agricultural UAV based on controller. J. Agricult. Mechaniz. Res. **44**(1), 218–222 (2022)
3. Wu, Y., Chen, Z., Su, L., et al.: Longitudinal attitude control of quad tilt-rotor UAV based on fuzzy LADRC. Flight Dynamics **38**(3), 28–33 (2020)
4. Yang, K., Yang, G.Y., Fu, S.: Research of control system for plant protection UAV based on Pixhawk - ScienceDirect. Procedia Comput. Sci. **166**, 371–375 (2020)
5. Shi, J., Pei, Z., Tang, Z., et al.: Design and realization of an improved active disturbance rejection quadrotor UAV control system. J. Beijing Univ. Aeronaut. Astronaut. **47**(09), 1823–1831 (2021)
6. Yao, H., Liu, Z., Zuo, H.: Research on control system of multi-functional UAV based on embedded system. J. Agricult. Mechaniz. Res. **44**(01), 142–145 (2022)
7. Pereira, L.G., Fernandez, P., Mourato, S., et al.: Quality control of outsourced LiDAR data acquired with a UAV: a case study. Remote Sens. **13**(3), 419 (2021)
8. Pang, Q., Wu, S., Niu, B., et al.: Design and implementation of micro air vehicle control system for IMAV 2019. Control Eng. China **28**(11), 2114–2122 (2021)
9. Zhu, H.: Design and experimental testing of safe flight control system for novel vertical take-off and landing aircraft. J. Vibroeng. **24**(3), 20 (2022)
10. Oh, D., Han, J.: Fisheye-based smart control system for autonomous UAV operation. Sensors (Basel, Switzerland) **20**(24), 7321 (2020)
11. Zhang, K., Bi, F.H., Li, K.L., et al.: Design and implementation of a dual-IP core UAV flight control system based on QSYS - ScienceDirect. Procedia Comput. Sci. **166**, 180–186 (2020)
12. Ding, D., Zhao, C., Jia, T.: Design of quadrotor unmanned aerial vehicle control system based on STM32. Mod. Electron. Techniq. **40**(01), 31–37+47 (2023)
13. Liu, G.: Application of genetic algorithm in plant protection UAV control system. J. Agricult. Mechaniz. Res. **45**(04), 221–224 (2023)
14. Qiu, J., Sun, Z.: RGPI observer-based control for wind disturbance rejection of quadrotor UAV. Comput. Simulat. **40**(01), 31–37+47 (2023)

Research on Railway Frequency Shift Signal Detection Based on Transient Electromagnetic Radar

Rong Zhang[✉], Hao Tang, and Qing Shi

Wuhan Railway Vocational College of Technology, Wuhan 430000, China
Zhangr133@163.com, ssl335@sina.com

Abstract. In response to the characteristics of railway frequency shift signals, transient electromagnetic radar is used to start from two aspects: carrier frequency and low frequency. The carrier frequency uses the same frequency signal detection principle, while the low frequency uses a step sequence that can cover the entire low frequency band to detect unknown low frequencies. The detection principle and steps of carrier frequency and low frequency are given, and a simulation model is built and verified. Finally, accurate frequency shift signals are detected. Finally, the accurate detection of railway frequency shift signal is also realized under the condition of in band harmonic interference and white noise interference, and the bit error rate is analyzed under different signal to noise ratios. Compared with the traditional railway frequency shift signal measurement methods, the threshold of signal to noise ratio based on transient electromagnetic radar can be lower when detecting railway frequency shift signal.

Keywords: Electromagnetic Radar · Frequency Shift Signal · Signal Detection

1 Introduction

As an important infrastructure in transportation, railway is known as the main artery of national economy due to its outstanding advantages of huge transportation volume. It plays a very important role in promoting the exchange of materials between different regions and the prosperity of social economy. In recent years, with the continuous acceleration of railway, especially the emergence and continuous development of high-speed railway, railway is playing an irreplaceable role in passenger transport, so it is of great practical significance to vigorously develop railway industry and improve railway transport capacity [1, 2]. However, in order to improve the railway transportation capacity, it is necessary to improve the train running speed, increase the traffic density, shorten the train running interval. But at the same time, the railway running safety will face great challenges, in order to ensure the running safety, we must vigorously develop the railway signal automation, to provide a strong safety guarantee for the railway system.

Railway frequency shift signal is a general term for signal, interlocking, blocking and other equipment used in railway systems. In the railway frequency shift signal system,

L. Yun et al. (Eds.): ADHIP 2023, LNICST 548, pp. 162–176, 2024.
https://doi.org/10.1007/978-3-031-50546-1_11

track circuit is a circuit composed of rail lines and rail insulation. It is an important basic equipment of railway frequency shift signal. Its function is to collect the real-time status of train operation, express the track occupancy, and transmit the frequency shift signal information to the signal display equipment through the way of electromagnetic radar [3–5]. With the continuous improvement of modern digital signal processing theory and the continuous improvement of digital signal processor performance, as well as the rapid development of digital demodulation technology, there are more implementation methods for the detection of railway frequency shift signal parameters, thus gradually solving various problems in the measurement of railway frequency shift signal parameters. For example, in reference [6], a parameter estimation and signal detection algorithm based on adaptive capture is proposed. First, the correlation between the signal and the leading sequence is used as the basis for frequency capture. Secondly, the frequency is estimated accurately based on interpolation algorithm. Finally, according to the characteristics of gradual frequency change, the phase-locked loop structure is used to track the frequency, and the frequency offset of the signal is eliminated in the digital down conversion stage, and the signal detection is realized. This algorithm consumes less resources and has low complexity. However, in the case of low signal-to-noise ratio, the detection accuracy will decrease and the bit error rate will be high.

Therefore, in order to solve the problem of low detection accuracy and high bit error rate under the condition of low SNR, a railway frequency-shift signal detection method based on transient electromagnetic radar is proposed, which can effectively reduce the detection SNR threshold and achieve accurate detection under the condition of low SNR. Firstly, the ResNet network model and LSTM network model are combined, and a dual-parallel RLA (ResNet LSTM Attention) network structure is proposed by taking advantage of the complementary advantages of the two models. Secondly, the structure is introduced in layers, and the function of each layer is defined. Then, the network structure is used for acquisition, feature extraction and fusion, and the modulation mode of radar signal is classified and recognized by RLA network. Finally, the amplitude and phase of the original signal data are extracted directly, the hidden information contained in the signal is distinguished from the time domain, and the storage detection is carried out to realize the detection research of railway frequency shift signal based on transient electromagnetic radar.

2 Railway Frequency Shift Signal Detection

The network structure of railway frequency shift signal transient electromagnetic radar is introduced. Detection of railway frequency shift signal by network structure. In order to improve the effectiveness of the network model and achieve a higher recognition rate of transient electromagnetic radar signal modulation methods, this chapter combines the ResNet network model and the LSTM network model, and proposes a dual-parallel RLA (ResNet LSTM Attention) network structure. Among them, ResNet is a deep convolutional neural network designed to solve the problem of gradient disappearance in deep network training. It builds the network structure by introducing residual connections. Residual connections allow information to skip directly several layers in the network, making it easier for deeper networks to learn the identity map. This design avoids the problem of gradients fading away, making the network deeper and easier to train. The model can better capture the frequency domain and time domain characteristics of the frequency shift signal. LSTM is a type of recurrent neural network (RNN) specifically designed to process time series data. It comprises an input door, an oblivion door and an output door. These gating mechanisms control the flow of information and help LSTM networks capture and remember long-term dependencies. The LSTM network consists of a number of LSTM units, each of which has its own internal cell state, and determines how to update and pass the state and information through a gating mechanism to better model these relationships and improve the accuracy of detection. Therefore, the combination of ResNet and LSTM network model can improve the generalization ability of the model, with features with stronger expression ability, so that the model can process complex data more accurately, and accurately extract the features of the data, improve the modeling and detection ability of the timing characteristics of the frequency shift signal, and thus improve the accuracy and stability of detection. Achieve accurate railway frequency shift signal detection. The RLA network structure mainly consists of four parts: input layer, parallel layer, feature fusion layer, and output layer. The model structure is shown in Fig. 1.

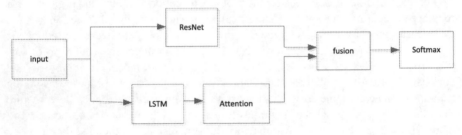

Fig. 1. RLA Network Model Diagram

The first part is the input part. The main function of the network input layer is to collect data. The data format of each input sample is a 2×2048 matrix. The upper half branch is a residual network, which takes advantage of its network advantages to complete the extraction of spatial features of radar modulated signal samples. ResNet adopts a residual unit composed of two convolution layers. Since the dimension of the data matrix input in this paper is not a traditional square, the convolution kernel size is set as $[3 \times 1]$, $[5 \times 1]$, which can effectively reduce model parameters. To improve the convergence speed, ResNet is formed by four residual units; the residual block is set as [2] by referring to ResNet18 network proposed by He Keming; the Pooling method is selected as Max Pooling; the learning rate is set as 0.001[7].The lower branch is a long-term and short-term memory network, mainly used to extract temporal features of radar signals. In order to balance algorithm accuracy, runtime, and convergence speed, through literature review and experimental comparison, it was found that blindly increasing the number of LSTM layers is not conducive to model performance. Therefore, a double-layer LSTM network with a Batch size of 1024 and a learning rate of 0.001 was adopted, and an Attention mechanism was added to allocate attention weights to the features extracted from the network, Summarize the features that are more prominent in the network extraction process.

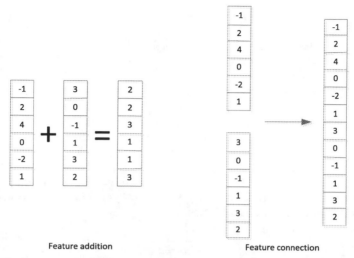

Feature addition   Feature connection

Fig. 2. Schematic diagram of feature fusion mode for the proposed method

The second layer is the feature fusion layer, whose main function is to integrate the features extracted from the double-channel parallel network that can identify and collect railway frequency shift signals. As shown in Fig. 2, there are generally two choices in the way of feature fusion. Due to the problem that two feature vectors cannot be added due to their different dimensions in the way of feature addition, or feature cancellation occurs after addition, the feature fusion mode in this section is selected as the feature connection mode, and features of the upper and lower branches are merged. That is to say, based on the spatial feature information mainly extracted by ResNet network and the temporal feature information mainly extracted by LSTM, new features are synthesized by the two features and sent to the next part [8]. Finally, the Softmax classifier classifies and outputs according to the features transmitted from the above part. This layer is an ordinary fully connected layer, including six nodes, which are used to output signals representing six different modulation modes respectively.

Using transient electromagnetic to identify and collect radar signal modulation methods often results in different feature extraction methods. At the same time, each modulated signal will have different time-frequency images and different feature attributes. When using time-frequency images as input, it is necessary to first convert the intercepted radar signal into a two-dimensional image before using a network model for recognition. Moreover, generating time-frequency images takes a long time, with a large number of sampling points for a single image and complex image processing, which is not conducive to the real-time requirements in radar processing [9]. Therefore, this chapter takes the time-domain amplitude and phase (AP) data features of the signal converted from the raw data of the transient electromagnetic radar signal as input, and uses the RLA network as the framework to classify and identify the modulation methods of the radar signal.

Generally speaking, modulated signals are represented as:

$$s(t) = A \exp\left[j(2\pi f_0 t + \varphi_0)\right] \tag{1}$$

In order to make the signal $s(t)$ collected by the transient electromagnetic radar receiver have better adaptability and simplicity in calculation, research and equipment processing, the amplitude and phase of the original signal data are extracted directly, and the hidden information contained in the signal is distinguished from the perspective of time domain.

In order to prevent overfitting of the model from being too complex, $L2$ norm naturalization was carried out on the extracted AP data, and $L2$ norm was defined as:

$$norm(x) = \sqrt{x_1^2 + x_2^2 + \ldots + x_n^2} \tag{2}$$

Among them, x_i represents the elements of the extracted amplitude data and phase data, and after normalization, each element is:

$$x_i = \frac{x_i}{norm(x)} \tag{3}$$

The processed data AP is stored in the matrix form of 2×2048 as the input form of the data set. Meanwhile, different modulation modes and corresponding signal-to-noise ratio information are also stored in the data set [10, 11].

Due to the frequent traffic of railway transmission vehicles, the site environment and transmitted information are becoming more and more complex, which affects the reception of railway frequency shift signals. Therefore, through the frequency shift signals collected in the data set, the transient electromagnetic radar can accurately detect railway frequency shift signals and ensure the normal transmission.

3 Experimental Analysis

The traditional railway frequency shift signal detection method has a good filtering effect on harmonic interference of traction current outside the frequency band of the railway frequency shift signal, but it cannot effectively filter out harmonic interference within the frequency band of 40 Hz. The following is an analysis of different proposed methods for carrier frequency detection and low-frequency detection, and the use of transient electromagnetic radar to analyze the harmonic interference in the traction current band, analyze the error rate, and verify the effectiveness of the proposed method in low-frequency band detection. And under white noise interference, the proposed method was used for ZPW-2000 railway frequency shift signal detection simulation and bit error rate analysis to verify the applicability of the proposed method. Finally, to further verify the superiority of the proposed method, under the same settings as the proposed method, the change in bit error rate between the proposed method and the traditional method was measured by changing the noise intensity. The dataset used for this test is the 'Railway Track Maintenance Data Set' dataset from the UCI Machine Learning Repository.

During the experiment, the amplitudes of railway frequency shift signals were added to be 0.01 V, 0.1 V, and 1 V, respectively. Then, traction current harmonics equivalent to noise were added, while other system parameters remained unchanged. In the process of judging the results, the detection statistics were calculated, and the phase diagram and domain output diagram were observed. The results obtained were consistent. In order to demonstrate more intuitive results, this article provides detection results for phase and domain diagrams in the following sections.

3.1 Carrier Frequency Detection

Firstly, analyze the railway frequency shift signal with an amplitude of 0.01 V and a carrier frequency of 2000–1 Hz. As the intensity of harmonic interference in the traction current in the band increases, calculate the signal-to-noise ratio and obtain that the desired frequency shift signal can still be detected through changes in the phase diagram when the signal-to-noise ratio is −24 dB. The results are shown in Fig. 3. The detection results of Fig. 3(a), Fig. 3(c), and Fig. 3(d) are in a chaotic state, that is, no co frequency signal with a carrier frequency of 2000–1 Hz was detected, while the oscillator with an internal driving force term set to 2000–1 Hz in Fig. 3(b) undergoes a phase transition, indicating that the co frequency signal was detected. When the amplitude of harmonic interference current is further increased to achieve a signal-to-noise ratio of −25 dB, the transient electromagnetic radar system cannot detect the same frequency signal.

Similarly, it can be concluded that with the increase of the amplitude of the in-band harmonic interference current, the frequency shift signal with a amplitude of 0.1 V can finally reach the detection SNR threshold of −19 dB; while with the increase of the amplitude of the in-band harmonic interference current, the frequency shift signal with a amplitude of 1 V can finally reach the detection SNR threshold of −17 dB. Therefore, it is found that with the increase of the amplitude of the signal to be measured, the higher the detection signal-to-noise ratio can be achieved, that is, compared with the noise, the weaker the signal of the detection system of the transient electromagnetic radar oscillator, the higher the detection sensitivity.

3.2 Low Frequency Detection

The most important aspect of detecting railway frequency shift signals is to obtain accurate low-frequency information. Similarly, when the amplitude of railway frequency shift signals is 0.01 V, 0.1 V, and 1 V, the minimum signal-to-noise ratio threshold that can be achieved by using transient electromagnetic radar is studied, that is, its anti-interference performance under the interference of harmonic interference of in band traction current.

Firstly, analyze the railway frequency shift signal with an amplitude of 0.01 V and a carrier frequency of 2000–1 Hz. After extensive simulation research, it can still detect low-frequency information using transient electromagnetic radar when the signal to noise ratio is −28 dB. The results are shown in Fig. 4, and Fig. 4(a) shows the waveform of the ZPW-2000 railway frequency shift signal under in band harmonic interference. At two consecutive steps, intermittent chaotic phenomena occurred, as shown in Fig. 4(b) and Fig. 4(c). When the signal-to-noise ratio is −29 dB, low-frequency information

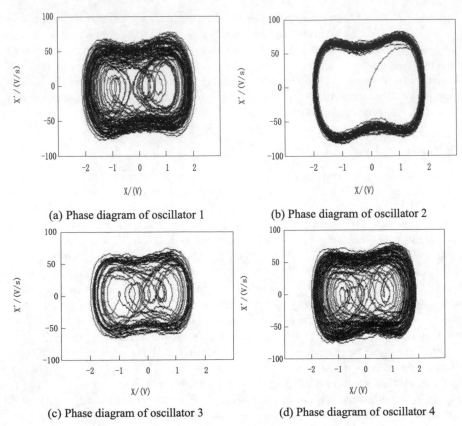

(a) Phase diagram of oscillator 1   (b) Phase diagram of oscillator 2

(c) Phase diagram of oscillator 3   (d) Phase diagram of oscillator 4

Fig. 3. Output Phase Diagram of Transient Electromagnetic Radar Detection System under 2600 Hz Intraband Harmonic Interference

cannot be detected. At the step size where intermittent chaos occurs, the zero crossing distance method is used to obtain a low frequency of 11.4254 Hz, which meets the requirement that the detection error should not exceed 0.03 Hz.

Change the amplitude of railway frequency shift signal. When the amplitude is 0.1 V, the threshold value of detected signal-to-noise ratio is -25 dB. When the amplitude is 1 V, the detection SNR threshold is -21 dB. After a large number of simulation experiments, the conclusion is that the use of transient electromagnetic radar detection system can accurately detect the low-frequency information under the condition of in-band harmonic interference, and the signal is weaker and can achieve a lower signal-to-noise ratio.

3.3   Bit Error Rate

In the analysis of bit error rate, two aspects can be approached: on the one hand, when the carrier frequency is the same but the low-frequency is not the same, and on the other hand, when the low-frequency is the same but the carrier frequency is not the same. Firstly, we study the bit error rate situation when the carrier frequency is not the

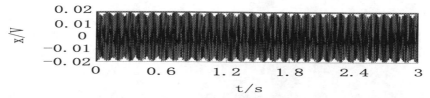

(a) Waveform diagram of ZPW-2000 frequency shift signal under harmonic interference in the 2000-1Hz band

(b) Time domain diagram of transient electromagnetic radar oscillator 2 output under

harmonic interference in the 2000-1Hz band with step size a_{143}

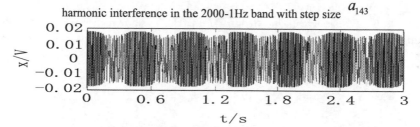

(c) Time domain diagram of transient electromagnetic radar oscillator 2 output under

harmonic interference in the 2000-1Hz band with step size a_{144}

Fig. 4. Low frequency detection results under harmonic interference in the 2000–1 Hz band

same, and select the low-frequency as 10.3 Hz and 29 Hz, respectively. The results are shown in Fig. 5. In Fig. 5(a), for a certain carrier frequency, the error rate increases as the signal-to-noise ratio decreases. When the carrier frequency is 1700 Hz and the signal-to-noise ratio is greater than or equal to −14 dB, the error rate is approximately considered 0. When the signal-to-noise ratio is around −20 dB, the error rate increases to around 0.5. And for different carrier frequencies, at the same signal-to-noise ratio, the bit error rate decreases as the carrier frequency decreases. For the highest carrier frequency of 2600 Hz, when the signal-to-noise ratio is greater than −9 dB, the bit error rate is basically considered to be 0.

Then analyze the bit error rate when the carrier frequency is fixed but the low frequency is changed. The low frequencies of 10.3 Hz and 29 Hz at 1700 Hz and 2600 Hz are analyzed respectively, and the results obtained are shown in Fig. 6. It can be seen that in Fig. 6(a), when the carrier frequency is fixed at 1700 Hz and the low frequency is 10.3 Hz, the bit error rate will increase as the signal-to-noise ratio decreases. When the signal-to-noise ratio is greater than or equal to −14 dB, the bit error rate will be approximately 0, and when the signal-to-noise ratio is about −20 dB, the bit error rate

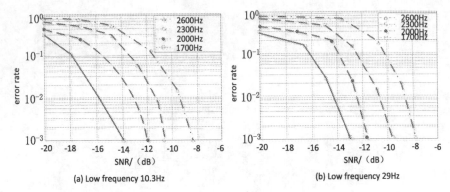

(a) Low frequency 10.3Hz   (b) Low frequency 29Hz

Fig. 5. Error rate of carrier frequency detection at different low frequencies

will be about 0.6. And with the increase of different low frequency value, the bit error rate is larger. When the carrier frequency is fixed at 2600 Hz, the variation curve of bit error rate is consistent with the trend of 1700 Hz, as shown in Fig. 6(b). For the most unfavorable 29 Hz, the bit error rate is basically 0 when the signal-to-noise ratio is greater than or equal to −9 dB.

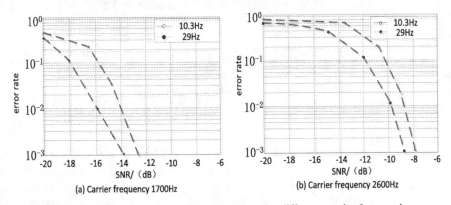

(a) Carrier frequency 1700Hz   (b) Carrier frequency 2600Hz

Fig. 6. Low frequency detection error rate under different carrier frequencies

The following conclusions can be drawn from the above detection of railway frequency shift signals under the interference of in-band traction current harmonics: The railway frequency shift signals under the interference of in-band harmonics can be detected by using the transient electromagnetic radar detection system.

3.4 Detection Simulation and Bit Error Rate Analysis of ZPW-2000 Railway Frequency Shift Signal Under White Noise Interference

In this section, the amplitude of the railway frequency shift signal is also set to 0.01 V, 0.1 V, and 1 V, respectively, and white noise is added, while the other system parameters remain unchanged. The frequency shift signal and bit error rate analysis of the transient electromagnetic radar oscillator detection under white noise interference are carried out.

(1) Carrier frequency detection

First, analyze the railway frequency shift signal with amplitude of 0.01 V and carrier frequency of 1700–1 Hz. As the strength of white noise increases, when the signal-to-noise ratio is −23 dB, the results are shown in Fig. 7. The detection results in Fig. 7(b), Fig. 7(c), and Fig. 7(d) are in a chaotic state, while the oscillator shown in Fig. 7(a) with the internal driving force term set to 1700-Hz undergoes a phase transition. Increase the strength of white noise. When the signal-to-noise ratio is −24 dB, the transient electromagnetic radar system cannot detect the same frequency signal.

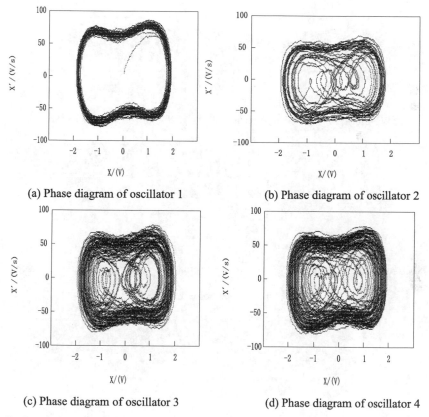

(a) Phase diagram of oscillator 1

(b) Phase diagram of oscillator 2

(c) Phase diagram of oscillator 3

(d) Phase diagram of oscillator 4

Fig. 7. Output × time domain waveform of transient electromagnetic radar detection system under white noise interference

Similarly, it can be obtained that the threshold value of signal to noise ratio for the frequency shift signal with a amplitude of 0.1 V is −18 dB, while the threshold value of signal to noise ratio for the frequency shift signal with a amplitude of 1 V is −16 dB.

(2) Low frequency detection

In the low frequency detection, the lowest SNR threshold that can be reached by the transient electromagnetic radar oscillator is studied when the amplitude of the frequency shift signal is 0.01 V, 0.1 V and 1 V respectively, that is, its anti-interference performance in the case of white noise interference.

Firstly, the frequency shift signal with a amplitude of 0.01 V was analyzed. After a large number of simulation studies, array chaos was detected at two consecutive steps, and the low-frequency information could still be detected by using the transient electromagnetic radar oscillator when the signal-to-noise ratio was −27 dB, as shown in Fig. 8. At the step of array chaos, the low frequency of 11.3722 Hz is obtained by using the zero crossing distance method, which meets the requirement that the detection error should not be greater than 0.03 Hz.

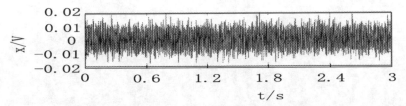

(a) Waveform of ZPW-2000 frequency shift signal with white noise interference

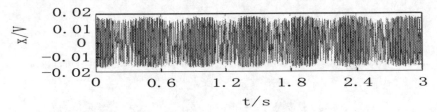

(b) Output time domain diagram of transient electromagnetic radar oscillator 1 when step size

is a_{143} under white noise interference

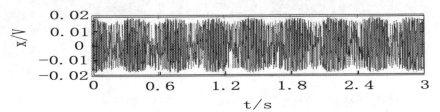

(c) Output time domain diagram of transient electromagnetic radar oscillator 1 when step size

is a_{144} under white noise interference

Fig. 8. Low frequency signal detection results under white noise interference

Similarly, the amplitude of railway frequency shift signal is changed. When the amplitude is 0.1 V, the threshold value of signal to noise ratio detected is −24 dB. When the amplitude is 1 V, the detection SNR threshold obtained is −20 dB. After a large number of simulation experiments, the conclusion is that the use of transient electromagnetic radar oscillator measurement system can accurately detect the low-frequency information in the case of white noise interference, and the signal is weaker and can achieve a lower signal-to-noise ratio.

(3) Error rate analysis

Starting from fixing the carrier frequency and low frequency separately, consider the error rate situation. Firstly, fix the carrier frequency, and the error rate results obtained as the low frequency changes are shown in Fig. 9. In Fig. 9(a), for a certain carrier frequency, the bit error rate increases as the signal-to-noise ratio decreases. Taking the 2600 Hz carrier frequency as an example, when the signal-to-noise ratio is greater than or equal to −8 dB, the bit error rate is so small that it can be approximately considered 0. When the signal-to-noise ratio is around −20 dB, the bit error rate increases to around 0.9. And for different carrier frequencies, at the same signal-to-noise ratio, the bit error rate decreases as the carrier frequency decreases. Therefore, for the lowest carrier frequency of 1700 Hz, when the signal-to-noise ratio is greater than −14 dB, the bit error rate is basically considered to be 0.

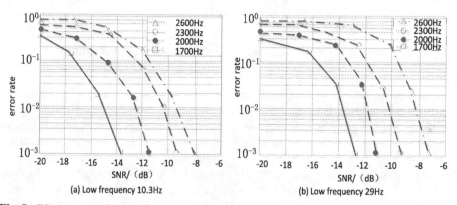

Fig. 9. Bit error rate of carrier frequency detection at different low frequencies under white noise interference

Next, analyze the error rate when the low-frequency is fixed but the carrier frequency changes. The results obtained by analyzing the low frequencies of 10.3 Hz and 29 Hz at 1700 Hz and 2600 Hz are shown in Fig. 10. In Fig. 10(a), when the carrier frequency is fixed at 1700 Hz and the low frequency is 10.3 Hz, the error rate increases as the signal-to-noise ratio decreases. When the signal-to-noise ratio is greater than or equal to −13 dB, the error rate is considered to be approximately 0, and when the signal-to-noise ratio is around −20 dB, the error rate is about 0.5. And as the values of different low-frequency frequencies increase, the error rate increases. The error rate variation curve when the carrier frequency is fixed at 2600 Hz is consistent with the trend at 1700 Hz, as shown in Fig. 10(b). For the most unfavorable 29 Hz, when the signal-to-noise ratio is greater than or equal to −8 dB, the error rate is basically 0. Compared with the analysis of bit error rate in the case of intra band harmonic interference, the values have slightly changed, but the trend is consistent.

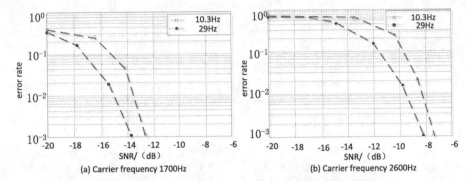

(a) Carrier frequency 1700Hz   (b) Carrier frequency 2600Hz

Fig. 10. Low frequency detection error rate under different carrier frequencies

After the above analysis, the frequency shift signal of railway can still be detected by the transient electromagnetic radar under the condition of offline noise interference of pantograph catenary equivalent to white noise.

3.5   Comparison of Bit Error Rate with Traditional Detection Methods

The above studies respectively aim at the bit error rate of railway frequency shift signal detection under different noise conditions. To demonstrate the superiority of the improved detection method, the bit error rates of the proposed method and different methods were measured by changing the noise intensity, as shown in Fig. 11. It can be seen from the figure that although the Bit error rate of the two methods decreases with the increase of the signal to noise ratio, compared with the curve trend of the two methods, the bit error rate of the proposed method is far lower than that of the traditional method under the same signal to noise ratio. Especially when the signal-to-noise ratio is about − 20 decibels, the error rate of the proposed method is 0.1, while the error rate of traditional methods is as high as 1. From this, it can be concluded that the new detection method based on transient electromagnetic radar discussed in this article can accurately detect ZPW-2000 railway frequency shift signals under low signal-to-noise ratio conditions, while also having a low bit error rate.

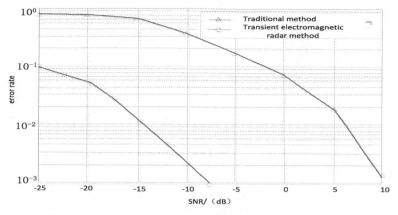

Fig. 11. Bit error rate curve of ZPW-2000 frequency shift signal

4 Conclusion

In summary, this article has completed the corresponding technical research on railway frequency shift signal detection and parameter estimation of transient electromagnetic radar. However, there are still many problems that need to be solved in the research content, mainly including the following parts: In the railway frequency shift signal detection method of transient electromagnetic radar, the test threshold is reduced through the method of equal division replacement, and there may be certain false alarms in the actual environment. In response to this problem, The algorithm in this article still needs further improvement; In the implementation of the transient electromagnetic radar detection method for modulation recognition, the types of modulation recognition for transient electromagnetic radar signals should be added in subsequent work to enhance the breadth of research. Further research is needed in this regard.

References

1. Li, B., Li, J., Xu, X., et al.: Research on corrosion detection method of tower grounding body based on transient electromagnetic method. In: 2020 IEEE 5th International Conference on Signal and Image Processing (ICSIP), pp. 056–062. IEEE (2020)
2. Liang, Z., Tian, B., Zhang, L.: Research on apparent resistivity imaging of transient electromagnetic method for oil and gas pipelines based on GA-BP neural network. Math. Prob. Eng. 011–015 (2019)
3. Xie, H., Truong, C.V., Wang, Y., et al.: Research on frequency parameter detection of frequency shifted track circuit based on nonlinear algorithm. Nonl. Eng. **10**(1), 592–599 (2022)
4. Xu, P., Zeng, H., Qian, T., et al.: Research on defect detection of high-speed rail based on multi-frequency excitation composite electromagnetic method. Measurement **187**, 110351 (2022)
5. Zyab, C., Czb, C., Ying, Y.D.: Principle of a low-frequency transient electromagnetic radar system and its application in the detection of underground pipelines and voids. Tunn. Undergr. Space Technol. **122**, 104392–104514 (2022)

6. He, Z., Sun, P., Gong, K., et al.: Parameter estimation and signal detection algorithm based on adaptive capture in non-cooperative communication. MATEC Web Conf. **336**(6), 4–12 (2021)
7. Li, J., Zhang, X., Zhang, C., et al.: Simulation research on high-speed railway dropper fault detection and location based on time-frequency analysis. J. Phys.: Conf. Ser. **1631**, 012100 (2020)
8. Lu, W., Duan, J., Li, Z., et al.: Radiation magnetic field research of primary/secondary fusion switch cabinet based on frequency scaling approach. In: Conference on Frontier Academic Forum of Electrical Engineering, pp. 1185–1189 (2020)
9. Xia, Q., He, C., Liu, Y., et al.: Research on pollution flashover detection of power insulator based on high frequency electromagnetic wave. In: AIAM2020: 2nd International Conference on Artificial Intelligence and Advanced Manufacture, pp. 162–172 (2020)
10. Wang, Z., Guo, L., Jiangting, L.I.: Research on phase shift characteristics of electromagnetic wave in plasma. Plasma Sci. Technol. **23**(7), 075001 (2021) (7pp)
11. Qian, W., Wang, H., Wu, M., et al.: Research on noise suppression method for transient electromagnetic signal. In: 2018 2nd IEEE Advanced Information Management, Communicates, Electronic and Automation Control Conference (IMCEC), pp. 083–096. IEEE (2018)

Multi Target Tracking Method for Rail Transit Crossing Based on Transient Electromagnetic Radar

Qing Shi$^{(\boxtimes)}$ and Jian Nie

Wuhan Railway Vocational College of Technology, Wuhan 430000, China
ssl335@sina.com

Abstract. Conventional multi-target tracking methods for rail transit crossings mainly use the Deep SORT tracking detector to measure the tracking Mahalanobis distance, which is vulnerable to the dynamic change of target tracking state, resulting in low accuracy of multi-target tracking. Therefore, a new multi-target tracking method for rail transit crossings needs to be designed based on transient electromagnetic radar. That is to say, the transient electromagnetic radar is used to collect the multi-target tracking data of rail transit crossings, build the multi-target tracking model of rail transit crossings, and design the multi-target tracking algorithm of rail transit crossings, thus realizing the multi-target tracking of rail transit crossings. The experimental results show that the designed multi-target tracking method based on transient electromagnetic radar has high accuracy, which proves that the designed multi-target tracking method for rail transit crossings has good tracking effect, accuracy, and certain application value, and has made certain contributions to improving the safety of rail transit crossings.

Keywords: Transient electromagnetic radar · Track · Traffic lane · Fork · Multi target tracking

1 Introduction

In today's society, rail transit system, as one of the important urban transportation modes, its safety and operating efficiency are crucial to ensure passenger travel and smooth urban traffic. As a key component of the rail transit system, the crossing is responsible for scheduling the direction of train travel, and its safe and smooth operation plays a vital role in the operation of the whole system [1, 2]. Therefore, the research on multi-target tracking of rail transit crossings has important background and significance. Through this research, the safety, operating efficiency and intelligent level of rail transit system can be improved, providing better security and services for urban traffic development and passenger travel [3].

The commonly used multi objective tracking methods for rail transit intersections include traditional computer vision based multi-objective tracking methods, deep

L. Yun et al. (Eds.): ADHIP 2023, LNICST 548, pp. 177–189, 2024.
https://doi.org/10.1007/978-3-031-50546-1_12

learning based multi-objective tracking methods, and multi-sensor fusion based multi-objective tracking methods. The multi object tracking method based on traditional computer vision methods utilizes traditional computer vision algorithms to track target objects at intersections, and detects and tracks targets based on their appearance, motion features, and trajectory information. The deep learning based multi-objective tracking method uses a deep neural network model to achieve target detection and tracking by learning a large amount of data, which to some extent improves the robustness of tracking. The multi target tracking method based on multi-sensor fusion can capture the information of target objects more comprehensively by integrating and fusing data from multiple sensors. Although the above methods can achieve target tracking, they usually require a large amount of annotated data for training, and for efficient target detection and tracking, they require huge computational resources to support, which limits the feasibility and scalability of these methods in practical applications, and the tracking accuracy is low [4–6].

In response to the above issues, this article proposes a multi target tracking method for rail transit intersections based on transient electromagnetic radar. Transient electromagnetic radar can detect and track targets from interference clutter, which can be converted into the precise distance between the radar and the target, which is beneficial for improving the accuracy of target tracking. A multi-objective tracking model for rail transit intersections has been constructed, which utilizes a state transition mechanism to handle the common phenomenon of random appearance and disappearance of targets in multi-objective tracking. At the same time, it can also utilize the performance advantages of existing single target tracking methods, which is beneficial for improving target tracking performance. By using Markov decision processes to model multi-objective tracking problems, the entire tracking process can be represented as a decision sequence. In modeling, it is necessary to define the state of the system, including information such as target position and speed, as well as other environmental states related to tracking tasks. Then, define the set of actions, which are selectable tracking operations in each state. Next, the reward function is defined to evaluate the pros and cons of each state and action to guide the decision-making process. Finally, using MDP solving algorithms such as value iteration or policy iteration, find the optimal strategy to maximize long-term cumulative rewards and achieve effective tracking of multiple objectives. This modeling method can provide systematic decision support and help optimize the performance and effectiveness of multi-objective tracking algorithms.

2   Design of Multi-target Tracking Method for Rail Transit Crossing Based on Transient Electromagnetic Radar

2.1   Acquisition of Multi-target Tracking Data of Rail Transit Crossings Based on Transient Electromagnetic Radar

Transient electromagnetic radar (TEM) is a kind of wireless detection device, which can detect targets and determine their spatial positions by radio. When there is relative movement between targets, frequency Doppler effect will be generated. Transient electromagnetic radar can detect and track targets from interference clutter and convert them into accurate distance between radar and target [7]. According to the principle of multi-target tracking at rail transit crossings, this paper selects 24GHz transient electromagnetic radar to transmit modulated signals of specific frequencies. The distance and speed of the target can be obtained by comparing the frequency difference between the transmitted signal at the moment and the echo signal at any moment. The composition block diagram of the radar is shown in Fig. 1 below.

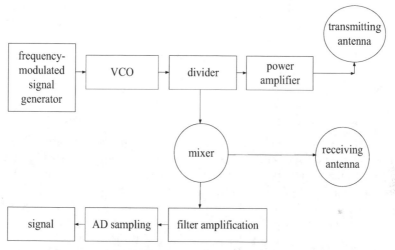

Fig. 1. Composition block diagram of 24GHz transient electromagnetic radar

It can be seen from Fig. 1 that the radar is mainly composed of three parts: antenna part, RF front-end and signal processing back-end. The antenna part is divided into two parts: transmitting antenna and receiving antenna; RF front-end mainly includes: power divider and mixer; The rest of the modulation signal generators and AD sampling constitute the back-end of signal processing. The modulation signal generator provides the modulation signal required by the radar, and generates a continuous signal [8] of a specific frequency through the VCO control of the voltage controlled oscillator. One part is transmitted from the transmitting antenna end through the power amplifier, and the other part is used as the local oscillator signal. In the process of outward transmission of the emitted electromagnetic wave, once encountering the measured target, it will return to the receiving antenna. At this time, the frequency difference between the echo signal and the non transmitted LO signal will occur. After passing through the mixer, the beat signal [9] will be obtained. The frequency of beat signal determines the distance and speed of the target. After the signal processing system calculates, the real distance and speed of the target can be obtained.

The innovative points of collecting multi target tracking data at rail transit intersections based on transient electromagnetic radar are as follows:

(1) Non visual perception: Traditional multi target tracking methods mainly rely on image or video data collected by visual sensors (such as cameras). The innovation based on transient electromagnetic radar lies in the use of non visual perception

to obtain information about the target object. Transient electromagnetic radar can detect and track target objects by measuring the propagation characteristics of electromagnetic waves between the target object and the radar, without being affected by factors such as lighting and occlusion.

(2) Distance and velocity information acquisition: Transient electromagnetic radar can provide relevant information such as distance and velocity of target objects, which is crucial for target tracking. By collecting and processing data from transient electromagnetic radar, real-time information such as the position, velocity, and motion trajectory of the target object can be obtained, thereby achieving accurate tracking of multiple target objects at intersections.

(3) Strong robustness: Transient electromagnetic radar has strong robustness in complex environments, unaffected by weather conditions such as light, rain, and snow, as well as changes in the appearance and occlusion of target objects. This enables the multi target tracking method based on transient electromagnetic radar to operate stably in various complex environments, providing reliable target tracking results.

In summary, the innovation of using transient electromagnetic radar to collect multi target tracking data at rail transit intersections lies in utilizing non visual perception to obtain information about target objects, and achieving accurate tracking of targets by providing data on key physical characteristics such as distance and speed.

There are also many kinds of frequency modulation modes for transient electromagnetic radar, such as linear frequency modulation continuous wave system, stepped frequency modulation continuous wave system and frequency shift keying system. Taking LFMCW sawtooth FM continuous wave as an example, this paper briefly introduces the principle of distance and speed measurement of traffic radar. Assume that the transmitted signal is sawtooth FM continuous wave, that is, it is composed of each sawtooth continuous linear FM pulse string. At this time, the frequency of the interpolated beat signal in the effective section f_b as shown in (1) below.

$$f_b = \tau K \tag{1}$$

In Formula (1), τ represents the transmitted signal, K represents the signal slope. When different targets are distributed at different distances, the beat signal obtained is the result of the linear addition of the beat signals of each target [10], but the frequency is separate, so the range information of the observed target can be obtained from the beat signal of a single target or multiple targets, and the wave path difference of the radar plane signal at this time φ as shown in (2) below.

$$\varphi = \frac{2\pi}{\lambda} d \sin \theta \tag{2}$$

In Formula (2), d represents the fixed distance difference of antenna, λ represents the length of electromagnetic wave emitted by radar, θ represents the angle between the target and the normal direction of the radar.

When collecting multi-target tracking data, it is necessary to identify and confirm multiple targets detected by radar, and associate, estimate and form tracks of the identified motion state information from the same target. The whole multi-target tracking process is a recursive process. Assuming that each target track has been successfully initiated during the tracking process, the newly received measurement information is first used to update the established target track, and the tracking gate is used to judge whether the newly received measurement data is reasonable. Data association is used to determine which track the measurement matches with the started track, and then the real parameters of the target track can be estimated according to the motion state, filtering and prediction. In the tracking process, the measured values that differ greatly from the established track may be clutter, and their effectiveness can be further judged according to the motion law or starting rules, and a new track can be started according to the appropriate situation; When some tracks do not match many new measurements, the tracks can be destroyed to reduce the amount of calculation; Finally, before the arrival of the new measurement, the center and size of the tracking gate at the next time can be appropriately adjusted according to the target prediction state, and a new round of recurrence cycle can be started again. According to the above reasoning, the data acquisition steps for multi-target tracking of rail transit crossings based on transient electromagnetic radar can be designed, as shown in Fig. 2 below.

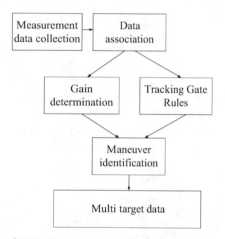

Fig. 2. Multi target tracking data acquisition steps

It can be seen from Fig. 2 that the resolution unit for detecting distance and speed of the above radar is small, so the receiving sensitivity and resolution are high. The size of the target will not cause too much impact when detecting long-distance targets. A target is usually detected to a reflection point. However, as the measured target is getting closer to the traffic radar, the same target will generate multiple reflection points due to different target shapes and large volumes. In the process of traffic radar detection, the transmitted signal returns after encountering multiple reflection points, generating multiple spectral lines in the spectrum, so that spectrum broadening occupies multiple resolution units, and each horizontal and vertical intersection resolution unit corresponds to a reflection point status information. As a result, the same target detected with multiple reflection points will generate multiple plot information, which actually corresponds to the same target. So when tracking multiple targets on the traffic road, it is the key to distinguish which targets belong to different targets from a large number of original plot information and which are from the same target, and to cohere multiple plot information from the same target well to ensure the subsequent high-precision and stable tracking of multiple targets. Therefore, the requirements for the original data preprocessing phase of the traffic radar multi-target tracking are high, especially the plot condensation part is the basis of the whole tracking process.

2.2 Build a Multi-target Tracking Model for Rail Transit Crossings

After collecting the multi-target tracking data of rail transit crossings, in order to effectively track the flow, it is necessary to build a multi-target tracking model of rail transit crossings. In this paper, the Markov decision process is used to model the multi-target tracking problem. The four states of active state, tracking state, lost state and inactive state are used to describe the different states of single target tracking. The whole life cycle of a target is modeled as a Markov decision process. At the same time, the tracking problem is also transformed into the learning problem of the transition strategy between different states of MDP. With the help of the state transition mechanism in MDP, this model structure can deal with the phenomenon of random appearance and disappearance of targets commonly seen in multi-target tracking, and can also take advantage of the performance advantages of existing single target tracking methods. This paper defines an action set composed of seven actions to describe the transition behavior between different states. The reward function is learned by training samples. The learning of the state transition strategy is realized through the reinforcement learning method, so the advantages of online learning and offline learning can be simultaneously used to realize data association. At this time, the multi-target tracking relationship is shown in Fig. 3 below.

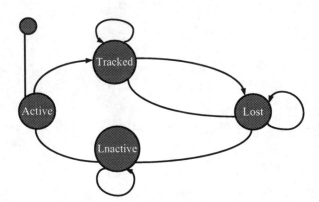

Fig. 3. Multi target tracking relationship

It can be seen from Fig. 3 that the Active state is to preliminarily screen the noisy detection frame obtained from the detector, and directly transfer the unreasonable detection frame to the Inactive state without target tracking. In the Active state, the MDP transfers the target detection frame to the Tracked state and Inactive state. In MDP tracking l31, in the Active state, support vector machine 431 is used to classify the detection frame. SVM uses the 5-dimensional feature vector of the detection frame to extract the 2-dimensional coordinates, width, height, and detection score of the detection frame. Then, the image width and height, maximum area detection frame width and height, and the highest detection score extracted from the training samples are used to calculate the 5-dimensional feature vector as the input of SVM,Take the detection box in the training sample as the parameter of training SVM, and the reward function $R_{Active}(s, a)$ as shown in (3) below.

$$R_{Active}(s, a) = y(a)(w_{Active}^T \phi_{Active}(s) + b_{Active}) \tag{3}$$

In Formula (3), $y(a)$ represents tracking 2D coordinates, w_{Active}^T represents the highest test score, $\phi_{Active}(s)$ represents a super parameter, b_{Active} represents the state transfer coefficient.

In the Tracked state, you need to decide whether to keep the target in the Tracked state or transfer the target state to the Lost state. As long as the target is not blocked and within the camera's field of view, the target needs to remain Tracked, otherwise the target will turn to Lost. In MDP tracking, an online tracking model based on appearance is established in Tracked to track a single target. The specific method is based on TLD.During initialization, K templates will be set for each target to represent the target's appearance characteristics. During the tracking process, LK tracking used in TLD will be used to calculate the target position through the template and perform single target tracking. The template update adopts the lazy update mechanism, and only one of the K templates is updated each time, so that the K templates retain the historical information of the target, which is the basis for data association in the Lost state. At this time, the rail transit intersection multi-target tracking model is built $R_{Tracked}(s, a)$ as shown in (4) below.

$$R_{Tracked}(s, a) = R_{Active}(s, a) \cdot y(a) \backslash otherwise \tag{4}$$

In Formula (4), *otherwise* represents the continuous detection value of multi-target tracking. The conversion of the Tracked state Lost state is based on two parameters. First, use the value measuring the tracking quality in LK tracking to calculate the median forward and backward optical flow. When the monthly marker encounters occlusion, the tracking quality becomes worse and the emedFB value rises; The second is the average overlap rate of K target templates and detection frames, which is used to determine whether the tracking target is a detector false detection target. If the current tracking target is a detection frame false detection result, the target will not be continuously detected by the detector, and the Omean value will continue to decrease. Finally, the state transition is carried out by setting the threshold value.

When the target is in the Lost state for more than a certain time threshold, the target will be directly transferred to the Inactive state to end tracking. Whether the target will remain in the Lost state or be transferred to the Tracked state is essentially a data association problem. If the target can match a detection frame in the current frame, it will be transferred to the Tracked state, otherwise it will remain in the Lost state.When tracking fails, the tracking target has saved K templates representing appearance information, and 12 dimensional feature similarity coding can be designed based on templates and detection frames *MDP*, , as shown in (5) below.

$$MDP = \frac{y(a)}{R_{Tracked}(s, a)}(W_{\max}(t) + b)$$ (5)

In Formula (5), $W_{\max}(t)$ represents the number of corresponding detection frames when the target realizes data association, b represents training initialization SVM parameters. According to the above feature similarity coding and multi-target tracking model, linear regression solution can be carried out $F(Z)$ as shown in (6) below.

$$F(Z) = \min \sum_{i} (f(x) - y_i)^2 + \lambda \|w\|^2$$ (6)

In Formula (6), $f(x)$ represents the target input function, y_i represents the weight matrix, w represents the ridge regression learning parameter matrix. According to the above model solution values, the sample cyclic shift parameters can be obtained, and the tracking transformation eigenvalue can be obtained to maximize the accuracy of multi-target tracking.

2.3 Design Multi-target Tracking Algorithm for Rail Transit Crossings

The most original data received by the 24 GHz transient electromagnetic radar used in this paper is in hexadecimal form, and each frame counts as a packet of data. The refresh rate of the radar is 50 ms, so the radar receives 20 frames of data per second, and each frame of data can detect up to 160 targets, that is, each frame packet data can contain up to 160 target motion status information in addition to the relevant information such as the unified header data format of each frame of data. Therefore, it is the first and key step to accurately analyze the original frame packet data in hexadecimal form. When parsing the original packet data of a frame, because the frame header identifier of each frame of data is the same, first judge and identify the frame header information of each frame of data, until the complete frame header data is detected and identified, then proceed to the next step, and convert the hexadecimal data after the frame header information into decimal data according to the corresponding bytes occupied by each part.

The received original frame packet data may contain some errors or receive incomplete frame header data due to various reasons of the radar itself or data transmission, which is called outlier in practical engineering applications. They generally differ greatly from the real measured values, or do not contain the effective motion state information of the detected target. Therefore, in the process of original data analysis, outlier removal is also included to reduce the calculation amount for the next step of plot condensation preprocessing.

In the process of radar data acquisition or transmission, some received frame packet data may have errors or incomplete data. The most obvious is that the frame header information is incomplete, which leads to great errors in the target motion parameters after the frame header information. Therefore, such frame packet data with incomplete header data will also be eliminated in the original plot pre-processing stage. The specific method is to first detect whether the header format of each received frame of data is consistent, directly eliminate the frame packet data with incorrect or incomplete header data format, do not convert from hexadecimal to decimal data, and then detect the next frame of data.

According to the performance indicators of the radar, this radar has a certain detection range. The measurement data outside the detection range itself is not meaningful for subsequent algorithm processing, and has a large difference from the measured value of the actual observation target. Therefore, these outliers can be eliminated in the pre-processing phase of the original plot. The radar detection range of this paper is shown in Fig. 4 below.

Longitudinal distance (m)	0~400
Angle (°)	-60~60
Speed (m/s)	-70~70

Fig. 4. Detection range of this radar

It can be seen from Fig. 4 that in the original data analysis, not only useful target motion parameters such as frame number, radial distance, speed and angle of the target detected by the radar are obtained, but also invalid values are eliminated, which reduces the amount of calculation for the following target tracking algorithm and improves the operation speed. In addition, the original data analysis also includes the calculation of the coordinates under the traffic radar coordinate system where the target is located. Through the radial distance and angle of each target plot, the coordinate value of the target plot is calculated, as shown in (7) below.

$$\begin{cases} x = d \cdot \sin\theta_a \\ y = d \cdot \cos\theta_a \end{cases} \quad (7)$$

In Formula (7), d represents the radial distance of the target, θ_a represents the target angle. After analyzing the original data, the multi-target tracking equation can be generated, as shown in (8) below.

$$\begin{cases} y_l = -a \\ y_r = W - a \end{cases} \quad (8)$$

In Formula (8), y_l, y_r represents the radar detection range, W represents the width of the traffic road, a represents the distance between the radar and the edge. Generally speaking, the traffic radar is installed on the road perpendicular to the driving direction of the road target. It is easy to know the width of the road where the traffic radar is installed and the specific position of the radar in the direction of the road width. The traffic radar uses the radar coordinate system in the target tracking process, so according to the above road information. The edges of traffic roads can be identified in the radar coordinate system. Since the installation position of the radar is perpendicular to the driving direction of the traffic road vehicles, the edge marking line of the traffic road is parallel to the radar normal. The method designed in this paper comprehensively compares and analyzes the distance, angle and speed, sets a comprehensive threshold value. The original plots from the same frame with similar angles and within the comprehensive threshold value are marked as points from the same target, and then the information of these marked plots is condensed D as shown in (9) below.

$$D = \sum_{i=1}^{m} \frac{p_i d_j}{\sum\limits_{i=1}^{m} (p_i)} \tag{9}$$

In Formula (9), p_i represents the tracking amplitude, d_j represents the radial distance. Using the above multi-target tracking algorithm for rail transit crossings, multi-target tracking can be carried out quickly to solve the tracking problem caused by the dynamic change of tracking state.

3   Experiment

In order to verify the tracking performance of the designed transient electromagnetic radar based multi target tracking method for rail transit intersections, this paper selected experimental datasets that meet the experimental requirements and compared the proposed method with conventional deep learning based multi target tracking methods and multi-sensor fusion based multi target tracking methods.

3.1   Experimental Software and Hardware Environment Settings

The experimental software and hardware environment is shown in Table 1.

Table 1. Experimental hardware and software environment

experimental environment	ambient condition
CPUmodel	Inter(R) Xeon E5-2697 v3
GPU	NVIDIA Tesla k40c Video storage 11 GB
	NVIDIA Quadro k420 Video storage 2 GB
Running memory (RAM)	192 GB
Programming Language	Python 3.5
Deep learning framework	Keras 2.1.5

From Table 1, it can be seen that the above software and hardware environments meet the training requirements of the experiment, improve the efficiency of the experiment, and ensure the effectiveness of the experimental results. At this point, subsequent multi target tracking experiments at rail transit intersections can be conducted.

3.2 Experimental Sample and Indicator Settings

Based on the experimental requirements, this article selects the MOT16 dataset as the experimental dataset. The MOT16 dataset includes 11 annotated training video sequences and 11 unlabeled test video sequences with only detection results. Figure 5 shows some experimental sample images.

Scenario 1   Scenario 2

Scenario 3   Scenario 4

Scenario 5   Scenario 6

Fig. 5. Experimental Scenario

After the dataset is selected, experimental indicators are set. In this paper, multi-objective tracking accuracy is selected as the experimental indicator, and the calculation formula is as follows (10):

$$MOTA = 1 - \frac{\sum_t m_t + fp_1 + mme_t}{\sum_t g_t} \quad (10)$$

In Formula (10), m_t represents the number of missed detections in t frame, fp_1 represents the number of false detections in t frame, mme_t represents the number of mismatches in t frame, g_t represents all the matching quantity in t frame. The higher the MOTA multi-target tracking accuracy is, the better the multi-target tracking effect is. On the contrary, the tracking effect is relatively poor.

3.3 Experimental Results and Discussion

Based on the above experimental preparation, a multi target tracking experiment for rail transit intersections was conducted. The designed multi target tracking method for rail transit intersections based on transient electromagnetic radar, conventional deep learning based multi target tracking method, and multi-sensor fusion based multi target tracking method were used for multi target tracking. Formula (10) was used to calculate the tracking accuracy of the three methods for different tracking targets, The experimental results are shown in Table 2 below.

Table 2. Experimental Results

Tracking target	Multi target tracking method for rail transit crossing based on transient electromagnetic radar (%)	Multi object tracking method based on deep learning (%)	Multi target tracking method based on multi-sensor fusion (%)
C1#a	95.44	75.36	65.65
C2#a	93.89	61.45	73.34
C3#a	96.32	62.28	62.53
C4#a	95.48	63.54	63.96
C5#a	94.14	76.12	66.45
C6#a	92.52	69.69	75.28
C7#a	96.95	65.85	74.66
C8#a	99.47	56.74	72.85
C9#a	98.58	64.26	66.42
C10#a	95.23	65.84	79.39

From Table 2, it can be seen that the multi target tracking method for rail transit intersections based on transient electromagnetic radar designed in this article has high accuracy in tracking different targets, while conventional deep learning based multi target tracking methods and multi-sensor fusion based multi target tracking methods have relatively low accuracy in tracking different targets. The above experimental results prove that the multi target tracking method designed in this paper has good tracking performance, high tracking accuracy, and effectiveness, with certain application value.

4 Conclusion

With the rapid development of road traffic, the total mileage of roads in China is getting higher and higher, resulting in many problems such as traffic congestion, mixed traffic, frequent accidents, low level of traffic management, among which the problem of frequent traffic accidents is the most serious. According to statistics, about 200000 road traffic accidents occur every year in China, resulting in more than 60000 deaths and about 210000 injuries. According to the investigation, most accidents are caused by drivers' illegal driving and abnormal parking without supervision; If the road management party can strengthen the monitoring and timely find out the violations of vehicle drivers or give early warning guidance to other vehicles after the accident, the occurrence of traffic accidents can be reduced. In this context, in order to deal with the severe traffic safety situation, it is necessary to carry out multi-target tracking for vehicles. In this paper, based on the characteristics of the current traffic, a multi-target tracking method for rail transit intersections is designed based on the transient electromagnetic radar. The experimental results show that the designed track crossing multi-target tracking method has good tracking effect, accuracy and certain application value, and has made certain contributions to improving traffic safety.

References

1. Merkert, M., Sorgatz, S., Le, D.D., et al.: Autonomous traffic at intersections: an optimization-based analysis of possible time, energy, and CO2 savings. Networks **79**(3), 338–363 (2022)
2. Hu, Z., Huang, J., Yang, D., et al.: Constraint-tree-driven modeling and distributed robust control for multi-vehicle cooperation at unsignalized intersections. Transp. Res. Part C: Emerg. Technol. **131**(10), 103353.1–103353.17 (2021)
3. Song, S., Yeom, C.: Reduction of plastic deformation in heavy traffic intersections in urban areas. Sustainability **13**(7), 4002 (2021)
4. You, S.: Optimization driven cellular automata for traffic flow prediction at signalized intersections. J. Intell. Fuzzy Syst.: Appl. Eng. Technol. **40**(1), 1547–1566 (2021)
5. Li, Y., Wang, B.: Multi-extended target tracking algorithm based on VBEM-CPHD. Int. J. Pattern Recogn. Artif. Intell. **36**(06), 2250026.1–2250026.15 (2022)
6. Li, X., Yu, N., Li, J.W., Jiang, J.W.: A moving target tracking method with overlapping horizons and multi-camera coordination. Comput. Simulat. **38**(11), 162–167 (2021)
7. Chand, S.: Modeling predictability of traffic counts at signalised intersections using Hurst exponent. Entropy **23**(2), 188 (2021)
8. Karnati, Y., Sengupta, R., Rangarajan, A., et al.: Subcycle waveform modeling of traffic intersections using recurrent attention networks. IEEE Trans. Intell. Transp. Syst. **3**, 2538–2548 (2022)
9. Maode, Y., Zhang, Y., Yang, P., et al.: Intelligent connected vehicle queue lane changing method based on multiobjective optimization. Comput. Simulat. **39**(3), 145–149 (2022)
10. Sallard, A., Bala, M.: Modeling crossroads in MATSim: the case of traffic-signaled intersections. Procedia Comput. Sci. **184**, 642–649 (2021)

A Data Mining and Processing Method for E-Commerce Potential Customers Based on Apriori Association Rules Algorithm

Xian Zhou[1(✉)] and Hai Huang[2]

[1] Institute of Economics and Management, Zhi Xing College of Hubei University, Wuhan 430011, China
zhou_xian2001@163.com
[2] Shanghai Dong Hai Vocational and Technical College, Shanghai 200241, China

Abstract. In order to improve the effectiveness of e-commerce potential customer data mining and processing, a method based on Apriori association rule algorithm for e-commerce potential customer data mining and processing is proposed. Innovatively adopting a multidimensional tree structure to improve the Apriori association rule algorithm, using frequent itemsets as candidate itemsets, and further expanding on this basis by adding judgment conditions to reduce the frequency of scanning the database; The Vector space model is used to calculate the similarity between e-commerce potential customers, and the similarity is used as a scalar value to complete the accurate calculation. The e-commerce potential customers at different levels in customer transaction data are divided. Obtain a sticky evaluation system for potential e-commerce customers from the perspectives of perceived usefulness, perceived ease of use, perceived service, perceived security, and perceived interest, as the basic indicators for subsequent mining and processing. The Quicksort method is used to sort each data dimension in the e-commerce customer data set, and the improved Apriori association rule algorithm is used to realize data mining and processing of e-commerce potential customers through high-density grid. The experimental results demonstrate that the method innovatively utilizes the improved Apriori association rule algorithm to mine three types of customer behavior data with an accuracy of over 80%, which is in line with the actual situation. It improves the effectiveness of e-commerce potential customer data mining and processing, effectively mining e-commerce potential customers, and providing good basic data for e-commerce platforms to adjust marketing strategies.

Keywords: Apriori Algorithm · E-Commerce Potential Customers · Data Mining

1 Introduction

At present, China's e-commerce is booming, with various e-commerce platforms constantly emerging, while explosive data growth is occurring. There is valuable information hidden in this data that needs to be mined, such as e-commerce potential customer data.

L. Yun et al. (Eds.): ADHIP 2023, LNICST 548, pp. 190–202, 2024.
https://doi.org/10.1007/978-3-031-50546-1_13

Data mining is the discovery of knowledge. Mining e-commerce potential customer data from the accumulated transaction data of e-commerce can find targeted and effective customer information in the massive information, which has important strategic significance for the development of modern enterprises [1–3]. E-commerce is a product of the rapid development of the internet industry, and it is a new type of business operation model. In an open online environment, buyers and sellers can engage in product selection, payment, and other trading activities on the Internet without meeting.

According to the survey report, e-commerce can be roughly divided into four types according to transaction partners: Firstly, B2C (Business to Customer): refers to e-commerce activities between enterprises and consumers, where consumers directly engage in commodity trading activities on the Internet. Secondly, B2B (Business to Business): refers to the business activities between enterprises. Can companies find suitable partners on the Internet? M line transaction activity. Thirdly, C2C (Consumer to Consumer): refers to e-commerce activities between consumers. Consumers can engage in sales activities in the form of retail investors on e-commerce platforms, such as Xianyu on the Alibaba platform. Fourthly, C2B (customer to business): refers to the business activities between consumers and enterprises. Consumers targeting the same consumer goods can purchase from enterprises in the form of teams, generate precise orders and produce goods, but this model is not yet mature.

Data mining can first provide customers with more comprehensive personalized services. By conducting in-depth mining and analysis of customer access information data, we can obtain customer purchasing behavior characteristics and preferences, understand customer habits, interests, potential needs, and loyalty issues to profile customer characteristics and reduce unnecessary content push. Taking Taobao as an example, it digs data information about customers' access methods and orders related information in a specific period of time, so as to understand buyers' needs or predict their potential for money, and designs the content and structure of the web page in a targeted way, and assigns each customer a personalized service package or a preferential combination based on appropriate marketing strategies, thus bringing profits to shops on e-commerce platforms [4, 5]. Secondly, data mining can optimize the design of website content. Over the years, e-commerce companies have accumulated a large amount of historical data. How to better utilize this data and explore valuable internal laws and potential customers has become an important means for e-commerce companies to survive and develop.

In order to meet the computational requirements of many data mining systems, technology needs to be adopted in hardware, operating system software, and database systems. These resources greatly increase costs and strain the information technology resources composed of technologists. Not only do distributed and non memory versions of current data mining algorithms need to be developed, but new algorithms also need to be developed, which is a new challenge faced by data mining technology. In order to improve the accuracy and efficiency of e-commerce potential customer data mining, a method for e-commerce potential customer data mining processing based on Apriori association rule algorithm is proposed. This article innovatively adopts a multidimensional tree structure to improve the Apriori association rule algorithm to reduce the frequency of scanning the database; The Vector space model is used to calculate the similarity between e-commerce potential customers and divide e-commerce potential customers at different levels in customer transaction data. Establish a stickiness evaluation system for potential e-commerce customers as a basic indicator for subsequent mining and processing. Utilize the improved Apriori association rule algorithm to complete data mining and processing of e-commerce potential customers through high-density grids. The study flow module of the method design is shown in Fig. 1:

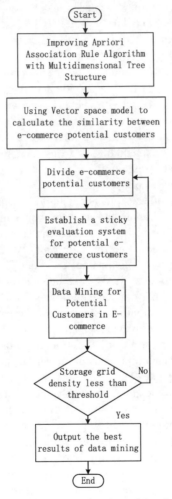

Fig. 1. Research process of each module of the article

2   Improvement of Apriori Association Rule Algorithm

The core idea of the Apriori association rule algorithm is to find all itemsets, ensure that their occurrence times are consistent with the defined minimum support, and then generate strong association rules that can meet the minimum credibility and support.

Then, based on the rules generated from the initial frequency set, generate all rules that only contain items with a set. Once generated, only rules with greater confidence than the given minimum can be left behind. Based on the above characteristics of the Apriori association rule algorithm, it is utilized for e-commerce potential customer data mining and processing [6, 7]. However, during use, this algorithm requires multiple scans of the original database, resulting in a large number of candidate sets being generated, resulting in many redundant rules and low efficiency. This requires further improvements to the algorithm.

The principle of the improved algorithm is to use frequent 1-item sets as candidate item sets, and further expand on this basis by adding a judgment condition Lk. count $\geq$ min sup port to obtain frequent k + 1-item sets. Then, use this as a candidate project set and continuously generate the next frequent project set according to the above steps, ultimately obtaining all project sets. Its advantage is that the frequency of scanning the database decreases, and the number of candidate sets generated continues to decrease. It is more suitable in large-scale databases and can effectively improve the efficiency of extracting frequent itemsets and discovering association rules [8]. To achieve this goal, consider the following principle: assuming there is a data item A, add it to I, and AI, the frequency of the result set (A $\div$ I) is lower than that of I. If I cannot meet the minimum support threshold, then the latter will also not meet, meaning that a set will not pass the test. The following points can be inferred from it.

Improved Apriori association rule algorithm based on multidimensional tree structure. Assuming that an association coefficient is the result of a mapping solution from the Initialization vector to the target vector, multidimensional association rules can be understood as a unified set space composed of multiple association coefficients.

The multidimensional tree structure is used as the basic application structure for large-scale data mining processing, consisting of multiple associated nodes. However, according to the different tasks performed, the corresponding data information objects of each node organization also vary. A Schematic representation of the structure of the association tree organization as shown in Fig. 2.

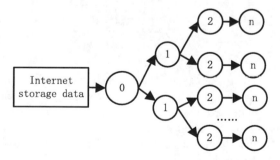

Fig. 2. Schematic representation of the structure of the association tree organization

In Fig. 2, the "0" node serves as the initial structure, responsible for docking with internet stored data, and can directly feed back the information parameters to be mined to the subordinate node structure. As a subordinate structure of node '0', node '1' has certain data classification capabilities and can be fed back to different storage units according to different encoding forms of data parameters. The "2" node - "n" node serves as the core processing structure of the association tree organization, directly executing data mining instructions, and displaying real-time transmission positions of data information parameters according to the processing results [9]. The actual value of the coefficient "n" varies depending on the length of the association tree organization connection.

Let c denote a randomly selected RFM value definition index, and the inequality condition of the coefficient $c \neq 0$ is constant, and β represent the node definition coefficient in the correlation tree structure. Combining the above physical quantities, the RFM value calculation expression based on the multidimensional association rule can be defined as:

$$z_c = \frac{\beta \cdot x_c}{(\alpha_c + \delta_c)^2 + 1} \tag{1}$$

where x_c represents the scale characteristic value of the e-commerce customer data, α_c and δ_c represent the two unequal multidimensional vector assignment calculation coefficients.

When solving the expression of the RFM value, it is required that the value of the coefficient x_c must belong to the physical interval of $[1, e]$.

In the organization of association trees, the arrangement of feedback nodes not only affects the calculation results of RFM values, but also changes the ability of multidimensional algorithms [10, 11].

Let χ denote the initial assignment of the feedback node distribution coefficient, whose minimum value can only be equal to the natural number "1". ϕ represents the feedback parameter of the data to be mined based on the multidimensional association rules, which is influenced by the solution expression of RFM value. The larger the calculated value of RFM value index is, the larger the actual value of ϕ coefficient is. With the support of the above physical quantities, the joint Formula (1), the multidimensional algorithm expression can be defined as:

$$V = \sum_{\chi=1}^{+\infty} \phi \cdot z_c \cdot \frac{\sqrt{b_1^2 + b_2^2}}{\gamma \cdot |\bar{b}|} \tag{2}$$

where b_1 and b_2 represent two unequal data information operation features, $\bar{b}$ represents the average value of coefficient b_1 and coefficient b_2, and γ represents the amount of data information extraction parameters based on multidimensional association rules. When constructing the multi-dimensional association rule algorithm, the calculation expression of RFM index should be highly unified. This completes the improvement of the Apriori association rule algorithm.

3 Characteristic Similarity Calculation of E-Commerce Prospect Data

For the behavior data of e-commerce potential customers, it has the characteristics of complex types and large amount of data. In order to accurately and quickly mine the behavior data of target customers, it is necessary to calculate the similarity of the behavior characteristics in advance. The similarity is taken as the scalar value to complete the accurate calculation, and the customers of the e-commerce potential customers at different levels in the customer transaction data are divided. The division structure is shown in Fig. 3:

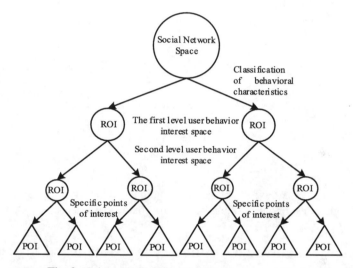

Fig. 3. Schematic diagram of different customer levels

This paper uses the vector space model to calculate the similarity between e-commerce potential customers, each customer is ROI (Region Of Interest region of interest) and POI (Point Of Interest point of interest) vector labeling,$R = [r_1, r_2, \ldots, r_n]$. In order to accurately calculate the increase the number of customer visits, the more similar the behavior expression, a_i is used to indicate the number of the i interest visits. Matrix $V_{m \times n}$ of all customers with the formula:

$$
V_{m \times n} = \begin{bmatrix}
v_{1,1} & v_{1,2} & \cdots & v_{1,n-1} & v_{1,n} \\
v_{1,2} & v_{2,2} & \cdots & v_{2,n-1} & v_{2,n} \\
\vdots & \vdots & \ddots & \vdots & \vdots \\
v_{m,1} & v_{m,2} & \cdots & v_{m,n-1} & v_{m,n}
\end{bmatrix}
\tag{3}
$$

In the formula, m represents the number of real-time customers; $v_{m,n}$ represents the number of interests of customers in a specific community space; and $v_{m,n-1}$ represents the specific number of interests of customers in the next community space. Take the community network as a space and the customer as a vector value in a dimension in the space. Let the vectors of customers α and β in n dimensions be expressed as U_α and U_β, then the similarity calculation formula between customer α and β is as follows:

$$sim(\alpha, \beta) = \cos(U_\alpha, U_\beta) = \frac{U_\alpha \cdot U_\beta}{\|U_\alpha\|\|U_\beta\|} \tag{4}$$

Since different customer interest points exist in different spatial neighborhoods, before calculating the similarity of customers under different spatial neighborhoods, it is necessary to cluster the similarity between the two different neighbors:

$$sim_{oreall} = \sum_{i=1}^{H} \mu sim_i \tag{5}$$

In the formula, μ represents the similarity weight, namely:

$$\mu = \frac{\chi_i}{\sum_{i=1}^{H} \chi_i} \tag{6}$$

In the formula, H is the total hierarchy of neighborhood division; χ_i is the weight coefficient. Thus, the similarity of customer data characteristics is calculated.

4   E-Commerce Potential Customer Data Mining and Processing

Based on the similarity results of e-commerce customer data features calculated in the previous section, considering the economic law of e-commerce potential customer stickiness, a e-commerce potential customer stickiness evaluation index system is constructed from aspects: perceived usefulness, perceived ease of use, perceived service, perceived security, perceived interest, and perceived cost performance, as the basic indicators for mining and processing. The evaluation index system of e-commerce customer stickiness is shown in Table 1.

Perceived usefulness represents the customer's belief that using e-commerce platforms is helpful for self-improvement; Perceived ease of use represents the customer's belief that using an e-commerce platform is easy to operate; Perceived service represents that customers believe that using e-commerce platforms has higher service quality [12]; Perceived security represents that customers believe that using e-commerce platforms has a high level of security; Perceived interest represents that customers believe that using an e-commerce platform has a certain level of interest and will increase their interest in using the platform; Perceived cost-effectiveness represents that customers believe that using e-commerce platforms to purchase goods has a higher cost-effectiveness.

Table 1. Evaluation index system of e-commerce customer stickiness

Target layer	The standard layer	Index layer
E-commerce customer stickiness evaluation index	Perceived usefulness	Information channels
		Source of knowledge
		Social help
	Perceived ease of use	Easy to operate
		Retrieve tool efficiency
		The purchase process is simple
		Convenient payment
	Perceived service	Pre-sale seller communication initiative
		Seller delivery timeliness
		Logistics delivery efficiency
		External packaging of goods
		Effectiveness of problem solving
	Perceived safety	Security of website registration and login
		Pay security
		Transaction information security
		Express safety
	Feel fun	Personalized recommendation
		Functional innovation
		The content is rich and interesting

In the improved Apriori association rule algorithm, when dividing the grid of e-commerce customer datasets, the interval segments with consistent boundaries are adjacent interval segments. Make the k-dimensional e-commerce customer dataset $D = \{d_{i1}, d_{i2}, \cdots, d_{iN}\}$, and D have N data points, and divide the i-dimensional e-commerce customer dataset into q intervals, that is, there are $[N/q]$ data points within each interval, that is, equal depth partitioning. Make the data points within each interval consistent. The i-th dimension and jth interval are $H_{ij} = \left(d_{(i)([N/q]*(j-1)+1)}, d_{(i)([N/q]*j)} \right)$, and because the number of data points within each interval is consistent, use the length of the interval to describe the density of the interval, that is $|H_{ij}| = \left(d_{(i)([N/q]*j)} - d_{(i)([N/q]*(i-1)+1)} \right)$. If the $|H_{ij}|$ of a certain grid exceeds the corresponding density threshold, then the grid is a high-density grid, and vice versa, it is a low-density grid. If a certain grid and its neighboring grids are both low-density grids, then the grid is treated as noisy data. Let $|H_{ij}|$ and $|H_{i(j+1)}|$ in the j + 1 interval be known.

If $|H_{ij}|$ does not obtain a reference source, then the density similarity of adjacent interval segments is $y = \frac{|H_{ij}|}{|H_{i(j+1)}|}$. If $|H_{ij}|$ has obtained a reference source, then $y = \frac{|H_{i(j+1)}|}{|H_{ij}|}$. There are m grids in D memory, and the α The density of each grid is x_α, and the density threshold of the grid is as follows:

$$\varepsilon = \frac{\sum\limits_{\alpha=1}^{m} x_\alpha \lambda}{N} \tag{7}$$

In the formula, the constant is the λ.

The mining result corresponding to the maximum ε value is, and the calculation formula is as follows:

$$k' = \arg\max \left\{ \frac{\sum\limits_{a=1}^{k} \sum\limits_{b=1}^{N} BWP(a, b)}{N} \right\} \tag{8}$$

The sample number of e-commerce customer data is b; the sample number of e-commerce customer data is N; and the mining class number is a.

The principle of improved Apriori correlation rules algorithm is through variable grid segmentation electricity customer data set, comparative analysis of k' and ε, choose more than ε grid, avoid isolated point on the influence of electricity customer viscosity data mining, through the improved Apriori association rules algorithm mining more than ε grid, get the best electricity customer viscosity related data mining results. The input of this algorithm is the data set D of e-commerce customer with N data points, the number of clusters k to expected segmentation, and the similarity threshold v; the output is the data mining result related to stickiness prediction of e-commerce customer. The specific steps of the algorithm to mine the e-commerce customer data set are as follows:

Step 1: Sort each dimension of data in the e-commerce customer dataset using a quick sorting method;

Step 2: Divide each dimension of e-commerce customer data into equal depths;

Step 3: Solve $|H_{ij}|$;

Step 4: Solve the neighboring interval segments of each dimension of e-commerce customer data ρ;

Step 5: Merge adjacent interval segments that meet the conditions;

Step 6: Cycle through the various dimensions of e-commerce customer data in D and merge the eligible interval segments;

Step 7: Calculate the density of the merged mesh;

Step 8: Store the solution record in grid set c;

Step 9: Solve ε;

Step 10: Store grid density exceeding in set d ε Class cluster of;

Step 11: Using the improved Apriori association rule algorithm, divide the high-density grid within d, and output the best results of k e-commerce customer stickiness related data mining.

This algorithm improves the representativeness of the selection of the initial mining center point. It removes the grid density lower than ε in the pruning mode, solves the problem of noise interference, and mines the e-commerce customer data set in the form of grid division, which can handle clusters of different shapes. Thus, the data mining processing of e-commerce prospects based on the Apriori association rules algorithm is completed.

5 Experiments and Analysis

5.1 Experimental Preparation

Taking an e-commerce platform as the experimental object, the public data set of the e-commerce platform contains more than 100,000 data records, which contains three dimensions, respectively, 1 dimension, 2 dimension, 3 dimension, 4 dimension, and 5 dimension, the data set is divided into three subsets according to the data dimension, recorded as subset 1, subset 2, subset 3, subset 4, and subset 5; mining relevant data of the e-commerce platform to improve the economic benefit of the e-commerce platform.

5.2 Comparative Analysis of the Mining Effect

According to the current situation that the data of different dimensions have influence on the mining effect of e-commerce customer behavior data, different dimension data sets are used to mine the user behavior data, the mined information is integrated, and the mining effect is determined through accuracy. Compared with the traditional data mining method of implicit user behavior and the user behavior data mining method based on big data generation, the experimental results are shown in Figs. 4, 5 and 6.

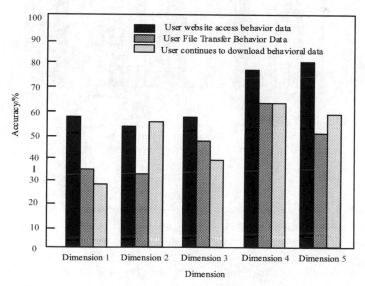

Fig. 4. Mining results of the implicit user behavior mining method

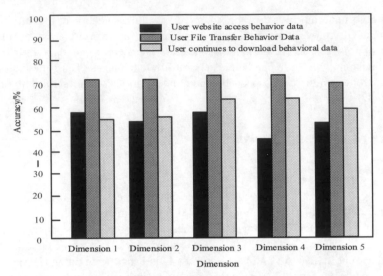

Fig. 5. Mining results of the Big data generation Methods

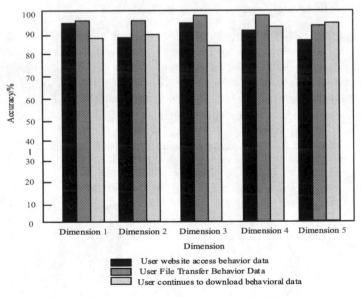

Fig. 6. User behavior mining results used in this paper

As can be seen from Figs. 4, 5 and 6, the user behavior data mined by the method in this paper has the highest accuracy, which is the most consistent with the actual situation. Taking data set 1 as an example, the data mining results are highly consistent with the actual situation, which thus accurately mines the user behavior data during this period; and then observe the other two methods have low accuracy, which is inconsistent with the actual situation. By comparison, only the method in this paper has a high accuracy of user behavior mining, and the mining accuracy of the other two methods is not high.

6 Conclusion

The data mining method of e-commerce potential customers based on Apriori association rule algorithm is proposed. Improvement of the Apriori association rule algorithm based on the multidimensional tree structure. Calculate the characteristic similarity of the e-commerce potential customer data, and divide the customers among the e-commerce potential customers at different levels in the customer transaction data. From the perspectives of perceived usefulness, perceived ease of use, perceived service, perceived security, perceived interest, the stickiness evaluation index system of e-commerce potential customers is obtained as the basic index of mining and processing. Using the improved Apriori correlation rule algorithm, the data related to e-commerce customer stickiness is mined in the mass e-commerce customer data. Enhance the mining quality of multidimensional data with a high-density grid. The test results prove that this method has high mining accuracy and can be used as the basic data for the e-commerce platform to adjust the marketing strategy. However, due to limitations in research conditions, this method only improves mining accuracy and does not significantly improve mining efficiency. Future research will not only improve mining accuracy, but also improve mining efficiency.

References

1. Zhou, S., He, J., Yang, H., Chen, D., Zhang, R.: Big data-driven abnormal behavior detection in healthcare based on association rules. IEEE Access **8**, 129002–129011 (2020)
2. Menghan, J., Feiteng, L., Zhijian, C., Yu, P.: Robust QRS detection using high-resolution wavelet packet decomposition and time-attention convolutional neural network. IEEE Access **8**(1), 16979–16988 (2020)
3. Qiu, K., Jiang, Y.: A service running data anomaly detection method based on weighted LOF and context judgment in cloud environment. Comput. Eng. Sci. **42**(3), 951–958 (2020)
4. Shuai, L., et al.: Human memory update strategy: a multi-layer template update mechanism for remote visual monitoring. IEEE Trans. Multim. **4**(23), 2188–2198 (2021)
5. Ghosh, D.: Novel trends in resilience assessment of a distribution system using synchrophasor application: a literature review. Int. Trans. Electric. Energy Sys. **31**(5), 129–134 (2020)
6. Jing, L., Siyu, F., Ya-u, Z.: Model predictive control of the fuel cell cathode system based on state quantity estimation. Comput. Simulat. **37**(3), 119–122 (2020)
7. Liu, S., Li, Y., Fu, W.: Human-centered attention-aware networks for action recognition. Int. J. Intell. Syst. **37**(12), 10968–10987 (2022)
8. Du, J., Han, G., Lin, C., Martínez-García, M.: ITrust: an anomaly-resilient trust model based on isolation forest for underwater acoustic sensor networks. IEEE Trans. Mobile Comput. **21**(7), 1684–1696 (2020)
9. Enron, F., Xuren, W., Qiuyun, W., Mengbo, X.: Database anomaly access detection based on principal component analysis and random tree. Comput. Sci. **47**(5), 94–98 (2020)
10. Yanan, S., Xuejing, Z.: An improved outlier detection algorithm and robust estimation. Chin. J. Appl. Prob. Statist. **37**(6), 136–154 (2021)

11. Xianhao, S., Chi, L., Qiong, G., Shaohua, N.: A method for detecting abnormal data of network nodes based on convolutional neural network. Mach. Tool Hydraul. **48**(8), 18–23 (2020)
12. Shiwei, W., Xiaobin, X., Zhongjun, L.: Hierarchical filtering algorithm for distributed abnomaly data based on urban computing. Comput. Integrat. Manuf. Syst. **27**(2), 2525–2531 (2021)

Design of English Mobile Online Education Platform Based on GPRS/CDMA and Internet

Bo Jiang[✉]

Navigation Technology Department, Tianjin Maritime College, Tianjin 300000, China
LX190625@126.com

Abstract. In order to increase the number of real-time online people on the education platform and realize the comprehensive promotion of English mobile online education, an English mobile online education platform based on GPRS/CDMA and the Internet is designed. Set up online education resource push module, mobile partner search module and English vocabulary learning module respectively to improve the hardware design of English mobile online education platform. On this basis, configure the GPRS/CDMA model, analyze the system's education needs by driving the Internet serial port, realize online education services at all levels, and combine relevant application components to complete the design of English mobile online education platform based on GPRS/CDMA and the Internet. The experimental results show that under the influence of GPRS/CDMA and the Internet system, the number of real-time online people on the education platform has significantly increased, which is in line with the practical application needs of comprehensively promoting English mobile online education.

Keywords: GPRS/CDMA model · Internet · Mobile online education · Resource push · Partner search · Network serial port · Number of people online · Educational services

1 Introduction

Learning, as a way to acquire knowledge and exchange emotions, has become an indispensable and important content in people's daily life [1]. With the gradual increase of Internet penetration, the mobile Internet era has given birth to mobile education based on personal electronic devices such as mobile phones and tablets. The network economy goes up, and the networked and mobile life habits are formed; With the rise of the middle class, people's awareness of paying has been awakened, and education demand and consumption have been upgraded; Lifelong learning and the increasing demand for knowledge in cross cutting fields promote the continuous extension of the time users need to learn. The scale of online education users has continued to expand, the market recognition of online education has gradually improved, and the requirements for the breadth and depth of online courses have increased, Users attach importance to the effectiveness of learning and constantly put forward new requirements for online learning experience [2]. Language, as an important tool for communication and understanding,

L. Yun et al. (Eds.): ADHIP 2023, LNICST 548, pp. 203–218, 2024.
https://doi.org/10.1007/978-3-031-50546-1_14

is the prerequisite for learning the latest information and advanced knowledge from countries around the world. College English education products have a strong momentum of development in China, but the research on product forms is still in its infancy. Therefore, this research takes English online education platform as a case to explore the methodology of mobile online education platform design.

Throughout the history of English education, English classes in schools are centered on teachers, textbooks and classrooms, which belong to the initial form of English education. VIPKID education system adopts a class switching teaching mode, which has the characteristics of strong feedback before class, full interaction in class, and focus on review after class. The courses include preview videos before class, summaries of important knowledge points, online real-time communication and interaction with teachers, homework completion after class, and weekly Q&A, which comprehensively plan students' preview and review, and more attract students' interest. The foreign teachers of VIPKID English are pure North American foreign teachers, which can let students cultivate in the original pure English environment for a long time, gradually complete the digestion of teaching knowledge and practice pure English pronunciation and thinking. The education system from the perspective of flow first conducted a research on the flow dimension of students' online education platform [3]. Then, comparative analysis of typical cases and user research were carried out to understand the learning needs and pain points of students' online learning, and parents' attitudes and needs to children's online learning through user interviews; Understand students' learning habits and problems in online learning through natural observation; Collect more sample data through questionnaire survey, further analyze students' online learning habits and learning experiences, and build an impact model of students' online learning willingness. However, the above two types of education systems can only carry a limited number of online people, which does not meet the practical application needs of comprehensively promoting English mobile online education.

GPRS is the English abbreviation of General Packet Radio Service. It is a new bearer service developed from the existing GSM system. Its purpose is to provide GSM users with packet data services. GPRS allows users to send and receive data in the end-to-end packet transfer mode without using the network resources of circuit switching mode. Thus, it provides an efficient and low-cost wireless packet data service. The theoretical bandwidth of CDMA can reach 300Kb/s, and the current actual application bandwidth is about 100Kb/s (bidirectional symmetric transmission). TCP/IP connection is provided on this channel, which can be used for Internet connection, data transmission and other applications [4]. CDMA wireless data communication system provides users with a high-speed, always on, transparent data transmission virtual private data communication network. Mainly aimed at the application of power system automation, industrial monitoring, traffic management, finance, securities and other departments, the CDMA network platform is used to realize the transparent transmission of data information. At the same time, considering the networking needs of various industry departments, the virtual data dedicated Internet is realized on the network structure. The current CDMA mobile network can support a variety of colorful data communication services, so the mobile data communication system based on CDMA network can fully meet the communication requirements of various data applications. In order to solve the problem of fewer people online in the VIPKID education system and the education system from the perspective of flow, an English mobile online education platform based on GPRS/CDMA and the Internet is designed. According to the online education resource push module, mobile partner search module and English vocabulary learning module, the application module of English mobile online education platform is constructed to improve the learning interest of learners. On this basis, configure a GPRS/CDMA model, combine it with internet serial port driver to achieve interconnection with the internet, analyze learners' educational needs, and complete the design of an English mobile online education platform.

2 Application Module Design of English Mobile Online Education Platform

The design of the application module of the English mobile online education platform needs to improve the real-time connection between the online education resource push module, the mobile partner search module, and the English vocabulary learning module. This chapter will study the specific design method of each component structure.

2.1 Online Education Resource Push Module

The online education resource push module is responsible for extracting English mobile online education information and normalizing it. Through the mining and analysis of learners' learning data, such as learning duration, frequency, learning motivation, viewing the detailed usage proportion of words and other data as input, the education platform host can locate learners' learning style through clustering algorithm.

The definition of student learning style feature vector in the online education resource push module meets the principles shown in Table 1.

Table 1. Characteristic vector of students' learning style

Index	Interpretation	Value type
Learning duration	Average browsing time per word in the past week	Float
Number of words	Average number of words learned per login in the past week	Int
Browsing method	In the past week, expand to view the proportion of detailed usage of words	Float
Login frequency	Number of user logins in the past week	Int
Learning motivation	User's motivation to use the selected software	Int

When different features are arranged together, due to different expressions of features, the online education resource push module needs to normalize the data and map it to the range of - 1–1, so that indicators of different units or scales can be compared and weighted.

For example, Formula (1) traverses every data in the feature vector of English mobile online education, and subtracts the sample mean first $\overline{W}$, divide the difference by the sample variance E, you can easily standardize data through formulas. The essence of data normalization is a linear transformation, which will not change the numerical ranking of the original data, but can improve the performance of English mobile online education data.

$$q = \frac{W' - \overline{W}}{E} \tag{1}$$

W' Represents the standard value of the English mobile online education feature vector provided by the online education resource push module.

After the online education resource push module obtains the user's learning style feature vector, it can use GPRS/CDMA and the Internet to divide the data points with "similar" characteristics in the data set into a unified category, and finally generate multiple learning styles. Using GPRS/CDMA and Internet, the steps are to select randomly χ As the initial cluster center, the distance between each educational resource object and each sub cluster center is calculated (usually using Euclidean distance), and each object is allocated to the nearest cluster center [5]. Cluster centers and objects assigned to them represent a cluster. Each time a sample is allocated, the cluster center of the cluster will be recalculated according to the existing objects in the cluster. This process will be repeated until a certain termination condition is met - when no objects are reallocated to different clusters, or the cluster center of no cluster changes again, the sum of error squares is locally minimum.

The solution of the cluster center coefficient of the online education resource push module meets the following expression:

$$R = \sum_{\alpha=1}^{+\infty} \frac{|\beta q - \overline{y}|^2}{(\delta - 1)^2 \times \sqrt{\chi}} \tag{2}$$

Among them, α cluster parameters representing English mobile online education information, β represents the recalculation coefficient of online education information samples, $\overline{y}$ represents the average value of educational information samples, δ indicates the cluster information marker coefficient of English mobile education data.

In the English mobile online education platform, the application idea of the online education resource push module is to divide the given data information sample set into multiple cluster organizations, so that the points in the cluster are as close as possible, and the distance between clusters is as large as possible.

It can be seen from the above contents that the module can acquire and push the Learning styles feature vector of learners through GPRS/CDMA and Internet technology. This mobile network based technology can achieve fast and real-time data transmission and push, and ensure that learners can obtain educational resources suitable for their Learning styles in time.

2.2 Mobile Partner Search Module

The realization of the mobile partner search function of English mobile online education platform is an extremely complex achievement. It needs to customize a unique learning scheme for learners, and constantly modify it according to the changes of users' internal dynamic development factors, such as the improvement of learners' cognitive level, so that it can adapt to learners' personalized needs. At present, common search modes include association rule based search, collaborative filtering search, hybrid search, etc.

Among them, collaborative filtering search is the most classic type of search algorithm, which is mainly divided into three types. The first is user based collaborative search, the second is item based collaborative search, and the third is model based collaborative search. The first two algorithms are similar. The first is to find the similarity between users and find out the resources learned by similar learners to recommend to target learners, which can help learners with different professional backgrounds and even different abilities to recommend learning resources [6]. The second algorithm calculates the similarity between different resources, predicts the score of similar resources with high similarity according to the target learner's score of existing resources, and recommends several similar resources with the highest score to users. However, the diversity of this recommendation is obviously poor, and it is generally applicable to small recommendation systems. The third type, model-based collaborative search algorithm is the most mainstream collaborative search type at present. It uses the idea of machine learning to model and solve the problem of association between learner characteristics and resource characteristics, mainly using clustering algorithms, classification algorithms, matrix decomposition, neural networks, and cryptic semantic models.

The complete mobile partner search module structure is shown in Fig. 1.

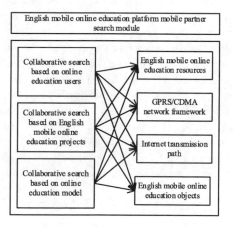

Fig. 1. Structure of mobile partner search module

The essence of the collaborative search mechanism of the mobile partner search module is "collaboration and mutual assistance". The system takes other users' English education data information as a reference, and uses it as a judgment basis to infer the new users' state and learning tendency. Judging from real life experience, the platform host can give the assumption that similar high learners who collect new word lists have similar English vocabulary and learning ability [7]. So for mobile online education system, the application value of collaborative search mechanism is reflected: mining learning partners for users can recommend the new words collected by learning partners to each other, and improve the learning efficiency of users.

With the increase of the number of people using the education platform and the accumulation of the data volume of English mobile online education resources, a well-designed mobile partner search module can not only save unnecessary storage space, help managers effectively store and manage data, but also help improve the software computing speed and reduce the occurrence of software failures. In the process of database design, it is necessary to fully understand the inevitable relationship between entities and follow the one-to-one design principle to reduce the occurrence of maintenance problems.

The mobile partner search module can discover learning partners with similar learning interests and tendencies to learners through collaborative search mechanisms, promote mutual assistance and cooperation among learners, provide functions such as recommending new words among learning partners, and improve learning efficiency and interactive experience.

2.3 English Vocabulary Learning Module

In the English mobile online education platform, the English vocabulary learning module will generate more meaningless symbols and function words for the processing of education resource information, so it is necessary to remove stop words, stem extraction, word form restoration and other operations first. First, use Internet functions to segment English mobile education text data, and process the information into the smallest unit that can be processed by computers. Secondly, the NLTK package is also used, in which the list of stop words is used to remove function words (conjunctions, prepositions, articles, etc.) and punctuation marks from the corpus. The implementation logic is very simple. Define an empty list and traverse the text list with good word segmentation. If the word does not exist in the disabled word list, append is added to the empty list [8]. Then, the education platform host uses the Porter stemming tool to extract the same stem from different morphisms of English words. Finally, use the WordNetLemmatizer function to restore the morphology. Just make clear the data structure of the incoming function of the English vocabulary learning module. When these steps are implemented using GPRS/CDMA language, importing packages and calling functions can be easily implemented.

The specific structure model of English vocabulary learning module is shown in Fig. 2.

In order to fully promote English mobile online education, the design of English vocabulary learning module must also ensure the similarity between vocabulary information. Similarity is a basic calculation in mobile online education and teaching information promotion. The key technology is mainly composed of two parts.

Simultaneous Formula (2) can express the similarity relationship between lexical information as:

$$U = (\phi|\Delta O|)^2 - \frac{1}{\varphi^2}\sqrt{\gamma \times R} \tag{3}$$

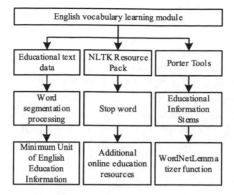

Fig. 2. Structure model of English vocabulary learning module

where, ΔO it represents the unit cumulative amount of English mobile online education information, ϕ indicates English vocabulary learning parameters, φ indicates the empty list definition coefficient in the module unit, γ represents the sampling coefficient of English mobile online education information.

Vocabulary is the brick and wood of a language and the most basic and key link of language learning. In the face of thousands of words, it is difficult to effectively master English words by rote without guidance of learning strategies. Referring to various theories of vocabulary memory, the contents that users may pay attention to when reciting vocabulary can be divided into: phonetic symbols, Chinese definitions, example sentences, phrase collocation, pictures, video clips, roots and affixes, similar words, etc.

This module processes and optimizes the information of educational resources by removing Stop word, Stemming, Lemmatisation and other operations to reduce the impact of meaningless symbols and function words. Extract the smallest computer processable vocabulary units, improve the quality and effectiveness of learning resources, and classify the content that learners are interested in. By providing diversified learning strategy guidance, help learners to master English words more effectively and improve vocabulary learning efficiency.

3   Realization of Online Education Service

On the basis of hardware modules at all levels, configure the GPRS/CDMA model, and complete the analysis of relevant education needs by driving the Internet serial port, so as to realize the application of English mobile online education platform based on GPRS/CDMA and the Internet.

3.1   GPRS/CDMA Model Configuration

The network header of the online education platform [9] GPRS/CDMA model follows all end-to-end headers (for example, the authentication header, if presented in clear text), which is followed by a networked IP packet. The sender encapsulates the original IP packet into the ESP, at least uses the sender's user ID and destination address to locate the correct security association, and then uses appropriate encryption transformation. If host based key generation mechanism is used, all sending users on a given system have the same security association for a specific destination address. If no key has been established, the key management mechanism will generate an encryption key for the connection session. This process will occur before the ESP is used. Thereafter, the (encrypted) GPRS/CDMA model ESP is encapsulated into an unencrypted IP packet as the final payload. If strict red black segmentation is performed, the optional payload and the address and other information in the plaintext IP header may be different from the values contained in the original packet (which has been encrypted and encapsulated now).

The configured English mobile online education platform GPRS/CDMA model is shown in Fig. 3.

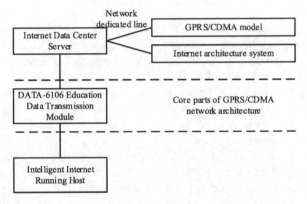

Fig. 3. GPRS/CDMA model

The configuration of GPRS/CDMA model meets the following process:
Step 1: Set the header of English mobile online education resources

$$p = i \times \sqrt{\frac{1}{U}} \tag{4}$$

i It represents the real-time transmission characteristics of English mobile online education resources.

Step 2: IP based processing of online education data

$$A = \frac{(\lambda - 1) \cdot p}{\sum\limits_{\varepsilon=1}^{+\infty} f \times |\hat{d}|} \tag{5}$$

λ Represents the IP distribution coefficient of English mobile online education information in the GPRS/CDMA network system, ε represents the IP processing parameters of online education resources, $\hat{d}$ the distribution characteristics of English mobile online education information in the GPRS/CDMA network system, f represents the transmission vector of online education resources in the GPRS/CDMA network.

Step 3: GPRS/CDMA model configuration conditions

$$F = \left| \frac{\iota_{max} - \iota_{min}}{\kappa} \right| \cdot \sqrt{(\tilde{g}^2 A) + \bar{h}^2} \tag{6}$$

ι_{max} Indicates the maximum value of key distribution parameters, ι_{min} represents the minimum value of key distribution parameters, κ ESP node application parameters representing English mobile online education information, $\tilde{g}$ it indicates the segmentation characteristics of English mobile online education information in GPRS/CDMA, $\bar{h}$ it represents the unit transmission average value of English mobile online education information in GPRS/CDMA.

The GPRS/CDMA model of online education platform maps an English mobile online education information IP address domain to another IP address domain when carrying out network address translation (NAT), usually to realize the mapping of an internal local IP address and an external global IP address. NAT is an important way to solve the IP address shortage of English mobile online education information and ensure the internal network security. To access the online education platform, you must have a legal IP address to realize the interconnection between GPRS/CDMA model and the Internet.

3.2 Internet Serial Port Driver

In order to meet the operation requirements of the GPRS/CDMA model, the English mobile online education platform sets two asynchronous serial ports at the same time when driving the Internet serial port. The serial port 1 is connected to the industrial control equipment, and the serial port 2 is directly connected to the wireless port in the English vocabulary learning module.

Serial port drive refers to the configuration of baud rate, data bit, stop bit, verification mode, working mode and selected clock of the serial port. The specific driving process is shown in Fig. 4.

When the Internet sender is ready to output English mobile online education data to the receiver, it activates the RTS signal. If the receiving end is ready to receive data, it will activate the CTS signal. Before receiving the signal that CTS is activated, the sending end cannot send data. In this way, the receiver actively controls the reception of data to prevent buffer overflow.

The main idea of Internet serial port driver is that "the receiver" controls "the sender" to prevent the sender from sending data too fast to receive. In order to facilitate the promotion of English mobile online education resources in a wide environment, the Internet serial port provides LCP [10]. LCP is used to automatically reach an agreement on encapsulation format options, handle changes in packet size, detect looped back links and other common configuration errors, and terminate links. Other optional equipment provided are: authentication of the same unit ID in the link, and decision when the link function is normal or the link fails.

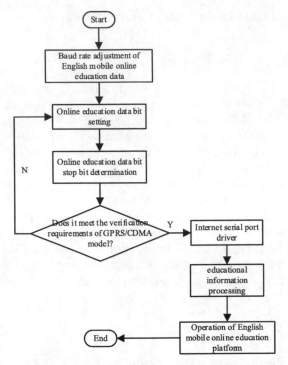

Fig. 4. Flow Chart of Internet Serial Port Driver

The solution of the Internet serial port driver expression meets the conditions shown in Formula (7).

$$G = F \times \left. \frac{\sum\limits_{\mu=1}^{+\infty} \vec{k} \cdot \vec{j}}{\sum\limits_{-\infty}^{+\infty} v \cdot L'} \right|_{v \neq 0} \tag{7}$$

Among them, μ indicates the education link connection parameters in the Internet environment, $\vec{k}$ represents the English mobile online education information transmission vector in the receiving end of the education platform, $\vec{j}$ represents the English mobile online education information transmission vector in the education platform sender, L' represent the response characteristics of English mobile online education information in LCP nodes, v represents the driving coefficient of the serial port of the education platform.

The English mobile online education platform Internet establishes communication through point-to-point links. Each end of the PPP link must first send LCP data packets to set and test the data link. The peer can be authenticated only after the link is established. Then, PPP must send NCP packets to select and set one or more English mobile online education information network layer protocols. Once each selected network layer protocol is set, data packets from each network layer protocol can be sent on the Internet link.

3.3 Education Demand Analysis

With the support of GPRS/CDMA and the Internet, the demand analysis of English mobile online education platform consists of functional requirements and feasibility requirements.

(1) Analysis of functional requirements of education platform

The user roles of English mobile online education platform are divided into learners and managers. The main functions of learners include: personal information maintenance, uploading corpus, completing tests, vocabulary learning, vocabulary comments and likes. The main functions of the administrator are to review the corpus and user comments, update the thesaurus, publish tests, recommend learning partners, recommend new words, etc. [11].

The function demand analysis expression of the education platform is:

$$Z_1 = \left(\theta_1^2 \times \left| \frac{G}{l_1} \right|^2 \right) - \varpi_1 \hat{X}^2 \tag{8}$$

θ_1 Represents the user role definition coefficient of English mobile online education platform, l_1 represent the functional response characteristics of the education platform related to user roles, ϖ_1 represents the real-time upload parameters of English mobile online education information, $\hat{X}$ represents the functional evaluation vector of the online education system.

(2) Feasibility demand analysis of education platform

At the application level, using GPRS/CDMA and the Internet to learn words at any time and anywhere has become a recognized way for the young generation. However, the new word recommendation function in the current mainstream mobile phone software for memorizing words is virtually non-existent, the language data is old, and the source is single. Therefore, the market for memorizing words has been unsatisfied by the old user needs, and the application prospect is broad. At the technical level, with the development of natural language processing technology, with the help of python toolkits such as Gensim and NLTK, you can easily achieve functions such as word segmentation, keyword extraction, word vector training [12]. At the same time, the improvement of computer computing power also provides hardware support for the use of machine learning or even deep learning algorithms to achieve the recommendation function.

The feasibility demand analysis expression of the education platform is:

$$Z_2 = \varpi_2 \hat{X} \cdot \left(\frac{l_2}{\theta_2} G \right) \tag{9}$$

θ_2 Indicates the promotion demand parameters of English mobile online education platform, l_2 refers to the keyword extraction coefficient of English mobile online education data, ϖ_2 represents the word vector training coefficient of English mobile online education data.

Formulas (8) and (9) are used to derive the operation expression of English mobile online education platform based on GPRS/CDMA and the Internet as follows:

$$B = \frac{\sqrt{\vartheta \cdot \left(\frac{Z_1}{c_1} \cdot \frac{Z_2}{c_2}\right)}}{\tilde{m}^2 \cdot (\sigma^2 - 1)} \tag{10}$$

where, ϑ represents the upload parameters of English mobile online education information corpus based on GPRS/CDMA and the Internet, c_1 represent the response coefficient of education service related to functional requirements, c_2 represent the response coefficient of education service related to the feasibility demand, $\tilde{m}$ it represents the response characteristics of English mobile online education services in PRS/CDMA and Internet systems, σ represents the real-time operation vector of the online education platform.

In the process of learner learning, the education platform continuously collects learner behavior data, converts these text and numerical real-time data into learner characteristics and resource characteristics [13], and sends them to the server to calculate the similarity between feature vectors. When helping learners begin to memorize words, the system calculates the similarity between user characteristics according to the learner capability model, matches learning partners for users, and recommends the vocabulary in the learning partner's vocabulary list to new users. Improve the efficiency of memorizing words by predicting which words are unfamiliar to the user in advance.

4 Example Analysis

In order to highlight the practical differences of English mobile online education platform based on GPRS/CDMA and the Internet, VIPKID education system, and education system from the perspective of flow, the following comparative experiment is designed.

4.1 Experimental Process

The specific implementation process of this experiment is as follows:

- Connect the Windows host and the 12th generation 12490F processor on demand, and build the experimental environment required for the connection of the education platform.
- Input the executive program of the English mobile online education platform based on GPRS/CDMA and the Internet into the Windows host, record the specific number of real-time online people in the network system under the effect of the platform, and the results are experimental group variables.
- Input the executive program of the VIPKID education system into the Windows host, record the specific value of the number of real-time online people in the network system under the effect of the platform, and the results are the first control group variables.

- Input the executive program of the education system from the perspective of heart flow into the Windows host, record the specific value of the number of real-time online people in the network system under the effect of the platform, and the results are the variables of the second control group.
- Statistic the variable data obtained, and summarize the experimental rules.

4.2 Principle and Preliminary Preparation

The real-time online number can be used to describe the promotion ability of the network host to English mobile online education. Without considering other interference conditions, the more real-time online number, the stronger the promotion ability of the network host to English mobile online education.

The following table records the specific models of the selected equipment components in this experiment (Table 2).

Table 2. Selection of Experimental Equipment

Project	Experimental equipment	Model and name
1	Teaching promotion host	Windows host
2	Teaching data processor	12th generation 12490F processor
3	Response chip	DS3231MZ + SOIC-8 real-time clock chip
4	Teaching task server	LM393 DIP8 inline server
5	Response device	AP LME49720 AD827/712
6	Client terminal	STM32F103 LQFP100

In order to ensure the fairness of the experimental results, the connection forms of equipment components in the experimental group and the control group are always consistent during the experiment.

4.3 Data Processing

This experiment takes 6 h as the total duration of the experiment, and records the real-time number of online people every 1 h. The specific experiment is shown in Fig. 5.

It can be seen from the analysis of Fig. 5 that the number of real-time online people in the experimental group kept a numerical trend of first rising and then stabilizing, and by the end of the experiment, the maximum number reached 225. The number of real-time online people in the first control group kept increasing, and by the end of the experiment, the maximum number reached 105, a decrease of 120 people compared with the maximum number of the experimental group. The number of real-time online people in the second control group remained stable first, and then increased. By the end of the experiment, the maximum number reached 75 people, which was 150 people lower than the maximum number of the experimental group. This is because the design platform pushes education information in real time through the online education resource

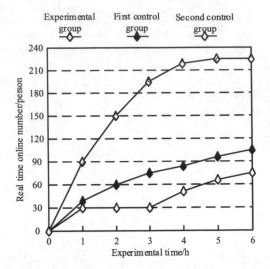

Fig. 5. Real time online population

push module, uses the mobile partner search module to ensure the learning efficiency of learners, and improves the quality of learning resources based on the English vocabulary learning module.

To sum up, the conclusion of this experiment is:

(1) The application of the VIPKID education system and the education system from the perspective of flow is not enough to improve the real-time online population of the education platform, so these two types of systems cannot effectively guarantee the ability of the network host to promote English mobile online education.

(2) Compared with the VIPKID education system and the education system from the perspective of flow, the application of the English mobile online education platform based on GPRS/CDMA and the Internet has achieved a significant increase in the number of real-time online people, which meets the actual application needs of improving the network host's ability to promote English mobile online education.

5   Conclusion

The above research aims to solve the defects of traditional word memorizing software such as outdated language data, low accuracy of personalized recommended words, and low sense of social participation when users recite words, assist users in professional language learning, rapidly expand vocabulary, and more importantly, learn professional related practical vocabulary, so that users can apply what they have learned. Aiming at the problems of poor learning efficiency and low interest of learners, the application module of English mobile online education platform is designed, which is composed of online education resource push module, mobile partner search module and English vocabulary learning module. By combining the GPRS/CDMA model and internet serial port driver to achieve interconnection, and analyzing the teaching needs of learners, the design of an English mobile online education platform is completed. The main work and research results of this paper include:

(1) Sort out the development history and current situation of online education and intelligent recommendation technology at home and abroad, sort out the functions of mainstream English word reciting software, and investigate the market share of word reciting software.
(2) Based on the support of learning theory, from concrete to abstract, from low-level to high-level teaching objectives, build a learner capability model. Based on the classification of learning styles in the existing literature, a new vocabulary learning learning style model is proposed according to the results of questionnaires, interviews and market research. This research enriches the existing relevant theories, provides theoretical support for the user modeling process of the online education platform, and improves the recommendation effect of the word reciting software.
(3) Based on the demand survey and previous literature research, an intelligent recommendation model is built for English vocabulary learning, which fully describes the process design of each function module of the word reciting software, and builds the design framework of the platform system.

In addition, there are many directions that can be improved in the research, including:

(1) The preliminary investigation and experimental sample data are insufficient, so large-scale data testing cannot be carried out, and the reliability and validity of the experimental conclusions are difficult to guarantee. It can further expand the scope of the experiment, collect data to verify the accuracy of the system's recommended new words and the effectiveness of the learning style classification model.
(2) The functional design needs to be gradually improved or updated in practice. For example, the word message function belongs to but one-way, other tourists/users can only praise high-quality comments, and can add the mutual evaluation function; At present, the review form of words is only designed to read the new word book, and the review form is mainly browsing. You can refer to the review function of the existing word reciting software to improve it.

References

1. Liu, S., Gao, P., Li, Y., et al.: Multi-modal fusion network with complementarity and importance for emotion recognition. Inf. Sci. **619**, 679–694 (2023)
2. Yang, F., Ma, M.: Identification method of internet link traffic based on load randomness. Comput. Simulat. **38**(11), 331–334+339 (2021)
3. Cheng, Z., Wang, Z., Luo, Y., et al.: Development and application of an online learning platform for nursing ethics: a teaching practice research. Nurse Educ. Today **112**, 105336 (2022)
4. Oliveira, A., Capovilla, C.E., Casella, I., et al.: Co-simulation of an srg wind turbine control and gprs/egprs wireless standards in smart grids. IEEE/CAA J. Automat. Sinica **8**(3), 656–663 (2022)
5. Zhou, L., Xue, S., Li, R.: Extending the technology acceptance model to explore students' intention to use an online education platform at a university in China. SAGE Open **12**(1), 75–90 (2022)

6. Bag, S., Aich, P., Islam, M.A.: Behavioral intention of "digital natives" toward adapting the online education system in higher education. J. Appl. Res. High. Educ. **14**(1), 16–40 (2022)
7. Borkotoky, D.K., Borah, G.: The impact of online education on the university students of Assam in COVID times. Indian J. Sci. Technol. **14**(13), 1028–1035 (2021)
8. Ciorua, B.V., Lauran, M., Mesaro, M., et al.: 2021 (124) - Perceptions of students from northwestern Romania on online education during the pandemic COVID-19. Asian J. Educ. Soc. Stud. **17**(4), 11–18 (2021)
9. Sigmon, A.J., Bodek, M.J.: Use of an online social annotation platform to enhance a flipped organic chemistry course. J. Chem. Educ. **99**(2), 538–545 (2022)
10. Gong, Y., He, Y., Xie, D., et al.: Research on path selection of digital governance of online education. Mod. Econ. **12**(3), 469–476 (2021)
11. Khan, M.M., Rahman, S., Islam, S.: Online education system in Bangladesh during COVID-19 pandemic. Creat. Educ. **12**(2), 441–452 (2021)
12. Abu-Bajeh, Z.Y., Rabab, A.: Justifications for using interactive online education. Int. J. Eng. Technol. **10**(1), 8–13 (2021)
13. Kassawat, M., Cervera, E., Pobil, A.P.: An omnidirectional platform for education and research in cooperative robotics. Electronics **11**(3), 499 (2022)

Application of Artificial Intelligence Technology on Online Cultural Education Mobile Terminal

Qiao Wu[1]([✉]) and Xiaoxian Xu[2]

[1] Changchun University of Architecture and Engineering, Changchun 130119, China
13304306690@163.com
[2] Physical Education Department, Xi'an Shiyou University, Xi'an 710065, China

Abstract. In order to meet the functional and performance requirements of users for online cultural education mobile terminals, the application research of artificial intelligence technology on online cultural education mobile terminals is proposed. The overall architecture of online culture and education mobile terminal is designed using artificial intelligence technology. Through communication connection design and identity verification design, the communication function of mobile terminal is designed. Combining the curriculum center module, evaluation module, terminal user information acquisition module and mobile terminal load balance module, the functional module of online culture and education mobile terminal is designed. Realize the application of artificial intelligence technology online cultural and educational mobile terminals. The test results show that the authentication module, course center module and evaluation module of the mobile terminal in this paper can meet the needs of users. In terms of user information acquisition delay and load balancing, they can also meet the performance requirements of users for mobile terminals.

Keywords: Artificial Intelligence Technology · Cultural Education · Mobile Terminal · Function Module

1 Introduction

The birth of distance online education has brought revolutionary innovation and development to China's education industry. This high-quality and low-cost learning method enables students to enjoy the most cost-effective and high-quality online cultural and educational resources [1] through the online education platform. With the deepening of education, students' demand for learning conditions is also constantly improving. Compared with the traditional online education, which has a fixed learning environment in time and space, the birth of mobile terminals has become an innovative breakthrough. Through the special carrier of mobile phones, the online cultural and educational resources of non academic and non academic degrees are promoted to the people, breaking the limits of learning time and space, greatly expanding the space and objects of online cultural education, and making the perfect combination of mobile networks and high-quality educational resources. The mobile communication terminal provides

L. Yun et al. (Eds.): ADHIP 2023, LNICST 548, pp. 219–234, 2024.
https://doi.org/10.1007/978-3-031-50546-1_15

real-time, convenience and multimedia, which can easily communicate with training institutions, students and lecturers and share information anytime, anywhere.

In domestic research, Liu Weijun et al. [2] considered the slow speed of traditional terminal online education data mining technology. In order to solve this problem, they proposed a model driven terminal online education data mining technology. Use the data transformation method of association analysis to transform data, and then use the model driven crowd behavior modeling method to design the task flow of terminal online education data mining. After the above work is completed, the key technologies of model driven data mining are optimized through screening, selecting data subsets, coding, setting thresholds, and evolutionary steps to achieve efficient data mining for terminal online education. The experimental results show that the mining speed of the proposed technology is similar when the data set size is small, and the mining speed increases gradually when the data set size is large. However, the mining speed of the traditional technology using a small dataset is basically similar to that using a large dataset, which proves that the proposed technology is faster. Kang Mengjie [3] designed an integrated student management terminal based on the Internet of Things technology to improve the informatization level of student management. The RFID module is designed based on the Internet of Things technology to realize the management of students' location information and teaching information. The embedded microprocessor is designed in the processor module to realize the control and data transmission of the terminal. In the tag anti-collision module, combining the dynamic frame slot algorithm and CGCT algorithm, the anti-collision algorithm of RF tags is designed to solve the tag collision problem in identification. In the database module, student information table, college information table and other data tables are designed to store terminal information. Design a variety of management procedures, including attendance management procedures, user information management procedures, etc., to achieve application module design. The terminal test results show that the multi label reading time of the design terminal is relatively short, the memory utilization rate is always lower than 50%, the CPU utilization rate is always lower than 55%, and the resources accessed by users match their own permissions relatively well.

In foreign studies, Zhang J et al. [4] took college English courses as an example to study the mobile terminal system of intelligent college English teaching and training mode. Based on the introduction of the concept, architecture, characteristics of WAP technology and the WAP application software development tools needed in the mobile terminal system teaching mode, this paper attempts to build a mobile terminal system teaching mode. This paper describes the model, proposes the collaborative filtering algorithm, and discusses how to use the collaborative filtering algorithm to construct the matrix and calculate the similarity. In order to verify the effectiveness of the system model, the test level comparison and questionnaire survey were conducted on the teaching model. The experimental results of this paper show that 91.8% of the students said that their comprehensive English level has been improved in the process of applying the mobile terminal system teaching mode. In addition, 87.5% of the students said it was necessary or basically necessary to use English for professional teaching, and 76.2% of the students were very satisfied with this model.

Distance education itself covers a wide range of markets, especially China's vast population base, the number of colleges and universities can not meet the needs of most people to receive higher education, which means that China's distance education has a huge space for development. At the same time, with the progress of the times, knowledge and technology, lifelong education has become the future trend. People from all walks of life need to involve different levels and different kinds of knowledge. The distance education model can successfully solve this problem in the information age today. On the other hand, statistics show that the use rate of smart phones has reached 70% worldwide, and the use rate of mobile terminals in China is also in a booming period. Therefore, the combination of distance education and mobile Internet technology and the design of online cultural education mobile terminals have important theoretical significance and practical application value. Artificial intelligence technology can provide personalized recommendations and customize learning content based on users' learning behavior and preferences. By analyzing user data, provide cultural and educational resources and learning paths suitable for each student's learning level, and provide a more personalized learning experience. Therefore, this article proposes the application of artificial intelligence technology on online cultural education mobile terminals. Adopt Natural language processing technology in artificial intelligence technology and Mina network framework to establish the overall architecture of online cultural education mobile terminal. Based on the overall architecture of the terminal, design online cultural education mobile terminal functions, including mobile terminal communication function, course center module, evaluation module, end user information acquisition module, and mobile terminal load balancing module. In the design of mobile terminal communication functions, an identity verification module is added to ensure user security. Students can register for courses online and participate in online assessments through this terminal. Online cultural education mobile terminals based on artificial intelligence technology can quickly obtain user information, have a small load, and have good application performance.

2 Design the Overall Architecture of Online Cultural Education Mobile Terminal

The online cultural education mobile terminal designed with AI technology mainly meets the needs of student users. Using Natural language processing technology in artificial intelligence technology, online cultural education mobile terminals can process and analyze user information. Natural language processing technology provides corresponding online cultural and educational information by identifying the user's purpose. Users can download the client software and register. After logging in, users can see the online culture and other information that has been opened and can be learned through intelligent terminals, and use various business modules through the Internet under the Mina framework [5]. Using the Mina framework for data interaction with servers can achieve reliable data transmission and ensure the normal operation of mobile terminals under artificial intelligence technology. The server mainly uses SQLite for data storage, capable of efficiently managing and storing a large amount of user data. The client uses JSON, XML and other network interfaces provided by the server-side Web system to realize data exchange. The overall architecture of online cultural education mobile terminal is shown in Fig. 1.

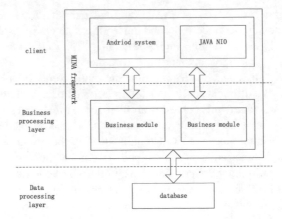

Fig. 1. Overall Structure of Online Culture and Education Mobile Terminal

According to the above demand analysis, the online cultural education mobile terminal based on artificial intelligence technology mainly meets the needs of students and the design and implementation of the C/S mode of the system, the design and implementation using C/S mode can enable students to easily access the required educational content and services on mobile terminals. The physical architecture of the mobile terminal is shown in Fig. 2.

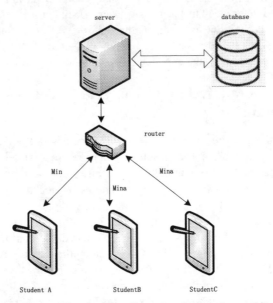

Fig. 2. Physical architecture of online cultural education mobile terminal

The client mainly interacts with the server through Mina network framework, and each mobile terminal is connected to the server through a router. The server is used to manage each business module of the system, store all user information, and ensure the normal operation of the mobile terminal under artificial intelligence technology.

3   Functional Design of Online Cultural Education Mobile Terminal

3.1   Mobile Terminal Communication Function Design

3.1.1   Communication Connection Design

The online cultural education learning terminal, as the client itself, does not directly process the user's request for service, but sends the user's request information to the server. The application calls the interface processing information provided by the server, and carries out relevant operations by calling the server API [6]. After receiving the request, the server confirms the information and returns it to the client, and the client will feed back the received results to the user, The system involves the main APIs, including Validate, Selectcourse, CheckExam, Notices, UserInfo, and SystemSetting APIs. The above interfaces are designed according to different functions and encapsulated in different Servlets for storage in the business logic layer for centralized processing. The interface calls are shown in Fig. 3.

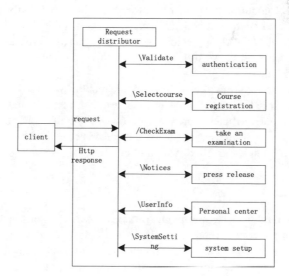

Fig. 3. Server side interface call diagram

As an online cultural education mobile terminal, it is mainly used to realize communication, data interaction, interface interaction and other functions. The interface is the activity involved in the terminal, which is used to realize two functions, namely, to obtain data from the server and display the space used for interface design for user operation and input the data that users need to operate and bind them to the processing module to read and write the data, finally upload to the server, where the data interaction format between the client and the server is JSON [7]. The client architecture is shown in Fig. 4.

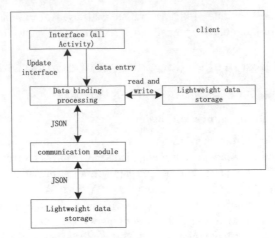

Fig. 4. Client architecture

3.1.2 Identity Authentication Design

The authentication module is a key for users to enter the client. Only legitimate users who have passed the system authentication can successfully log in to the client and use the relevant functions of the client, which also indicates that the communication between the client and the server is successful. The specific flow chart of the client login module is shown in Fig. 5 below.

It can be seen from the flow chart of the module that the user does not need to submit the data to the server before entering the mobile terminal. First, the network availability check and data format verification are carried out locally on the client. When these are verified, the data is sent to the server to verify the user's effectiveness. After successful login, the user can jump to the main system interface.

3.2 Course Center Module

The course center module is a core module of the entire mobile terminal. The following is the workflow of the course center module:

Step 1: Students log in to the client, and all course information is initialized.

Step 2: After the system reads the student login information, it reads the major, and displays the optional courses according to the major. Students can register online or cancel the registered courses.

Step 3: Students can choose to play videos online or download and watch the learning courses. Students can evaluate the courses during the learning process [8], and can also communicate online.

The flow chart of online student enrollment is shown in Fig. 6.

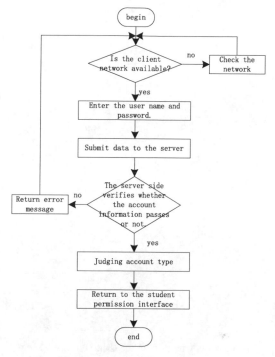

Fig. 5. Login module flow chart

3.3 Evaluation Module

The main function of the evaluation module is to carry out exercises for course learning. The training methods include basic training, simulated examination and online examination [9]. If students choose the basic training mode, they can select the question type for training. During the question making process, they can collect the wrong questions stored in the system by default, and the collected questions and wrong questions can be viewed in the personal center. At the same time, students can take a mock exam on the eve of the exam to get familiar with the exam question type and time. The mock exam completely simulates the online exam process and adopts the real-time timing scoring method. When the system announces the exam information, students will take the exam at the corresponding time. The evaluation module flow chart is shown in Fig. 7.

3.4 Design the Terminal User Information Acquisition Module

When fast acquiring end-user information, first establish vector space, use idf algorithm to calculate the weight of end-user information, complete feature selection of user information through information gain method, and complete fast acquiring end-user information according to KNN acquisition method.

Vocabulary is the basic expression item in the end user information. It shows certain rules and has a high frequency of occurrence [10] in the end user information. User information is obtained through different feature words.

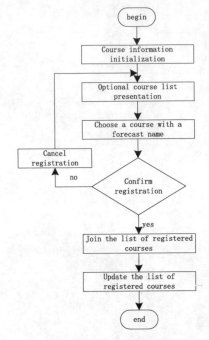

Fig. 6. Flow Chart of Online Course Registration

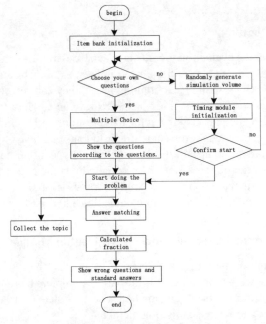

Fig. 7. Evaluation module flow chart

The vector space expression of end user information is:

$$v(d_i) = \left(w_1(d_i), w_2(d_2), \cdots, w_j(d_i)\right) \tag{1}$$

where, j it represents the total number of feature items used in the selection of end user information features, $w_j(d_i)$ on behalf of the j end user information feature items in the terminal document d_i weight value in.

df Represent the frequency of the end user information document, which means that the terminal contains $term_i$ user information of feature item, recorded as df_i. In the terminal, df_i the larger it is, the more characteristic it represents $term_i$ the lower the accuracy of measuring the similarity between user information in the terminal. idf it represents the inverted document frequency in the terminal, and represents the characteristics of user information feature items in the entire terminal document collection. Its role is to measure the distribution status of user information feature items in the entire terminal document collection. idf and df in general, it is inversely proportional, and its calculation formula is:

$$idf_i = \log\left(\frac{N}{df_i}\right) \tag{2}$$

where, N represents the total number of user information in the end user information document collection.

When idf when the value of is larger, it indicates the characteristic item $term_i$ it plays a greater role in the end user information differentiation. When there is only one characteristic item in one end user information, $idf_i = \log N$ when a feature item exists in all user information of the terminal, $idf = \log 1 = 0$.

set up $\{c_1, c_2, \cdots, c_n\}$ it is a collection of user information categories in the terminal target space. Characteristic value of user information $term_i$ the information gain formula of is:

$$G(T) = f(t) \sum_{i=1}^{n} f(c_i|t) \cdot \log f(c_i|t) \tag{3}$$

where, t characteristic items representing user information, c_i the category representing the end user information. For the end user information category c_i and user information characteristics t, in the information gain, by examining the user information category c_i whether the user information feature item appears t, measure user information characteristics t for end user information category c_i information gain of. When user information feature item t and user information category c_i when the information gain of is greater, the user information characteristics t for user information category c_i the greater the contribution, the more important.

Using machine learning algorithms in artificial intelligence technology to calculate user information similarity and information weights, and select the one with the highest similarity to user information in the training user information set k user information, set an initial value of user information h the initial value of user information is generally in the range of hundreds to thousands. At the end of the new user information h among the neighbors, the weight of each user information is calculated in order. The calculation formula is as follows:

$$W_{xj} = \sum_{i=1}^{k} \text{sim}_{xi} \cdot y_{ij} \tag{4}$$

where, W_{xj} information on behalf of new users x stay j weight in user information, sim_{xi} information on behalf of new users x and user information i similarity between, y_{ij} it is a constant 1 or 0. By comparing the weights of various user information, the end user information can be quickly acquired.

3.5 Design Mobile Terminal Load Balancing Module (2 Under Cloud Platform)

Hypothesis $K/K/1$ indicates the waiting value model of the wireless source for terminal information request, $I_{\max}$ indicates the total length of information queue of multi terminal system, I it indicates the length of the information queue of the multi terminal system. When the server load is higher, the remaining processing capacity of the terminal node will be very low. This indicates that the value of the processing capacity of the multi terminal node is determined by the number of loads on the multi terminal node [11]. The formula for calculating the processing capacity of the multi terminal node is as follows:

$$\vartheta = \left(1 - \frac{Q\_Load}{MaxLoad}\right) * M_0 \tag{5}$$

where, Q_Load represents the comprehensive load of multiple terminal nodes, $MaxLoad$ indicates the load value when the multi terminal node is fully loaded. When the CPU utilization and memory utilization are both 100%, the number of running programs is obtained by calculating the values of the maximum number of processes supported by the operating system, M_0 it indicates that the server node can complete tasks per second in the optimal state. According to the terminal information request radio source waiting value model [12], when the number of cloud service stations is equal to 1, the calculation formula of tasks completed by the multi terminal system is as follows:

$$W = \frac{o^*}{\vartheta} \tag{6}$$

where, o^* it indicates the amount of tasks completed by the information request. When the expected information queue that is not joined is an information queue that is not queued, the expression that can continue to hold the information processing capacity is as follows:

$$I_p = \frac{I_{\max}}{W} - I \tag{7}$$

According to the above formula φ it represents the average arrival rate of expected information. The queuing law is calculated as follows:

$$I_p = o\left(1 + \frac{w}{1-o}\right) = \frac{o}{1-o} = \frac{\varphi}{\vartheta - \varphi} \tag{8}$$

According to the above formula, the average arrival rate of expected information is calculated as follows:

$$\varphi = \frac{I_p}{1 + I_p} * \frac{(MaxLoad - Q\_Load) * M}{MaxLoad} \tag{9}$$

According to the above formula, φ the value of is Q_Load is a function of the Q_Load it shows that when the load of a terminal node is high, the amount of information required will decrease, which reflects the load balancing core. The expression of multi terminal adaptive load balancing information distribution is as follows:

$$A_\iota = \left[\frac{\varphi_\iota}{\sum\limits_{\tau=1}^{m} \varphi_\tau} \right] * I_{all} \tag{10}$$

where, I_{all} represents all information queues of the multi terminal load balancer port, A indicates the number of terminal nodes in the multi terminal system, φ_ι represents a multi terminal node ι average arrival rate of, A_ι indicates an ECS node ι the amount of tasks completed by the information request should be obtained. When using a server in the form of FIFO, if the load balancer will rotate to assign output or input information to each terminal node [13–15], if the amount of information of the terminal node meets the requirements, it will not be allocated.

The queuing theory is used to calculate the processing capacity of multiple terminal nodes, and the information handling capacity is calculated at the same time. According to the queuing law, the average arrival rate of expected information is obtained. When the load of a terminal node is high, the amount of information required will be reduced, and the adaptive load balancing of multiple terminals is realized.

4 Test Analysis

4.1 Test Environment

This article mainly introduces the design and development of the distance education learning terminal, so the running environment is mainly for the test of the client environment. The main test environment is as follows:

Test environment 1: Android PAD, 8-inch main screen, 1280×800 screen resolution, Android OS 4.3

Test environment 2: Android PAD, 7-inch main screen, 480×800 screen resolution, Android OS 4.1

Test whether different screen sizes and resolutions fit the application. In the development process, the mobile phone manufacturers and manufacturers in the market are different, so that the mobile phone or tablet has different specifications. The application developed in this paper will run on different mobile phones or tablets. The UI interface of the same application will display differently on different devices.

The development environment is set at 1280×800 screen resolution, $1280 \times$ The 800 screen resolution tablet can display a friendly user interface, while the system's UI interface is $800 \times$ The 480 screen resolution can also adapt to the device screen size.

4.2 Function Test

In order to ensure the smooth operation of the mobile terminal designed in this paper, the function of each module is tested. The detailed results are as follows. The terminal can be carried out smoothly as a whole, with good results. If there is a slight deficiency, modification suggestions will be proposed in the test analysis.

The authentication module can normally match the user information and load the corresponding main page. The authentication function adopts the threshold test, and the test results are shown in Table 1.

Table 1. Authentication module test table

step	Test module	Test content	Test method	test result
1	User Login	Whether the user can log in normally	Do not enter ID, password and verification code	Remind you to enter ID, password and verification code
2	User Login	Whether the user can log in normally	Enter wrong user name, password and verification code	Remind that the ID, password and verification code entered are wrong
3	User Login	Whether the user can log in normally	Close the client's network connection	Remind you to check the system network connection
4	User Login	Whether the user can log in normally	Shut down the server	Connection timeout, server side exception, please contact the administrator
5	User Login	Whether the user can log in normally	The information input is accurate and the network connection is normal	Successfully enter the main interface of mobile terminal

The course center module can obtain majors according to student login information and display the list of unselected courses according to majors. Students can also cancel the selected courses within the specified time. Courses can be played online and tracked to record the learning process. At the same time, courses can be downloaded. The default storage address is memory card. Students can evaluate courses. The system has no word limit, and will be improved later, with 140 characters as the limit. The test results are shown in Table 2.

The evaluation module can obtain majors and carry out targeted question type training according to the student login information. Students can start the timing function in the process of simulated examination, count down in seconds and score in real time. The test results are shown in Table 3.

Table 2. Module Test Table of Course Center

step	Test module	Test content	Test method	test result
1	Course Center	Show unregistered courses	Whether to display courses according to majors	Course information initialization and display according to specialty
2	Course Center	Cancel Enrollment	Can I cancel the course	Selected courses can be canceled and reloaded into unselected lines
3	Course Center	Online playing of courses	Whether it can be played online	Video and audio can be played online, and learning records can be displayed
4	Course Center	Course download	Can I download courseware	Video and audio files can be downloaded. The default storage address is the memory card
5	Course Center	Course evaluation	Whether the course can be evaluated	The number of words should be limited for evaluable courses

Table 3. Test Table of Evaluation Module

step	Test module	Test content	Test method	test result
1	Evaluation	Basic question type training	Whether the question type is displayed according to the specialty	Initialise question bank and display question types according to specialty
2	Evaluation	Timing function	Can you time	The entry test timing function is enabled, in seconds
3	Evaluation	Scoring function	Whether to score	Enter the exam scoring function to score in real time
4	Evaluation	Collection topic	Can I collect the topic	Click Collect to successfully collect the question
5	Evaluation	Collection error	Whether to collect wrong questions	Answer matching error, automatic collection of wrong questions

According to the above tests, the authentication module, course center module and evaluation module of the mobile terminal in this article can all pass the functional test to meet the needs of users.

4.3 Performance Test

In order to avoid the singleness of the experimental results, in the performance test, comparing mobile terminals based on model driven and intelligent training mode with those in the text, and the terminal user information acquisition delay and load balance are tested. The results are as follows.

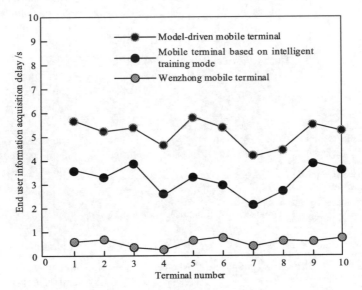

Fig. 8. End user information acquisition delay

According to the results in Fig. 8, among model driven mobile terminals, mobile terminals based on intelligent training mode, the delay of user information acquisition exceeds 2s, which cannot guarantee the real-time nature of user information acquisition. The mobile terminal based on artificial intelligence technology can control the delay of user information acquisition within 1s, improve the efficiency of user information acquisition, and meet the performance requirements of users for mobile terminals.

It can be seen from the results in Fig. 9 that the load balance of model driven mobile terminals, intelligent training mode based mobile terminals in operation is less than 80%, while the load balance of AI based mobile terminals in operation is more than 90%, indicating that the mobile terminals designed in this paper are more stable in operation.

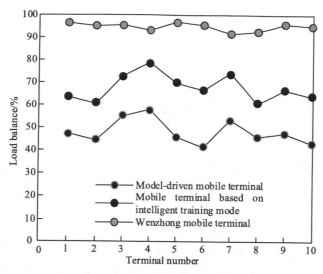

Fig. 9. Load balance

5 Conclusion

This paper proposes the application research of artificial intelligence technology on online cultural education mobile terminal. Through testing, it is found that the function and performance of the mobile terminal can meet the needs of users. Although the research in this paper has achieved certain results, there are still many shortcomings. For the development of Android technology, the current development of Android system is at its peak, and Android technology is mature and increasingly refined. In terms of its current market share, its dominant position is hard to shake, so it is a trend to develop application software on Android system devices. In the future, on the basis of completing the basic functions of software, it is necessary to improve the software performance.

References

1. Yang, J., Shi, Z., Liu, Z.: Simulation of rapid acquisition of terminal users' information under big data analysis. Comput. Simulat. **35**(2), 441–445 (2018)
2. Liu, W., Li, X. Research on terminal online education data mining technology based on model driving. Mod. Electron. Techniq. **43**(16), 112–114+118 (2020)
3. Kang, M.: Design of intelligent classroom integrated student management terminal based on Internet of Things technology. Mod. Electron. Techniq. **46**(3), 177–181 (2023)
4. Zhang, J., Feng, H.: Mobile terminal system of intelligent college English teaching and training mode. Mob. Inf. Syst. **8**, 1–9 (2021)
5. Zhang, X., Gao, X., Yi, H., et al.: Design of an intelligent virtual classroom platform for ideological and political education based on the mobile terminal app mode of the Internet of Things. Hindawi Limited, **2021**(Pt.23), 9914790.1–9914790.12 (2021)
6. Sun, X., Feng, L., Zhu, Z., et al.: Optimal design of terminal sliding mode controller for direct torque control of SRMs. IEEE Trans. Transp. Electrific. **8**(1), 1445–1453 (2021)

7. Huang, H., Jiang, W., Zhang, T., et al.: Shared radiator based high-isolated tri-port mobile terminal antenna group design. Int. J. RF Microw. Comput. Aided Eng. **7**, 32 (2022)
8. Liu, S., He, T., Li, J., et al.: An effective learning evaluation method based on text data with real-time attribution - a case study for Mathematical Class with Students of Junior Middle School in China. ACM Trans. Asian Low-Resour. Lang. Inf. Process. **22**(3), 63 (2023)
9. Zhang, J.: Research on classroom teaching evaluation and instruction system based on GIS mobile terminal. Mob. Inf. Syst. **2021**(11), 1–11 (2021)
10. Chen, X.: Application of the intelligent mobile terminal in the medical oral English teaching. Basic Clin. Pharmacol. Toxicol. **S1**, 124 (2019)
11. Wang, C.: Acute teaching method of college physical skills based on mobile intelligent terminal. J. Inter. Net. (2022)
12. Shankar, R., Ramana, T.V., Singh, P., et al.: Examination of the non-orthogonal multiple access system using long short memory based deep neural network. J. Mobile Multim. **18**(2), 451–473 (2022)
13. Ahmed, U., Lin, J.C., Srivastava, G.: A resource allocation deep active learning based on load balancer for network intrusion detection in SDN sensors. Comput. Commun. **184**, 56–63 (2022)
14. Rawls, C., Salehi, M.A.: Load balancer tuning: comparative analysis of HAProxy load balancing methods. arXiv preprint arXiv:2212.14198 (2022)
15. Finnerty, P., Kamada, T., Ohta, C.: A self-adjusting task granularity mechanism for the Java lifeline-based global load balancer library on many-core clusters. Concurr. Comput.: Pract. Exp. **34**(2), e6224 (2022)

College Psychological Mobile Education System Based on GPRS/CDMA and Internet

Zhang Liang[✉] and Zhao Yu

Changchun University of Finance and Economics, Changchun 130000, China
zliang1510@163.com

Abstract. In order to solve the problem of poor targeted teaching ability of college psychological mobile education and improve the mental health level of college students, a college psychological mobile education system based on GPRS/CDMA and the Internet was designed. On the basis of the distributed service framework, the IIS mechanism of the education system is set up, and then the mobile psychological learning module is combined to complete the hardware operation scheme design of the college psychological mobile education system. Perfect the VPN workflow in the GPRS/CDMA network, determine the layout form of the database organization with the help of the PPP Internet connection protocol, realize various technical functions in the mobile education system, and complete the design of the college psychological mobile education system based on GPRS/CDMA and the Internet in combination with the relevant hardware application structure. The experimental results show that the application of GPRS/CDMA and the Internet system can achieve accurate matching of college psychological mobile education in student terminals, effectively solve the problem of poor targeted teaching ability of psychological education, and can better improve the mental health level of college students, in line with the actual application needs.

Keywords: GPRS/CDMA Network · Internet · Psychological Mobility Education · Distributed Framework · VPN Process · PPP Connection Protocol · Database

1 Introduction

In today's society, with the rapid development of economy, the rapid change of science and technology, the complex and changeable interpersonal relationships, the pace of life in the whole society is getting faster and faster, and people are under increasing pressure in life. However, college students are generally between the ages of 18 and 25. The development of self-consciousness is not yet fully mature, and there are psychological contradictions between ideal and reality. They are a relatively special social group, and there are many problems unique to them, such as difficulties in adapting to the new environment, more interpersonal conflicts, greater emotional fluctuations, relatively immature personality and other psychological problems. How can we make the young students avoid or eliminate the psychological crisis caused by the pressure

L. Yun et al. (Eds.): ADHIP 2023, LNICST 548, pp. 235–250, 2024.
https://doi.org/10.1007/978-3-031-50546-1_16

of learning, making friends, and working, prevent the occurrence of psychosomatic diseases, face various psychological problems in a good psychological state, and then adapt to the complex social environment. How to better manage students' mental health has become a common concern and urgent problem for university administrators [1]. The online life has become an indispensable part of the life of young people at present. The characteristics of the network mean that they can express their opinions and feelings more boldly and truly on the network. Using the network platform to carry out mental health education for students can find out students' psychological problems in a timely manner and guide them. For the relevant administrators of colleges and universities, they can use information means to master students' mental health, more convenient and efficiently.

The application of time-series based education system and web-based model based education system is to install relevant software on the special client, and then publish the test on the web page. Students need to log in to the system to test, so as to replace the traditional paper and pencil tests, issue mapping machine card readers, retrieve questionnaires, and then analyze the statistical data. This process, to a large extent, liberates the staff of mental health education in colleges and universities from the complicated and transactional work when doing the psychological census of students. It saves resources and manpower, improves the coverage of students' psychological survey, and improves the timeliness of mental health education in colleges and universities. However, this system still has great defects. Only on the computer equipped with the relevant software system client, students can log in to the software system for testing and operation. Generally, it takes a lot of resources from the school to equip the corresponding hardware and software systems, and it requires special personnel to maintain and manage the equipment, which does not bring basic convenience to students.

The full name of Internet GPRS is General Packet Radio Service, which is the abbreviation of general packet radio service technology. It is a mobile data service available to GSM mobile users, and belongs to the data transmission technology in the second generation mobile communication. The full name of Internet CDMA is Code Division Multiple Access, which is the abbreviation of Code Division Multiple Access. It is a new and mature wireless communication technology developed from the spread spectrum communication technology, a branch of digital technology. GPRS/CDMA is different from traditional GSM circuit type data service in that GSM mobile users monopolize certain wireless resources for a long time. Under packet data service, all mobile users share wireless resources, and each user dynamically applies for and occupies wireless resources only when there is service data transmission. Therefore, packet data mode can achieve "always on" [2]. For example, the peak rate of GPRS is 115. 2kbit/s, and that of CDMA 1X system is 153. 6kbit/s. Aiming at the problem of limited application of conventional education system, based on GPRS/CDMA and the Internet, this paper designs a new college psychological mobile education system, and highlights the practical value of this system through comparative experiments.

2 Design Scheme of College Psychological Mobile Education System

For the improvement of the hardware design scheme of the college psychological mobile education system, it is carried out simultaneously from three aspects: the construction of the distributed service framework, the IIS setting of the education system, and the connection of the mobile psychological learning module. The overall framework of the psychological mobile education system in universities is shown in Fig. 1.

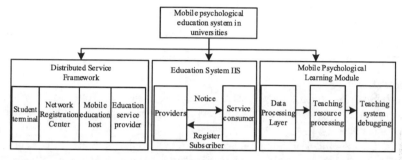

Fig. 1. Overall framework of mobile psychological education system in universities

This chapter will study its specific design methods.

2.1 Distributed Service Framework

The background server of college psychological mobile education system can be divided into an application server for processing logic and a database server for storing data. The application server here uses tomcat, while the database uses the commonly used MYSQL database.

The traditional background application architecture of the psychological mobile education system maintains a vertical layout. With the gradual expansion of the Internet platform business, the vertical architecture will make each education service quite bloated, and the coupling between modules is too high, making it difficult to modify and expand. In the long run, it will cause a lot of subsequent redundancy and insecurity [3]. Therefore, it is necessary to consider adopting a distributed service architecture. The Dubbo model came into being as the times require. As a distributed service framework, it can extract independent core businesses, which not only simplifies the background engineering structure, but also improves the performance, and greatly improves the response speed and stability of front-end applications.

The complete distributed service framework of college psychological mobile education system is shown in Fig. 2.

The Dubbo model is structured in a hierarchical manner to maximize the psychological education resources for disaster relief [4]. From the perspective of design pattern, Dubbo adopts the subscriber pattern and defines two main roles, one is the service provider and the other is the service consumer. Their interaction is conducted through

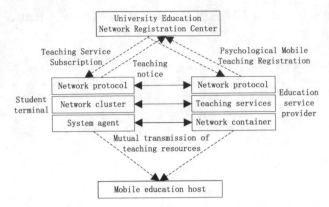

Fig. 2. Distributed service framework of psychological mobile education system

the mechanism of registration, subscription and notification. From the perspective of process operation, the overall operation is an asynchronous process, which is safer and more efficient.

In the Dubbo model of the mobile education system distributed service framework, the specific application capabilities of each structural component are as follows:

Student terminal: Dubbo defines the user who calls the service remotely as the service consumer, which is also different from most RPC (remote procedure call) frameworks.

Education service provider: provide specific services and realize the business logic of services. When using a registry, you need to expose your own services to the registry. When not using a registry, you can directly expose your own education services to consumers.

Mobile education host: monitor the call times and specific call duration of college psychological mobile education services.

College education network registration center: registration, subscription and search of college psychological mobile education services, not responsible for forwarding requests.

2.2 IIS Settings of Education System

Setting up IIS, a college psychological mobile education system, should start with the lower teaching characteristics of psychological education. Since IIS component is a mobile application component with autonomous operation capability, in order to ensure that the component configuration results can meet the scheduling requirements of college psychological mobile education resources in the Internet environment, the following expression conditions should be followed when selecting teaching information P_α:

$$P_\alpha = \frac{\alpha\left[o_{\max}\left(i_{\max}\right) - o_{\min}\left(i_{\min}\right)\right]^2}{\sum\limits_{\chi=1}^{+\infty} \beta \times |\Delta T|} \tag{1}$$

where, α represents the real-time transmission parameters of college psychological mobile education resources in IIS components, $o_{\max}$ represents the maximum value of teaching resource sharing vector, $\dot{I}_{max}$ represents the maximum value of the sharing characteristics of college psychological mobile education resources, o_{min} represents the minimum value of teaching resource sharing vector, $\dot{I}_{min}$ represents the minimum value of the sharing characteristics of college psychological mobile education resources, ΔT indicates the mobility transmission cycle of psychological education resources, β represents the scheduling coefficient of psychological education resources in IIS components, χ indicates the educational resource tag parameter.

The general understanding of psychology is that the relatively stable psychological characteristics of "traits" and a person's behavior, such as intelligence, interest, attitude and personality, can be regarded as characteristics. In the process of measurement, it is a very complex work to take them as operational measurement objects. Psychological measurement is an indirect measurement. Psychological characteristics are implicit. We can not directly measure the quantity of psychological characteristics as the weight or length of measurement, but infer a person's psychological characteristics by measuring his specific behavior in a specific situation [5]. Psychological measurement is a standardized test developed and gradually tested, revised and perfected by experts in this field. The preparation of the scale is a highly specialized systematic work.

The specific IIS component setting process of the education system is shown in Fig. 3.

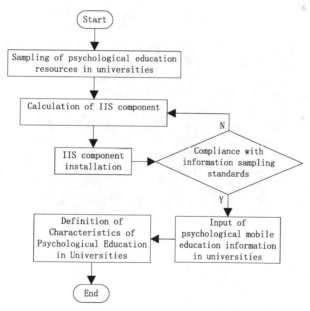

Fig. 3. Flow chart of IIS component setting of psychological mobile education system

The application ability of IIS components in setting psychological mobile education information in colleges and universities is mainly reflected in the following aspects. (1) The identification of the mentally handicapped and the college students with mental disorders is an important driving force to promote the development of psychological tests. Psychological test is still an important method to diagnose mental retardation, mental disease and brain dysfunction. (2) In psychological mobility education, measurement can be used to find the reasons for students' poor academic performance or social adaptation;It can be used to describe and evaluate people's strengths and weaknesses in intellectual, academic and personality characteristics, so that individuals can know their strengths and weaknesses; The data obtained from psychological measurement can be used as the basis of psychological consultation, such as comprehensive achievement test, intelligence test, ability test, vocational interest test and personality test. It can provide suggestions on a person's future career direction and help visitors make correct career choices. (3) The data of personality test and clinical psychological obstacle test help clients to improve their psychological adaptability, regulate their emotions and self-consciousness.

2.3 Mobile Psychological Learning Module

Mobile psychological learning function is a platform to release mental health knowledge. In this module, professional mental health knowledge can be published, which can be articles, music, pictures or videos. Let students learn to correctly deal with psychological problems and establish good learning habits and mentality through browsing these knowledge. Provide the database query function of psychological related information for students. The administrator can complete operations such as publishing, modifying and deleting data.

At the same time, in the function of online mobile psychological learning, according to the keyword association of the question, it is designed to be the form of automatic response, and the answer is automatically given according to the keyword of the student's question. Students can choose whether to view it, save consulting time, and can browse, view or download the psychological knowledge they are interested in.

The structural model of mobile psychological learning module is shown in Fig. 4.

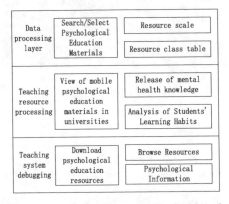

Fig. 4. Structure model of mobile psychological learning module

The mobile psychological learning module includes online consultation, consultation appointment and message consultation. For students who are clearly aware of their own psychological problems and want to seek psychological help, this module can directly make an appointment with a psychological consultant for consultation. Online consultation is the most important function of this module. Students voluntarily choose their own consultants to carry out online consultation [6]. At the same time, you can make an appointment or leave a message if the counselor you want is not online. You can communicate with the psychological counselor at the appointed time in a timely manner to improve the students' use experience. At the same time, you can improve the college psychological counseling manual reception of visiting students, arrange the counselor, time and venue, and avoid the cumbersome process of making an appointment. It is more efficient, fast, privacy and security.

For students who have special circumstances and need to be paid more attention by the department, on the premise of protecting students' privacy, they will summarize data according to different situations, such as academic problems, family problems, emotional problems, etc., and feed back to the department to remind student counselors and relevant staff to care for students. Psychological tutors regularly pay attention to students' learning and living conditions, and give real-time feedback on students' psychological conditions with special attention.

The specific learning mode is shown in Fig. 5.

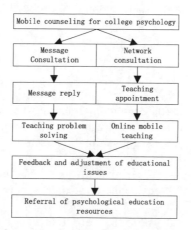

Fig. 5. Learning mode of mobile psychological learning module

The management of mobile psychological education is the main function of the learning module. It can carry out comprehensive management of relevant educational institutions, including adding the name of the institution, editing the category of the institution and setting the attributes of the institution. Permission management is a comprehensive management of the permissions of all users in the system, including the operation permissions of adding and deleting users. The system log records the operations of all users in the system, as well as the operation of the system. The mobile psychological learning module is the main functional module of the college psychological mobile education and comprehensive counseling management information system, that is, the operation authority and role management, system configuration, and database management of all users in the system. System users include adding, editing, modifying, and deleting users' personal basic information. Different roles and users have corresponding permissions. The permissions of each role can also be set according to the needs of the position.

3 Implementation of Functions Based on GPRS/CDMA and Internet

On the basis of application components at all levels, in order to realize various executive functions of the college psychological mobile education system, we should improve the VPN workflow in the GPRS/CDMA network, and then complete the design of the database system through the forwarding function of the PPP Internet connection protocol.

3.1 VPN Workflow in GPRS/CDMA Network

In the college psychological mobile education system, the VPN workflow of GPRS/CDMA network consists of the following steps.

Step 1: The student terminal sends an APN dedicated to the Internet through the TE, uses the APN to find the IP address of the GGSN at the access end of the education system, and the TE sends an AT command message to the MS, and transmits the APN information to the MS.

The calculation of APN lookup requirements U meets the following expression:

$$U = \dot{R} \cdot \left(1 + \frac{P_\alpha}{\delta \cdot \hat{y}}\right) \tag{2}$$

Among them, $\dot{R}$ represents the TE sending characteristics of psychological education information, $\hat{y}$ represents the forwarding vector of psychological education information in the APN cycle, δ indicates the forwarding parameters of psychological education information in GPRS/CDMA network.

Step 2: After receiving the AT command from the TE, the MS sends the SGSN the context request message of "Activate the Packet Data Protocol PDP", which contains the APN.

Step 3: SGSN searches the IP address of the access end of the education system through the DNS of the GPRS/CDMA network according to the APN.

The definition formula of IP address E in GPRS/CDMA network is:

$$E = \left[\frac{1}{(1 + \gamma)}\right]^2 \cdot P_\alpha \tag{3}$$

where, γ indicates the access coefficient of psychological education information in GPRS/CDMA network.

Step 4: GGSN sends the RADIUS authentication request message containing the student terminal information to the corresponding GPRS/CDMA server.

The service request W formula is:

$$W = \sum_{\varepsilon=1}^{+\infty} \phi^{-e} \cdot P_\alpha \tag{4}$$

where, ε represents the real-time transmission parameters of psychological educa-tion information, ϕ indicates the authentication permission of GPRS/CDMA server for psychological education information, e indicates the grid connection parameters of GPRS/CDMA network.

Step 5: After passing the authentication, assign the private IP address of the GPRS/CDMA intranet to the student terminal, and the RADIUS server sends back a RADIUS authentication permission message.

Step 6: Improve the VPN workflow.

The definition formula of VPN workflow Q in GPRS/CDMA network is:

$$Q = \frac{\bar{s}}{\vec{A}} \sum_{-\infty}^{+\infty} \left| \frac{1}{\varphi} \right| (U \cdot E \cdot W)^2 \tag{5}$$

where, $\vec{A}$ represents the transmission vector of psychological education information in GPRS/CDMA network, $\bar{s}$ represents the cumulative mean value of psychological education information, φ indicates the operation coefficient of psychological education information in VPN mode.

In the college psychological mobile education system, the packet data core network based on GPRS/CDMA network should include HA[7]in addition to PDSN and RADIUS servers. HA is responsible for assigning IP addresses to student terminals, sending packet data to underlying student users through tunnel technology, and realizing macro mobility management between PDSNs. At the same time, PDSN should also add the function of VPN operation, be responsible for providing tunnel exits, and send psychological education information to mobile terminals after unpacking.

3.2 PPP Internet Connection Protocol

PPP Internet Connection Protocol is the most important protocol file in GPRS/CDMA network, with the following characteristics.

(1) Be able to control the establishment of psychological education information link.
(2) It can allocate and use the IP address of the education system, and allows multiple network layer protocols to be used at the same time.
(3) It can configure and test the psychological education information link, conduct error detection, have negotiation options, and negotiate the address of the network layer and the mobile transmission behavior of psychological education information.

The complete PPP Internet connection protocol structure is shown in Fig. 6.

The college psychological mobile education system accepts the simultaneous adjust-ment of GPRS/CDMA network and Internet organization, so the PPP connection proto-col must include the following three components. (1) The method of compressing multi protocol self addressing psychological education information packet; (2) LCP used to establish, set and test the data link connection of psychological education information; (3) A family of NCP nodes used to establish and set different network layer protocols.

In order to be convenient enough to use in a wide environment, PPP provides LCP. LCP is used to automatically reach an agreement on encapsulation format options, handle

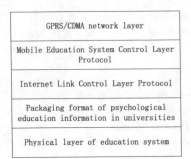

Fig. 6. PPP Internet Protocol Structure

changes in packet size, detect looped back links and other common configuration errors, and terminate links. Other optional equipment provided are: authentication of the same unit ID in the link, and decision when the link function is normal or the link fails [8].

The rating D of PPP Internet protocol connectivity follows the following expression:

$$D = \frac{Q \cdot \sqrt{\left|\frac{g_1}{d_1}\right|^2 + \left|\frac{g_2}{d_2}\right|^2 + \cdots + \left|\frac{g_n}{d_n}\right|^2}}{(\lambda - 1)^2} \tag{6}$$

where, $d_1 \text{、} d_2 \text{、} \cdots \text{、} d_n$ indicates GPRS/CDMA network n Psychological education information objects responding to PPP Internet Connection Protocol, and the inequality value condition of $d_1 \neq d_2 \neq \cdots \neq d_n$ is always true, $g_1 \text{、} g_2 \text{、} \cdots \text{、} g_n$ respectively represents the internet protocol link coefficients that match $d_1 \text{、} d_2 \text{、} \cdots \text{、} d_n$, λ represents the real-time rating coefficient of PPP Internet protocol in GPRS/CDMA networks.

Point to point connection may cause many problems with current network protocols. These problems are handled by a family of network control protocols (PPP), each of which manages the special requirements of its own network layer protocol [9]. To make the PPP link easy to configure, the standard default value can handle all configurations. The student terminal can improve the default configuration, which is automatically notified to its equivalent unit without the intervention of the teaching terminal. Finally, the teaching terminal can clearly set options for the link so that it can work normally.

3.3 Database Design

In the design of college psychological mobile education system, psychological education information is the core, and information is transferred and processed between layers in the form of data inside the system, and these data will eventually become the table structure and content of the database. Therefore, the mobile education system based on GPRS/CDMA and Internet architecture style takes the division of psychological education information and business logic as an important reference basis in database design [10]. The database of the system is designed according to the information division and business functional requirements of the system. The following will take part of the database table design of the system as an example to show the method of database design.

Regulations $\dot{j}$ it is the storage feature of psychological education information, and its solution $\dot{j}$ expression is:

$$\dot{j} = \sum_{-\infty}^{+\infty} |\kappa \times D|^{-\frac{1}{\iota}} \cdot \sqrt{\left(\frac{\eta \times \tilde{k}}{|\Delta h|}\right)} \tag{7}$$

where, κ indicates the classification parameters of psychological education information storage in the Internet environment, ι indicates the service logic connection parameters of GPRS/CDMA network, η indicates the transmission efficiency of psychological education information in GPRS/CDMA network [11], $\tilde{k}$ represents mobile transmission vector of psychological education information, Δh represents the unit cumulative amount of psychological education information in the Internet system.

Using formula (7), deduce the storage characteristics L based on $\dot{j}$ the coding conditions for the mobility of psychological education information of:

$$L = \left| \frac{\left| \sum_{-\infty}^{+\infty} |f \cdot \dot{j}|^{-1} \right|^2}{\mu \times \tilde{Z}} \right| \tag{8}$$

where, f represents the guidance amount of psychological education information transmission based on GPRS/CDMA and Internet architecture, $\tilde{Z}$ represents the transmission characteristics of psychological education information between database hierarchical organizations, μ represents the real-time recognition parameters of the mobile education host for the psychological education information.

On the basis of formula (8), the behavior X of psychological education information transmission between database hierarchical organizations can be expressed as:

$$X = \varsigma \times \int_{\varpi=1}^{+\infty} \vartheta^2 \left(\frac{L}{\tilde{C}}\right)^2 \tag{9}$$

where, ς represents the real-time display parameters of psychological education information in GPRS/CDMA network, ϖ refers to mobile education behavior rating parameters based on GPRS/CDMA and the Internet, ϑ represents the Internet connection coefficient, $\dot{C}$ represents the business logic connection characteristics of the database mechanism.

For the efficiency of database operation [12], each GPRS/CDMA network node maintains a mapping table of IP addresses and MAC addresses of nodes in the Internet. When a node has an address resolution requirement, it first looks up its own mapping table. If there is a corresponding item in the mapping table, it can directly look up the table to get the corresponding MAC address. Only when there is no corresponding item in the mapping table, the inquiry message of psychological mobile education will be displayed, and its own IP address and MAC address will be attached to it, so as to improve the address resolution efficiency of the entire network, and when the response message is received, the corresponding item will be filled in the mapping table.

The solution expression for the database connection response condition M of mobile education system is:

$$M = \left(X\sqrt{\psi} + 1 \right) \times \frac{V'}{\vec{b}} \tag{10}$$

where, ψ indicates the message inquiry coefficient of psychological mobile education information in the database mechanism, V' indicates the MAC address [13, 14] coding parameters based on GPRS/CDMA and the Internet, $\vec{b}$ represents the mapping vector of psychological mobile education information in the database system.

The database of the university psychological mobile education system takes the resource layer as the entrance of the client request, and passes through the business logic layer and the data access layer in turn. The main functions of the resource layer are to publish resource identification URIs externally[15], receive client requests, parse request data formats, call the business logic layer, encapsulate data and respond to clients, etc. It is the main feature of GPRS/CDMA and the Internet. The business logic layer is mainly responsible for encapsulating business function operations and providing support for the resource layer. The data access layer is mainly used for database access.

4 Example Analysis

In order to highlight the practical differences among the college psychological mobile education system based on GPRS/CDMA and the Internet, the time series based education system, and the WEB model based education system, the following comparative experiments are designed.

4.1 Principle and Steps

The targeted teaching ability of college psychological mobility education can be used to describe the mental health level of college students. Without considering other interference conditions, the stronger the targeted teaching ability of college psychological mobility education, the higher the mental health level of college students.

The specific implementation process of this experiment is as follows:

- Six different student groups were selected as the experimental research objects;
- The total amount of psychological resource information allocated by the education host for different students is counted, and the result is the standard value;
- Record the total amount of psychological resource information allocated by the education host to different students under the action of the college psychological mobile education system based on GPRS/CDMA and the Internet, and the results are experimental group variables;
- Record the total amount of psychological resource information allocated by the education host to different students under the action of the education system based on time series, and the results are the variables of control group A;
- Record the total amount of psychological resource information allocated by the education host to different students under the action of the education system based on the WEB model, and the results are the variables of control group B;
- Compare the experimental results of the experimental group, control group A and control group B with the standard values, and summarize the experimental rules.

4.2 Results and Discussion

The following table records the standard value of information allocation of psychological resources.

Table 1. Standard value of mental resource information allocation

Group	Student Category	Allocation of psychological resource information/ × 107Mb
1	First kind	7.4
2	Second kind	3.6
3	Third kind	5.1
4	Fourth kind	4.8
5	Fifth kind	6.2
6	Sixth kind	6.7

The following figure reflects the specific experimental values of information allocation of psychological resources under the action of different education systems (Fig. 7)

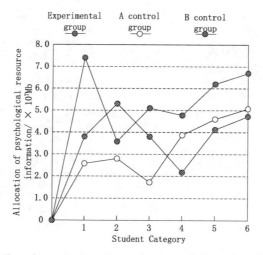

Fig. 7. Experimental value of mental resource information allocation

Combined with Table 1 and Fig. 6, the difference level between the experimental value of mental resource information allocation and its standard value is calculated. See Table 2 for details.

Analysis of Table 2 shows that the difference of information allocation of psychological resources in the experimental group is always zero throughout the experiment; Under the effect of the education system of control group A, for the first kind of students,

Table 2. Difference of mental resource information allocation

Student Category	Experimental group / × 107Mb	A control group / × 107Mb	B control group / × 107Mb
First kind	0	4.8	3.6
Second kind	0	0.9	1.6
Third kind	0	3.4	1.3
Fourth kind	0	0.9	2.7
Fifth kind	0	1.6	2.1
Sixth kind	0	1.7	2.0

the difference in the distribution of psychological resources information is the largest, reaching 4. 8/ × 107 Mb, and its average level was far higher than that of the experimental group during the whole experiment; Under the effect of the education system of control group B, for the first kind of students, the difference in the distribution of psychological resources information is the largest, reaching 3. 6/ × 107 Mb, although less than the maximum value of control group A, its average level is still far higher than the experimental group.

To sum up, the conclusion of this experiment is:

(1) The application of time series based education system is not enough to achieve accurate matching of college psychological mobile education, so it cannot achieve the purpose of improving college students' mental health.

(2) Although the application ability of the web-based model based education system is slightly stronger than that of the time-series based education system, it is still unable to effectively control the difference between the experimental value and the standard value of the psychological resource information allocation, so it cannot solve the problem of poor targeted teaching ability of college psychological mobile education.

(3) The application of college psychological mobile education system based on GPRS/CDMA and the Internet can solve the problem of poor targeted teaching ability of college psychological mobile education and improve the mental health level of college students. Compared with the time series based education system and the WEB model based education system, it is more consistent with the actual application needs.

In order to better validate the application effect of the mobile psychological education system in universities based on GPRS/CDMA and the Internet, 2000 students were selected to visit three systems and compare the response times of different systems, as shown in Fig. 8.

From Fig. 8, it can be seen that as the number of online users increases, the system response time also increases. When the number of online users reaches 2000, the response time of A control group is 550ms, and the response time of B control group is 480ms. The

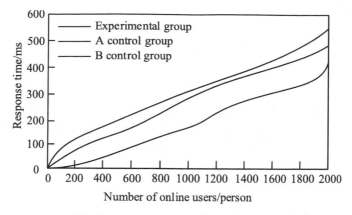

Fig. 8. Response time of different systems

response time of the experimental group is only 420ms, indica ting that the experimental group can quickly respond to student requests and the system has high efficiency.

5 Conclusion

The above research analyzes college psychological mobile education in detail from two perspectives of teaching environment and teaching philosophy. It attaches to mobile devices, gets rid of the constraints of time and space, makes full use of scattered time pieces, and improves learners' learning efficiency. Make use of rich online education resources to develop "suitable for people" teaching services for learners. "Personalized, timely and efficient" mobile teaching environment makes "mobile education" attract much attention. The teaching concept of "flipped classroom" is constantly integrated into the application of mobile teaching, which is used to reshape the teaching structure and provide new vitality for "mobile education".

The complexity of mobile applications is mainly due to the construction of heterogeneous platform systems and the limitations of resources. GPRS/CDMA and the Internet not only support the construction of heterogeneous platform network systems, but also have lightweight, especially resource oriented features compared with traditional Web Service, making GPRS/CDMA networks very suitable for the design and implementa tion of mobile application systems. In the detailed design of the system, the client uses the concept of prototype design to build a rapid prototype model. The server side takes resources as the center, and elaborates resource design and database design in detail. The system needs to build a distributed service framework, set up an education system, and connect mobile psychological learning modules to ensure that students can use the system efficiently and safely. On the basis of hardware, improve the VPN workflow in GPRS/CDMA networks, complete database design through the forwarding function of PPP internet connection protocol, improve system efficiency, and thus achieve vari ous execution functions of the college psychological mobile education system. Through experiments, it has been proven that the college psychological mobile education system based on GPRS/CDMA and the Internet has solved the problem of poor targeted teaching ability in college psychological mobile education, improved the mental health level of college students, and shortened response time to process student requests faster.

References

1. Oliveira, A., Capovilla, C.E., Casella, I., et al.: Co-simulation of an SRG wind turbine control and GPRS/EGPRS wireless standards in smart grids. IEEE/CAA J. Autom. Sinica (JAS) **8**(3), 656–663 (2022)
2. Ou, A.W., Hult, F.M., Gu, M.M.: Language policy and planning for English-medium instruction in higher education. J. English-Medium Instruct. **1**(1), 7–28 (2022)
3. Bas, M., Carot, J.M.: A Model for developing an academic activity index for higher education instructors based on composite indicators. Educ. Policy **36**(5), 1108–1134 (2022)
4. Wilkins, S., He, L.: Student mobility in transnational higher education: study abroad at international branch campuses. J. Stud. Int. Educ. **26**(1), 97–115 (2022)
5. Prendergast, C.O., Satkus, P., Alahmadi, S., et al.: Fostering the assessment processes of academic programs in higher education during COVID-19: an example from James Madison University. Assess. Update **34**(2), 1–16 (2022)
6. Mcclelland, R., Houdt, K.V., Akbari, M., et al.: Design, implementation and academic perspectives on authentic assessment for applied business higher education in a top performing Asian economy. Education + Training **64**(1), 69–88 (2022)
7. Nguyen, N.L.: The effects of leader expectation and coworker pressure on research engagement in higher education: the moderating role of achievement value. J. Appli. Res. Higher Educ. **14**(3), 1114–1126 (2022)
8. Vicente, B., Pérez, D: Keeping the score: Reflections on the (un)intended consequences of studying Latino male success in higher education. New Directions Student Serv. **2022**(177), 59–68 (2022)
9. Dhawan, S:. Higher Education Quality and Student Satisfaction: Meta-Analysis, Subgroup Analysis and Meta-Regression. Metamorphosis J. Manag. Res. **21**(1), 48–66 (2022)
10. Zhou, Y., Wang, G.: Research on security structure retrieval simulation of sensitive information in mobile internet. Comput. Simulat. **39**(09), 451–455 (2022)
11. Zhu, X.: Complex event detection for commodity distribution Internet of Things model incorporating radio frequency identification and wireless sensor network. Futur. Gener. Comput. Syst. **125**, 100–111 (2021)
12. Liu, S., He, T., Li, J., et al.: An effective learning evaluation method based on text data with real-time attribution - a case study for mathematical class with students of junior middle school in China. ACM Trans. Asian Low-Resour. Lang. Inform. Process. **22**(3), 63 (2023)
13. Pintor, L., Atzori, L.: A dataset of labelled device Wi-Fi probe requests for MAC address de-randomization. Comput. Netw. **205**, 108783 (2022)
14. Chen, K.: Backoff-toleration-based opportunistic MAC protocol for underwater acoustic sensor networks. IET Commun. **16**(12), 1382–1392 (2022)
15. Wan, S., Winiewski, R., Alexandropoulos, G., et al.: Special Issue on Optimization of Cross-layer Collaborative Resource Allocation for Mobile Edge Computing, Caching and Communication. Comput. Commun. **181**, 472–473 (2022)

Path Planning Method of Garbage Cleaning Robot Based on Mobile Communication Network

Xinyan Tan and Xiaoying Lv[✉]

Dalian University of Science and Technology, Dalian 116052, China
lvxy1986@126.com

Abstract. Aiming at many problems brought by the complex running environment, controller performance and obstacles of garbage cleaning robot, this paper puts forward a path planning method of garbage cleaning robot based on mobile communication network. Jud that initial signal rate of the mobile communication network, constructing a motion model of the clean robot, and planning the robot grasping trajectory according to the model; According to the straight path planning and turning path planning, the planning method is studied. The experimental results show that the navigation deviation of the proposed method is small, which can avoid obstacles and effectively plan the path of the garbage cleaning robot.

Keywords: Mobile Communication Network · Garbage Cleaning Robot · Path Planning · Initial Rate · Motion Model

1 Introduction

With the acceleration of urbanization, the amount of urban garbage is increasing, which poses a serious threat to the environment and public health. The traditional garbage cleaning method needs a lot of manpower, material resources and time cost, which is inefficient and has the problem of incomplete cleaning [1]. In this context, robots are widely used in industrial development to replace manual cleaning operations. With the development of society, people's lifestyles have undergone many changes, and modern scientific and technological means have been improved year by year [2]. At present, people have more and more strict requirements for the autonomy of garbage cleaning robots, and whether the path planning can be effectively completed has become one of the main identification conditions for robot autonomy detection, so the path planning of garbage cleaning robots has become the main research field of robotics [3, 4]. The garbage cleaning robot based on mobile communication network can realize remote control and data transmission through the Internet, and has the advantages of autonomous navigation and rapid cleaning. Path planning is the basis of robot's autonomous movement, so the rationality and efficiency of path planning method are directly related to the actual operation effect of garbage cleaning robot.

© ICST Institute for Computer Sciences, Social Informatics and Telecommunications Engineering 2024
Published by Springer Nature Switzerland AG 2024. All Rights Reserved
L. Yun et al. (Eds.): ADHIP 2023, LNICST 548, pp. 251–263, 2024.
https://doi.org/10.1007/978-3-031-50546-1_17

At present, reference [5] proposes a vision-based detection method for underwater garbage cleaning robot YOLOv4, and selects YOLOv4 algorithm as the basic neural network framework for target detection. In order to further improve the detection accuracy, YOLOv4 is converted into a four-scale detection method; In order to improve the detection speed, the new model is trimmed. Through the improved detection method, the robot can collect garbage on its own. The detection speed is as high as 66.67 frames per second, and the average accuracy is 95.099%. The experimental results show that the improved YOLOv4 has good detection speed and accuracy. Reference [6] puts forward a beach garbage collection robot based on wireless communication, which effectively uses the Internet of Things to keep the continuous connection between the central server and the garbage disposal and collection network, relies on the system to produce accurate results, and greatly reduces the cost, thus providing a feasible solution to minimize the manpower and cost in the garbage collection process. There is always a trade-off between the accuracy, efficiency and cost of garbage collection.

However, the above methods have some defects, such as high planning complexity and easy to be disturbed by the environment, and the intelligent algorithm needs a large number of sample data to support it. Therefore, how to establish an efficient and reliable path planning method has become the research focus of the practical application of garbage cleaning robots. Combining the characteristics of mobile communication network and garbage cleaning robot, this paper proposes a path planning method based on mobile communication network. This method evaluates the initial rate of mobile communication network signal, establishes the motion model of garbage cleaning robot, and plans the grabbing trajectory based on this model. Through the methods of linear path planning and curve path planning, the research is carried out. The experimental results show that the proposed method has a small deviation in the navigation process, can effectively avoid obstacles, and successfully plan the path of the garbage cleaning robot. The path planning method of garbage cleaning robot has the characteristics of high efficiency, intelligence, accuracy, reliability and expansibility, and can better meet the needs of users.

2   Path Planning of Garbage Cleaning Robot Under Mobile Communication Network

2.1   Mobile Communication Network Signal Initial Rate Determination

In the path planning of garbage cleaning robot, the initial speed of mobile internet communication network signal is judged [7]. In the robot multi-sensor information transmission, because the motion postures of the active and passive terminals change in real time, the sensor information can start to change before reaching the optimal predicted transmission rate, and the slow start mechanism will not quickly judge the correct network bandwidth in a short mobile communication period. In order to increase the end-to-end data throughput, all wired source nodes must choose the best expected rate for information transmission after the correct network connection is established, and must make full use of all available wireless broadband and select an appropriate transmission window to establish wireless connection, but at the same time, the transmission rate determined on the transmission window cannot exceed the transmission speed used in the whole transmission link [8, 9]. At the same time, it should also prevent the speed of the sender from changing too fast to cause congestion, which will greatly reduce the performance of the whole garbage cleaning robot network system.

The process of determining the initial signal rate of the mobile communication network is as follows: firstly, the starting frequency of the source node of the mobile communication network must meet the restriction requirements, that is, the forwarding frequency determined by the starting window does not exceed the data processing frequency of the mobile communication network, so as to prevent the occurrence of bottleneck link congestion. Through the data information transmission of the source node layer, the forwarding end timely adjusts and dynamically adjusts the forwarding frequency of data packets so as to approach the effective width of the wireless network. The initial rate of the mobile communication signal is determined as shown in Formula (1):

$$P_U = R_T (n_1 + n_2 + n_3 + n_4) \tag{1}$$

In formula (1), P_U represents the communication signal transmission rate; R_T represents the number of transmissions; n_1, n_2, n_3 and n_4 respectively represent the four processes of transmitting communication signals, receiving communication signals, sending data, and inputting to the source node. Through the above formula, the available wireless bandwidth can be estimated, connected to the transport layer and fed back to the source node, and the source node sets the corresponding initial rate according to the received feedback signal.

2.2   Constructing the Motion Model of Cleaning Robot

According to the initial signal rate of the mobile communication network, the state quantity of the robot at t moment is obtained, and the pose transformation of the cleaning robot is solved, as shown below:

$$M_t = F(A_t, B_B) + \lambda \times P_U \tag{2}$$

In formula (2), A_t represents the state of the cleaning robot at t, B_B represents the control quantity, $F(\cdot)$ represents the state transfer function, and λ represents the noise generated by the cleaning robot during the movement.

Using the way of mobile communication network, the trajectory of garbage cleaning robot is optimized to make it continuous. The path of the road section can be expressed as:

$$P(o) = \sum_{i=1}^{n} \frac{L_i}{t_i} \times M_t \tag{3}$$

In formula (3), $P(o)$ represents the polynomial programming result of the o-th path; L_i represents the curve of i short path, and t_i represents the time required for the i road section. In this section, you can get the starting point and ending point of locating the search node:

$$\begin{cases} q_1(t_1) = F_1 \times P(o) \\ q_2(t_2) = F_2 \times P(o) \end{cases} \tag{4}$$

In formula (4), $q_1(t_1)$ represents the road section at the starting point q_1 at t_1 time; F_1 indicates the distance traveled at this time; $q_2(t_2)$ represents the end point q_2 section of t_2 time; F_2 represents the distance traveled during this period. Combining the whole path, can get:

$$Q(t) = \begin{cases} q_1(t) = \sum_{i=1}^{n} \left(\frac{F_1}{t_1}\right)^2, t_1 \leq t \leq t_2 \\ q_2(t) = \sum_{i=1}^{n} \left(\frac{F_2}{t_2}\right)^2, t_2 \leq t \leq t_3 \\ \cdots \\ q_n(t) = \sum_{i=1}^{n} \left(\frac{F_n}{t_{n-1}}\right)^2, t_{n-1} \leq t \leq t_n \end{cases} \tag{5}$$

In formula (5), $Q(t)$ represents the function value represented by each endpoint in the piecewise function; t_1, t_2, t_{n-1} and t_n represent the time at each endpoint. Under this preset optimization problem, the time coefficient is obtained according to a certain proportion:

$$t_n = \left(\frac{F_{\max}(t)}{K_{FGH}}\right)^{\frac{1}{n}} \times Q(t) \tag{6}$$

In formula (6), t_n represents the time required for preset optimization under the n segment in a certain proportion; $F_{\max}(t)$ represents the maximum function value under piecewise function in different time periods; K_{FGH} represents the expected value. Combined with the above formula, the search node of garbage cleaning robot can be located, so as to find the planning path.

In the process of finding the path planning results, the extended nodes are used to explore the next step, so as to reduce the complexity of the mobile communication network and the number of iterations, thus reducing the running time of the method and improving the efficiency. At this time, it is necessary to locate the search nodes of mobile communication network equipment, and the actual cost of each node is:

$$R_{EU} = \sqrt{\frac{(E_x - U_x)^2}{(E_y + U_y)^2}} \times t_n \tag{7}$$

In formula (7), R_{EU} represents the actual cost of searching nodes for each mobile communication network device; E_x and E_y represent the vertical and horizontal coordinates of the starting point, and U_x and U_y represent the vertical and horizontal coordinates of the ending point. Combined with this cost function, the heuristic functions of intermediate nodes and target nodes can be redefined and the pheromone of mobile communication network can be initialized. In the mobile communication network, pheromone

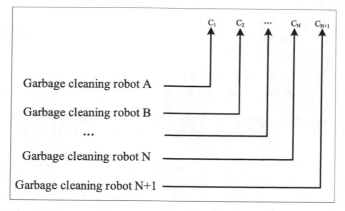

Fig. 1. Structure diagram of path information of garbage cleaning robot during driving

is generally used as the path information of the garbage cleaning robot during driving, and its structure is shown in Fig. 1.

In the path information structure of the garbage cleaning robot in the driving process as shown in Fig. 1, the kinematic constraints of the garbage cleaning robot of the mobile communication network equipment can be determined:

$$C_{N+1}(V_m, V_n) \geq 0 \tag{8}$$

In formula (8), $C_{N+1}(V_m, V_n)$ represents the constrained pose of the target area. Calculate the path planning evaluation function of the robot from this:

$$\Psi(\psi) = \sqrt{\frac{t(\psi)}{\zeta(\psi) + \mu(\psi)}} \times C_{N+1}(V_m, V_n) \tag{9}$$

In formula (9), $\Psi(\psi)$ represents the planning and evaluation function of the garbage cleaning robot within this path; $t(\psi)$ represents the time parameter; $\zeta(\psi)$ represents the minimum cost function; $\mu(\psi)$ represents the path search function. By combining the above formula, a motion model of the garbage cleaning robot can be established.

The posture transformation of the cleaning robot is related to the state quantity and the control quantity, so the motion model of the cleaning robot is constructed based on the linear state transition relation, which is expressed by the equation expression as follows:

$$\begin{bmatrix} a_l \\ b_l \\ c_l \\ d_l \end{bmatrix} = \begin{bmatrix} 1 & 0 & 0 & \Delta t & 0 & 0 \\ 0 & 1 & 0 & 0 & \Delta t & 0 \\ 0 & 0 & 1 & 0 & 0 & \Delta t \\ 0 & 0 & 0 & 1 & 0 & 0 \end{bmatrix} \times [M_t] + W \tag{10}$$

In formula (10), $X_t = (a_l, b_l, c_l, d_l)$ represents the attitude motion model of the garbage cleaning robot, and W represents the noise predicted by the model when the garbage cleaning robot is working.

Based on the mathematical model of the cleaning robot, the pose relationship of the robot components is obtained, and the dynamic analysis of the garbage cleaning robot is realized.

2.3 Trajectory Planning of Garbage Cleaning Robot

Because the garbage cleaning robot needs to plan the trajectory path on the basis of robot dynamics when it grabs foreign objects at work, the garbage cleaning robot can efficiently grab foreign objects [10, 11].

During the operation of the manipulator of the garbage cleaning robot, the position and posture of the end effector of the manipulator is set to K_i, and the manipulator of the cleaning robot will move smoothly from K_i to the position of the foreign object target point in the rectangular coordinate space, and its target position and posture is defined as K_{i+1}, so as to realize the planning of the trajectory target. Wherein the coordinate space is shown in Fig. 2.

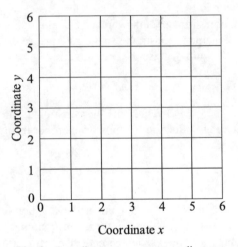

Fig. 2. Coordinate space structure diagram

According to the rectangular coordinate space structure and the kinematics analysis of the garbage cleaning robot [12], the current pose and target pose of the manipulator are defined respectively:

$$\begin{cases} T_i^0 = T_B^0 + K_i + T_B^{-1} \\ T_{i+1}^0 = T_{B+1}^0 + K_{i+1} + T_{B+1}^{-1} \end{cases} \tag{11}$$

In formula (11), T_B^0 represents the linkage system, T_{B+1}^0 represents the tool coordinate system, K_{i+1} represents the homogeneous transformation of the target pose, T_B^{-1} represents the end linkage transformation of the manipulator, and T_{B+1}^{-1} represents the homogeneous transformation of the current pose.

According to formula (11), the manipulator of cleaning robot will undergo driving transformation when it moves from K_i to K_{i+1}, that is $D(\lambda)$, which is expressed as:

$$D(\lambda) = T \times (K_i + K_{i+1}) \times \left(T_i^0 \times T_{i+1}^0\right) \times \chi \tag{12}$$

In formula (12), χ represents the time function, and T represents the total time of the robot's trajectory movement. Therefore, the formula for calculating the grasping trajectory planning of the garbage cleaning robot is:

$$D(\lambda) = L(\lambda) \times \left(T_i^0 \times T_{i+1}^0\right) \tag{13}$$

In the rectangular coordinate space, the position, speed and acceleration of the end effector of the garbage cleaning robot will be uniformly used as a time function. After kinematics analysis, the parameter values of the end effector of the robot manipulator in the current posture state will be solved, and then the parameter values of the end effector in the target posture state will be solved by inverse kinematics. Finally, the values of the two positions will be interpolated, from which the expected trajectory of the end effector of the garbage cleaning robot manipulator will be planned, and the grasping path planning of the robot manipulator will be realized.

2.4 Realize Path Planning

Based on the mobile communication network, the improved S-shaped trajectory planning method ensures the shortest planned path distance, and when the robot encounters obstacles, the software system can quickly plan the path to effectively avoid obstacles [13, 14], in which the path planning is divided into straight-line path planning and turning path planning. The details are as follows:

2.4.1 Linear Path Planning

Step 1: When the trajectory of the garbage cleaning robot from the origin to the garbage target point is a straight line, the planning value of the garbage cleaning robot's grasping trajectory is introduced, and the robot's origin O and destination G are expressed by vector equation. The formula is as follows:

$$\begin{cases} O = X_1 \times m + Y_1 \times n + Z_1 \times D(\lambda) \\ G = X_2 \times m + Y_2 \times n + Z_2 \times D(\lambda) \end{cases} \tag{14}$$

In formula (14), X, Y and Z respectively represent the coordinates of the origin and destination of the disinfection robot; m and n represent vector coefficients respectively.

Step 2: Calculate the position vector of the garbage cleaning robot at any time according to the running speed and total time of the intelligent robot by using the linear distance between the origin and the destination of the garbage cleaning robot based on the mobile communication network. The formula is as follows:

$$\begin{cases} X = \left(\dfrac{X_2 - X_1}{q}\right) \times v_X t + X_1 \\[3mm] Y = \left(\dfrac{Y_2 - Y_1}{q}\right) \times v_Y t + Y_1 \\[3mm] Z = \left(\dfrac{Z_2 - Z_1}{q}\right) \times v_Z t + Z_1 \end{cases} \tag{15}$$

In formula (15), q represents the linear distance between the origin and the destination of the intelligent robot; v represents the velocity vector of robot motion; t represents the running time of the robot in this linear distance.

Step 3: In the path navigation software system of the garbage cleaning robot, according to the position vector of the intelligent robot and the obtained garbage target point, draw the motion trajectory of the garbage cleaning robot, and then calculate the change process of the robot direction to complete the linear path navigation planning of the garbage cleaning robot based on the mobile communication network.

2.4.2 Turning Path Planning

Step 1: The garbage cleaning robot automatically navigates the planned turning path based on the mobile communication network. Firstly, the three-dimensional coordinate system of the garbage cleaning robot is converted into a two-dimensional coordinate system, and any three points that are not on the same straight line are selected to determine a turning plane, and then the plane is divided vertically [15]. The central angle of the path is determined by calculating the radian of the turning path. The coordinate system of turning path is shown in Fig. 3.

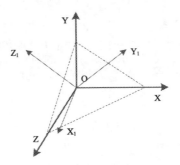

Fig. 3. Turning path coordinate system

Step 2: Introduce a transformation matrix to represent the transformation between the turning path coordinate system and the basic coordinate system. When the garbage cleaning robot based on mobile communication network moves, it will produce the basic conditions of relative coordinate system, so the transformation matrix can be obtained by rotating the coordinate axis, which is expressed by the following formula.

$$
Z_{HJZ} = \begin{bmatrix} \cos\alpha & -\cos\alpha\sin\alpha & \sin\alpha\cos\alpha & X \\ \sin\alpha & \cos\alpha\cos\beta & -\sin\beta\cos\alpha & Y \\ 0 & \sin\beta & \cos\beta & Z \\ 0 & 0 & 0 & 1 \end{bmatrix} \tag{16}
$$

In formula (16), Z_{HJZ} represents the transformation matrix; α represents the angle between the basic coordinate system and the X axis; β represents the angle between the basic coordinate system and the Y axis.

Step 3: There is a certain correspondence between the basic coordinate system and the turning path coordinate system in any time period, and the central angle $\Delta\gamma$ and the total step P corresponding to the trajectory period are calculated according to the following formula. Where M is the moving length of the basic coordinate axis and R is the length of the turning path.

$$
\begin{cases} Q = (X, Y, Z) \times M(X \times \alpha) \times M(Y \times \beta) \\ \Delta\gamma = \frac{v \times t}{R} \\ P = \frac{\gamma_1 + \gamma_2}{1 + \Delta\gamma} \end{cases} \tag{17}
$$

Step 4: Calculate the position coordinates of the i trajectory point in the turning path of the garbage cleaning robot based on the mobile communication network according to the central angle and the total step calculated by the formula (17).

Step 5: Based on the automatic path navigation software program of the garbage cleaning robot in the mobile communication network, combined with the main controller in the hardware platform, the pose angle of the end effector of the garbage cleaning robot in the position coordinates of all trajectory points can be calculated, and then the path planning of the garbage cleaning robot can be completed. The formula is as follows:

$$
G_H = Q \times P + \begin{bmatrix} X_i \\ Y_i \\ Z_i \end{bmatrix} \tag{18}
$$

In formula (18), G_H represents the trajectory of turning position. To sum up, the path planning method of garbage cleaning robot based on mobile communication network is completed.

3 Experimental Analysis

In order to verify the effect and effectiveness of the path planning method of garbage cleaning robot based on mobile communication network, experiments are carried out. Taking a garbage dump cleaning robot as an example, the path planning of this kind of robot is implemented by using this method, and the practical application performance of this method is tested by the planning results. Two robots with the same size and specifications were selected from the garbage cleaning robots in the garbage dump, named Robot A and Robot B respectively, and a virtual map was established, which was expressed in the form of a network. Then, according to the real-time transmitted garbage stacking information and robot position information, the path of the robot is planned by using the mobile communication network. Specifically, the goal of mobile communication network is to make robots clean up garbage as efficiently as possible, and at the same time avoid colliding with obstacles. In the process of mobile communication network, the fitness of candidate paths is determined by the length of paths and the collision situation on paths. The simulation experiment is carried out in Matlab2017 software environment, and the key parameters in the experiment are shown in Table 1.

Table 1. Setting of key experimental parameters

Serial number	Parameter name	Parameter value
1	Total number of grids in grid model/unit	900
2	Target point coordinates of robot a	(29.5,29.5)
3	Target point coordinates of robot b	(0.5,29.5)
4	Number of particles/piece	30
5	Maximum weighting coefficient	0.9
6	Minimum weighting coefficient	0.6
7	Weight coefficient $\omega 1$	1
8	Weight coefficient $\omega 2$	3
9	Normal number	0.5

Under the setting of the above experimental parameters, the robot path is planned by using this method. Navigation deviation is used as an index to evaluate the accuracy of path planning of each method. The larger the navigation deviation value, the lower the navigation accuracy of the method; The smaller the navigation deviation value, the higher the navigation accuracy of the method. The navigation deviation value consists of horizontal deviation and vertical deviation, and the calculation formula is as follows:

$$\begin{cases} H_1 = |C - c| \\ H_2 = |D - d| \end{cases} \tag{19}$$

In Formula (19), H_1 and H_2 respectively represent the lateral deviation value and the longitudinal deviation value; C and D respectively represent the horizontal and vertical values of the disinfection target point; c and d represent the horizontal and vertical coordinates of the robot terminal respectively.

Plot the navigation deviation test results of the proposed method, the method of reference [5] and the method of reference [6], and the results are shown in Fig. 4.

From the analysis results in Fig. 4, it can be known that the proposed method has high navigation accuracy and stability in the path planning of garbage cleaning robots, and its navigation deviation is smaller than that of the methods in Reference [5] and Reference [6]. The maximum navigation deviation values of the proposed method, Reference [5]

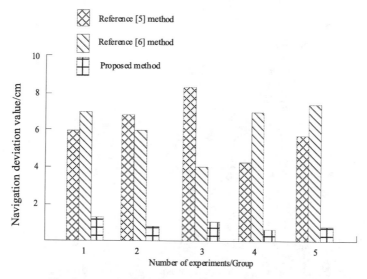

Fig. 4. Navigation deviation values of different methods

and Reference [6] are 1.2 cm, 8.2 cm and 7.8 cm respectively. With the increase of the number of experiments, the navigation deviation value of the proposed method presents a relatively stable trend. This shows that the proposed method can maintain consistent navigation performance in different scenarios and will not be greatly affected by environmental changes. However, the navigation deviation values of reference [5] method and reference [6] method fluctuate greatly and lack consistency. Therefore, for the path planning task of garbage cleaning robot, the proposed method shows a high level of navigation accuracy and stable performance. This will effectively improve the efficiency and accuracy of the garbage cleaning robot and help people better meet the needs of garbage cleaning.

Using the proposed method, the test robot can intelligently plan the obstacle avoidance path, and the result is shown in Fig. 5.

According to the results of Fig. 5, it can be confirmed that the proposed method has successfully designed and measured the intelligent obstacle avoidance path planning of the garbage cleaning robot. In the final planned cooperative path of robot, the robot can effectively avoid obstacles without changing the original optimal path. This means that the proposed method can simultaneously consider the requirements of avoiding obstacles and keeping the shortest path in the path planning of garbage cleaning robots. Through intelligent algorithm and technology, the robot can perceive and analyze obstacles in the environment in real time, and make intelligent planning according to their positions and attributes, thus realizing an efficient and safe garbage cleaning path. This intelligent planning method, which can avoid obstacles, is of great significance to the practical application of garbage cleaning robots. It not only improves the working efficiency of the robot, reduces the risk of collision and damage, but also can better adapt to the needs of garbage cleaning in different environments. To sum up, the method proposed in this paper successfully realizes the intelligent path planning of garbage cleaning robot, which

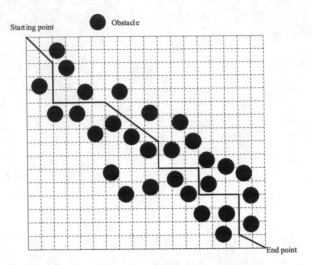

Fig. 5. Result diagram of intelligent path planning of garbage cleaning robot

can effectively avoid obstacles and maintain the optimal path, and bring higher benefits and reliability to the practical application of garbage cleaning robot.

4   Conclusion

The article proposes a path planning method for garbage cleaning robots based on mobile communication networks, and the following conclusions are obtained through research:

(1) The navigation deviation values of the proposed method are all smaller than those of the reference method, indicating high and stable navigation accuracy.
(2) The method proposed in this article can effectively plan the path of the garbage cleaning robot by avoiding obstacles in the final planned robot collaborative path.

The garbage cleaning robot studied in this paper can improve the efficiency of garbage cleaning, ensure the personal safety of operators, and achieve the basic Functional requirement of the robot. However, due to limited time, and currently only tested on the artificial layer, it has not yet been translated into actual industrial value, so further research on the garbage cleaning robot is needed. Future research work can be carried out in the following aspects:

(1) Add waterproof devices. Because the robot did not consider cleaning garbage on the water surface, water splashes may occur during the specific operation process. Therefore, it is necessary to improve the waterproof performance of the robot.
(2) Water surface path planning. Design more complex path planning schemes, calculate the most suitable route, and carry out garbage cleaning.
(3) Collect and treat according to the type of garbage. At present, the designed robot can recognize a small number of garbage types. In the future, the garbage types that can be identified by the target recognition algorithm should be added, and Waste

sorting devices should be added, so as to collect according to the types of garbage and protect the ecological environment.

Acknowledgements. 1. Research on small surface garbage cleaning Robot and target recognition algorithm-Innovation and Entrepreneurship Training Program for college students in 2022(202213207027)

2. Research on the Prediction method and system of base station network traffic based on Smart City-2021 Basic Scientific Research Project of Education Department of Liaoning Province (Youth Project).

References

1. Sivaram, K.M.: Design engineering smart river floating garbage cleaning robot using iot and embedded system. Design Eng. (Toronto) **2021**(5), 1455–1460 (2021)
2. Yan, Z., Ouyang, B., Li, D., et al.: Network intelligence empowered industrial robot control in the F-RAN environment. IEEE Wirel. Commun. **27**(2), 58–64 (2020)
3. Van, M., Ge, S.S.: Adaptive fuzzy integral sliding-mode control for robust fault-tolerant control of robot manipulators with disturbance observer. IEEE Trans. Fuzzy Syst. **29**(5), 1284–1296 (2020)
4. Ali, M., Atia, M.R.: A lead through approach for programming a welding arm robot using machine vision. Robotica **40**(3), 464–474 (2022)
5. Tian, M., Xiali, L.I., Kong, S., et al.: A modified YOLOv4 detection method for a vision-based underwater garbage cleaning robot. J. Zhejiang Univ. **23**(8), 1217–1228 (2022)
6. Mulani. A., Pirjade, K., Walhe, H., et al.: Wireless Communication based garbage collection robot on the beach. JETIR **8**(5), 633–643 (2021) https://www.jetir.org
7. Peng, H., Li, F., Liu, J., et al.: A symplectic instantaneous optimal control for robot trajectory tracking with differential-algebraic equation models. IEEE Trans. Industr. Electron. **67**(5), 3819–3829 (2020)
8. Vecchietti, L.F., Seo, M., Har, D.: Sampling rate decay in hindsight experience replay for robot control. IEEE Trans. Cybern. **52**(3), 1515–1526 (2020)
9. Garate, V.R., Gholami, S., Ajoudani, A.: A Scalable framework for multi-robot tele-impedance control. IEEE Trans. Rob. **37**(6), 2052–2066 (2021)
10. Liu, X.D., He, X.P., Hu, Y., et al.: Mobile robot path planning based on orientation-information strategy RRT algorithm. Comput. Simul. **39**(6), 444–495 (2022)
11. Sawadwuthikul, G., Tothong, T., Lodkaew, T., et al.: Visual goal human-robot communication framework with few-shot learning: a case study in robot waiter system. IEEE Trans. Industr. Inf. **18**(3), 1883–1891 (2022)
12. Pw, A., Bo, W.B.: Multi-sensor detection and control network technology based on parallel computing model in robot target detection and recognition - ScienceDirect. Comput. Commun. **159**, 215–221 (2020)
13. Liu, S., Li, Y., Fu, W.: Human-centered attention-aware networks for action recognition. Int. J. Intell. Syst. **37**(12), 10968–10987 (2022)
14. Juang, C.F., Bui, T.B.: Reinforcement neural fuzzy surrogate assisted multiobjective evolutionary fuzzy systems with robot learning control application. IEEE Trans. Fuzzy Syst. **28**(3), 434–446 (2020)
15. Kanagaraj, G., Masthan, S., Yu, V.F.: Inverse kinematic solution of obstacle avoidance redundant robot manipulator by bat algorithms. Int. J. Robot. Autom. **36**(1), 18–26 (2021)

Research on Electrical Equipment Status Monitoring Method Based on Wireless Communication Technology

Rong Zhu[1][✉] and Wenwei Li[2]

[1] Yili Xingtian Coal Chemistry Co., Ltd., Yili 835000, China
zzhappy888@163.com
[2] Qinzhou Power Supply Bureau of Guangxi Power Grid Co., Ltd., Qinzhou 535019, China

Abstract. The current methods for monitoring the status of electrical equipment are prone to interference from the external environment, resulting in low accuracy of monitoring results and longer monitoring time. Therefore, this study proposes a method for monitoring the status of electrical equipment based on wireless communication technology. Firstly, a low-power wireless transceiver module is established based on RF transceivers. After selecting a suitable wireless communication receiving device, a wireless monitoring framework is established. Then, Fourier transform technology is used to collect electrical equipment status monitoring signal data, and wavelet analysis technology is used to organize the collected signals. Finally, neural network technology is used to evaluate the real-time status of electrical equipment. Through data mining, conduct in-depth analysis of signal data to obtain the final monitoring results. The experimental results show that this method can effectively improve the accuracy of monitoring results and shorten the output time of monitoring results.

Keywords: Wireless communication technology · Electrical equipment · Status monitoring · Fourier transform

1 Introduction

With the continuous rise of the country's overall economic level, the power industry (especially the power generation enterprises) develops rapidly, showing an unparalleled prosperity. On the one hand, the rapid growth of electricity consumption makes the power system network increasingly large, the number of electrical equipment used is increasing, forming a unified organic whole in the process of power production, transmission and use, which is conducive to improving the efficiency of the system. On the other hand, the unsafe factors affecting the safe operation of the power system have also increased sharply. Any major failure or failure of the electrical equipment in the system will have a chain effect of system collapse, resulting in human casualties and huge economic losses, affecting the harmony and stability of the society. Therefore, ensuring the safe and healthy operation of electrical equipment has become the focus of power system workers [1–3].

© ICST Institute for Computer Sciences, Social Informatics and Telecommunications Engineering 2024
Published by Springer Nature Switzerland AG 2024. All Rights Reserved
L. Yun et al. (Eds.): ADHIP 2023, LNICST 548, pp. 264–276, 2024.
https://doi.org/10.1007/978-3-031-50546-1_18

Electrical equipment failure refers to the abnormal working conditions of the power system, partial functional failure of electrical equipment, or the performance indicators of electrical equipment exceeding its rated range, usually resulting in electrical equipment entering a fault state. The structure of electrical equipment is complex, and the system is prone to malfunctions during operation. The main cause of failure refers to the physical, chemical, biological or mechanical processes that cause the failure of electrical equipment under operating conditions, such as corrosion, creep, wear, heating, aging, etc. [4]. With the rise and development of science and technology such as microprocessors and new anti-interference transmission, the research on online status monitoring of power equipment has gradually deepened. The research on dissolved gas systems, online monitoring of partial discharge and leakage of human mouth power generation equipment, and other equipment has successfully enabled real-time monitoring of experiments that could only be conducted through equipment debugging, providing the possibility for online management of equipment status.

With the development of infrared and optical fiber technology, the research of on-line condition monitoring is gradually carried out. Online condition monitoring overcomes the drawbacks of regular maintenance, carries on the condition assessment of electrical equipment, and then carries on the necessary maintenance of electrical equipment on the premise and basis of the condition assessment, avoids the waste of human and financial resources, and effectively guarantees the reliability of power supply. The state assessment of electrical equipment needs to make comprehensive use of all state information such as operating conditions, temperature and electromagnetic, etc. The principle is to process, classify and evaluate the collected state information of the equipment, and the later maintenance of the equipment is based on this technical support [5, 6]. In the state assessment, the fault prediction and alarm of electrical equipment is the most widely used.

At present, there are a lot of problems in condition monitoring methods of electrical equipment, and it is urgent to optimize and perfect them. Therefore, a method of electrical equipment condition monitoring based on wireless communication technology is proposed in this study. On the basis of the current method, the work cost is saved, the service life of the equipment is extended, and the reliability of the power system is greatly improved. This method achieves signal data acquisition and organization through the use of Fourier transform technology and wavelet analysis technology, fundamentally improving data quality, improving the accuracy of later monitoring results, and shortening monitoring time by effectively avoiding environmental interference.

2 Construction of Wireless Communication Electrical Equipment Monitoring Framework

The communication system is an important component of the power system. In recent years, with the continuous deepening of China's power system informatization construction, the power communication network has preliminarily formed. At present, various and fully functional communication methods such as microwave, power line carrier, optical fiber, and wireless mobile communication have been formed, playing a huge role in power load management, distribution automation, and power system status monitoring.

With the continuous development of the wireless communication industry and break-throughs in second and third generation wireless communication technologies, more and more wireless communication technologies are being applied to power system communication, and wireless communication is also playing an increasingly important role in power communication [7]. Compared to various wireless communication methods, wireless radio frequency technology was selected in this study to construct a wireless communication electrical equipment monitoring framework.

The wireless data transmission module (RF transceiver) adopted in this study is a micro-power wireless transceiver module, which has the following characteristics: Receive and transmit in one, the working frequency is the international unified data transmission frequency of 433 MHz, FSK modulation, low transmission power, strong anti-interference ability, transmission range up to 450 m, suitable for the communication within the substation and other short distance power equipment. The performance parameters of wireless transceiver are shown in Table 1.

Table 1. Performance parameters of wireless communication transceiver

Parameter	Value result
Operating temperature	−35 °C–80 °C
Operating frequency	430 MHz
Transmitted power	10 dBm
Frequency modulation mode	FSK
Receiving sensitivity	−110 dB
Operating voltage	2.2–3.6 V
Operating frequency	1.2–1500 kbps
Communication distance	450 m
Emission current	60 Ma@10 dBm
Receiving current	10 mA

Based on the data in Table 1, set and adjust the signal receiving and transmitting device, and apply it to monitor the operating signals of electrical equipment. To obtain more reliable signal values, it is necessary to control the external interference suffered by the device. According to the power transmission model related to communication and interference links, the interference signal ratio of the target receiver can be obtained, and the transmission power (PTs) of the signal can be improved. GTs is the gain of the transmitting antenna, and GRs is the gain of the receiving antenna; Ls reduce the transmission path loss of signals, which can reduce the input interference to signal ratio of the receiver, maintain the system from unnecessary interference, and ultimately ensure the reliability of non information transmission.

Another anti-interference measure is to increase the anti-interference tolerance of the system. The receiver of the communication system can be represented as shown in Fig. 1.

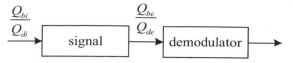

Fig. 1. Receiver signal ratio model

In Fig. 1, $\frac{Q_{bi}}{Q_{di}}$ represents the signal interference ratio at the input end of the receiver, $\frac{Q_{be}}{Q_{de}}$ represents the signal noise ratio at the input end of the information demodulator in the receiver, $M_Q = \frac{Q_{be}}{Q_{de}} - \frac{Q_{bi}}{Q_{di}}$ represents the signal interference ratio gain obtained from signal processing before the receiver has no demodulator, and T represents the loss of signal interference ratio after the completion of the gain processing. According to the content shown in Fig. 1, the above Settings are integrated, then:

$$R_j = (\frac{Q_{bi}}{Q_{di}})_{\max} = M_Q - \left[T + (\frac{Q_{bi}}{Q_{di}})_{\min} \right] \tag{1}$$

Therefore, by increasing M_Q and reducing T and $(\frac{Q_{bi}}{Q_{di}})_{\min}$, the interference tolerance of the system is increased, thereby improving the system's anti-interference ability against interference.

3   Electrical Equipment Condition Monitoring Method

The condition monitoring of the equipment is to evaluate the running condition of the equipment through various signal measurement, detection, processing and analysis methods, combined with the history and current situation of the system operation, and display, record and trend analysis of the equipment status, timely processing of abnormal conditions, and provide basic facts and data for the running condition analysis and equipment performance evaluation of the monitored equipment. According to the known structural characteristics, parameters and environmental conditions, and combined with the operation history of the equipment (including operation records and previous failure and maintenance records), the nature, degree, category and position of equipment failure are determined, the relationship between fault, symptom, cause and system is defined, and the development trend of fault is indicated [8, 9].

Therefore, based on the wireless communication framework, this study optimizes the current methods for monitoring the status of electrical equipment. The process of monitoring the status of electrical equipment is divided into three stages, corresponding to signal sorting, evaluation, and status diagnosis. The specific design concept is shown in Fig. 2.

3.1   Electrical Equipment Signal Acquisition and Sorting

Fourier transform frequency domain analysis is one of the most widely used signal analysis methods in the field of equipment operation status monitoring. The occurrence and development of faults usually cause changes in the frequency components of equipment

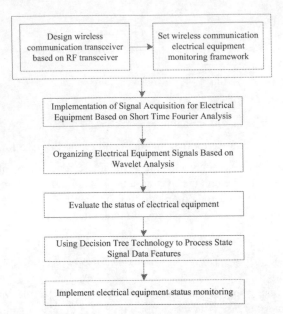

Fig. 2. Schematic diagram of electrical equipment status monitoring methods

vibration signals. The basis of frequency domain analysis is spectrum analysis, and the most commonly used method is Fourier transform, which decomposes complex signals into the sum of finite or infinite spectral components. The definition of Fourier transform is as follows:

If $f(t) \in Z^2(H)$ and $Z^2(H)$ represent square integrable real number spaces, that is, signal spaces with energy roots:

$$f(t)\frac{-\int_{-\infty}^{+\infty} F(\alpha)gd\alpha}{2\pi} \tag{2}$$

The traditional Fourier transform analysis method has made outstanding contributions to the development of signal processing technology. Fourier analysis is a global transformation, although it can connect the characteristics of the signal in the time domain and frequency domain, but only from the time domain and frequency domain observation, and can not combine the two organically. Therefore, this study optimized it and designed a short-term Fourier analysis process.

Short time Fourier analysis is a Time–frequency analysis method, also known as windowed Fourier analysis [10]. Its basic idea is to use the Fourier transform at the same time, before the basis function of the Fourier transform, multiply a time-limited function $k(t)g$ plays the role of frequency limit, $k(t)t$ plays the role of time limit, and then carry out time-domain localization analysis of signals through time-frequency double constraints. The short time Fourier transform is defined as:

$$K(\alpha, \delta) = \int_{-\infty}^{+\infty} k(t - \alpha)gd\alpha \tag{3}$$

After completing this operation, perform wavelet analysis on the signal. From the perspective of basis function, the characteristics of several corner basis (frequency analysis) in Fourier transform and time shifted window function in short-time Fourier transform are absorbed to form the basis function of oscillation and attenuation.

Wavelet analysis method is a time-frequency localization analysis method where the window size is fixed but its shape can be changed, and both the time and frequency windows can be changed. Wavelet transform observes signals at different scales (resolutions) and decomposes them into different frequency bands. It not only provides a comprehensive view of the signal, but also details of the signal. It has multi resolution ability, which means it has higher frequency resolution and lower time resolution in the low-frequency part, and higher time resolution and lower frequency resolution in the high-frequency part. The specific calculation process of wavelet analysis is set as follows:

$$\aleph_{c,\varepsilon}(t) = h^{-\frac{1}{2}} \varpi \left(\frac{t-u}{v} \right) \tag{4}$$

The wavelet transform of signal $x(t)$ is:

$$HT_x(c, l) = h^{-\frac{1}{2}} x(t) \varpi \left(\frac{t-u}{v} \right) dt \tag{5}$$

In the equation, c is referred to as the scale parameter, and l is referred to as the translation parameter. The scale parameter c changes the shape of the continuous wavelet, while the translation parameter l changes the displacement of the continuous wavelet. According to this section, organize the signals and store them in a suitable database for backup.

3.2 Assess the Status of Electrical Equipment

In this study, neural network technology is used to complete the real-time state assessment of electrical equipment. At present, the most commonly used neurons in typical examples of applied neural networks are perceptron and sigmoid unit. Generally speaking, the neural unit composed of linear activation function is usually called perceptron, and the neural unit composed of nonlinear continuous activation function such as S-shape function and bipolar S-shape function is called sigmoid unit. In this study, S-shape function and bipolar S-shape function are selected to complete the training and evaluation of the original signal.

A linear signal combination of a vector of real values is taken as the perceptron input, and some function is applied to calculate the input. If the output vector is larger than the threshold set in advance, the output is 1, otherwise the output is 1. Let a_1 to a_n be the input of a perceptron, then the output calculated by the perceptron is:

$$W(a_1, a_2, ..., a_n) = \begin{cases} 1, & if \ e_0 + e_1 a_1 + e_2 a_2 \mid ... \mid e_n a_n > 0 \\ -1, & otherwise \end{cases} \tag{6}$$

Among them, e_n is the weight value, which is a vector composed of a set of real number constants. It serves as a measure of the contribution of each input a_n to the output of the perceptron, commonly referred to as the contribution rate to the output. $-e_0$ is the pre-set threshold, which compares the output result with the size of the threshold. If the perceptron is to obtain a result with an output of 1, then the weighted sum of all input vectors $e_1a_1 + e_2a_2 + ... + e_na_n$ must be greater than the threshold $-e_0$. The decision plane equation of the perceptron is $\vec{e} * \vec{a} = 0$. Based on this perceptron, build sigmoid unit and set activation function. Based on the signal acquisition results, set the sigmoid unit structure as shown in Fig. 3.

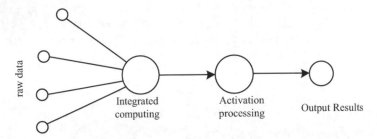

Fig. 3. Schematic diagram of sigmoid unit structure

Combined with the image, the signal training criterion is set. In this study, an error function is defined to determine the gap between the training output and the target output, so as to obtain the most appropriate weight vector. There are many ways to define the error function, and one of the most commonly used and convenient measurement criteria is:

$$U(\vec{e}) = \frac{\sum\limits_{c \in C} (z_c - g_c)^2}{2} \tag{7}$$

Among them, C is the overall set of training samples, z_c is the target output, and g_c is the actual output. The goal of using delta rule is to minimize the above error function. The process of finding the minimum value using the gradient descent search method is to start with an arbitrarily set initial weight value, take small steps along the steepest direction of the surface, continuously modify the weight vector according to the direction of advance, and repeat the above process until the point in the entire space that meets the minimum error is found. Gradient descent algorithms are divided into two types: standard gradient descent and random gradient descent. The idea of the standard gradient descent algorithm is to summarize the errors of all samples before the weight update, that is, to sum multiple samples at each step of the weight update. Its gradient descent update rule is as follows:

$$\triangle e_i = \alpha \sum_{c \in C} (z_c - g_c) a_{ic} \tag{8}$$

The main idea of stochastic gradient descent is to update weights by looking at each training instance. This approach can sometimes avoid falling into local minimums because it uses different derivatives to guide the search. Using the above Settings, the collected signals are trained and evaluated as the preliminary results of condition monitoring.

3.3 Realize Real-Time Monitoring of Electrical Equipment Status

Apply data mining technology to cluster and mine signals in the trained signal database, in order to obtain the final electrical equipment status monitoring results. Using decision tree technology to complete this step based on the characteristics of signal data. Decision trees belong to recursive tree building algorithms, where each leaf point represents a data classification, and each tree point is associated with an attribute with information and gain. The value of this attribute can indicate the branching rules from each leaf point to its child nodes. According to this principle, the information entropy gain of the signal is calculated.

Hypothesis: A represents the total set of training samples, in which the number of training samples is represented by $|A|$; If the classification type has v' different values, then the training sample is divided into v' categories by classification category, each category is represented by V_i', and the number of training samples in each category is represented by $|V_i'|$, then the probability that any training sample belongs to $|V_i'|$ is $P' = \frac{|V_i'|}{|A|}$, and the average information entropy $\eta(A)$ of a given sample classification can be expressed as:

$$\eta(A) = -\sum_{i=1}^{v} P' \lg_2 P_i'$$ (9)

Set attribute V to contain n different values, divide the sample into n subsets, represented by A_n, and each subset has the same value in the sample. Set $|A_{ik}|$ as the number of samples after classification in the subset. The conditional entropy of this data classification can be expressed as:

$$\eta(\frac{A}{\sigma}) = \sum_{k=1}^{n} \left[P_k'(-\sum P_i' \lg_2 P_k') \right]$$ (10)

where, $P_k' = \frac{|A_k|}{|A|}$ and $P_{ik}' = \frac{|A_{ik}|}{|A_k|}$ represent the probability that the sample belongs to the target class. Generally, $\eta(\frac{A}{\sigma})$ and $\eta(A)$ are not equal. Attribute V provides information for classification, and the information entropy value of training sample classification changes. The change of information entropy caused by attribute V is called its information gain $Gain(V)$ for classification, then:

$$Gain(V) = \eta(A) - \eta(\frac{A}{\sigma})$$ (11)

Use formula (11) to calculate the information entropy of the original signal data, and compare it with the historical data to determine the current operating state of electrical equipment. By organizing the content set in the previous text, the design of an electrical equipment status monitoring method based on wireless communication technology has been completed.

4 Experimental Demonstration and Analysis

In recent decades, the sudden failure of power plant equipment has caused great economic losses and casualties, so the fault diagnosis technology of power plant equipment is very necessary. Fault diagnosis is to analyze the real-time monitoring data of the unit and classify them, whether they belong to the fault state or normal state, and if they belong to the fault state, which kind of fault they belong to and so on. Thus, the problem of fault diagnosis is transformed into the problem of classifying monitoring data.

4.1 Equipment Data Samples

In the actual work of this research, a transformer fault sample was collected and sorted, which contains 10 sets of fault data, five typical faults: general overheating fault (fault 1), serious overheating fault (fault 2), partial discharge fault (fault 3), spark discharge fault (fault 4), arc discharge fault (fault 5). Since the actual measured gas content is a continuous value, in order to apply the decision tree method, we first need to discretization the data. For the convenience of calculation, it is divided into three levels based on its size: not high [0-5 ppm], high [0-5 ppm], and high ([0-5 ppm]). Based on the above settings, the device data sets used in this study were set.

Table 2. Equipment data sample

Signal group number	Total signal volume	Category of abnormal signals contained	Contains abnormal signal level
A1	10000	Fault 1	High
A2	10000	Fault 4, Fault 3	Higher
A3	10000	Fault 3	Not high
A4	10000	Fault 2	High
A5	10000	Fault 1, Fault 3, Fault 4	Not high
A6	10000	Fault 2	Higher
A7	10000	Fault 5, Fault 2	Not high
A8	10000	Fault 3, Fault 2, Fault 4	High
A9	10000	Fault 5, Fault 3, Fault 1	Not high
A10	10000	Fault 2	Not high

In this experiment, the data in Table 2 will be applied to analyze the application effects of the methods, basic methods (based on power hardware monitoring methods), and machine learning methods in this paper. In order to obtain more realistic experimental results, the experimental environment was set up in two parts: the laboratory and the factory. The specific experimental operation process and results are shown in the following text.

4.2 Abnormal State Recognition Rate Test Analysis

According to the preset experimental scheme, the following experimental results were obtained:

Table 3. Abnormal state recognition rate (unit: %)

Signal group number	Textual method		Basic method		Machine learning method	
Experimental environment	Laboratory	Factory	Laboratory	Factory	Laboratory	Factory
A1	90.39	90.53	90.77	80.7	90.83	85.12
A2	90.51	90.72	90.63	80.2	90.57	86.91
A3	90.3	90.26	90.8	81.19	90.94	86.58
A4	90.91	90.61	90.7	80.98	90.13	85.25
A5	90.01	90.15	90.79	81.34	90.48	85.29
A6	90.05	90.24	90.42	80.48	90.54	86.15
A7	90.73	90.17	90.58	80.84	90.42	85.64
A8	90.67	90.34	90.12	80.78	90.25	86.73
A9	90.66	90.67	90.44	80.81	90.2	85.73
A10	90.18	90.16	90.88	81.83	90.17	85.97

By analyzing the data in Table 3, it can be seen that the recognition abilities of the three methods are different in different experimental environments. Taking the laboratory environment as an example, in this environment, the abnormal state recognition ability of the three methods is roughly the same, and the whole is close, indicating that under the premise of no interference, the three methods have relatively close recognition results, and the overall recognition level is high. In the experimental environment of the factory, the recognition ability of the three methods changed greatly. The recognition rate of the proposed method does not fluctuate, while that of the other two methods decreases obviously. Based on the above experimental results, it can be determined that the proposed method has a strong ability of state recognition of electrical equipment.

4.3 Test and Analysis of Abnormal State Monitoring Accuracy

Based on the experimental results of the previous group, the accuracy of abnormal state monitoring for different methods is calculated using the following formula:

$$\partial = \frac{U'_i}{U'_{all}} * 100\% \tag{12}$$

where, U'_{all} represents all abnormal state data; U'_i indicates that abnormal status data is detected. According to this formula, the following experimental results are obtained.

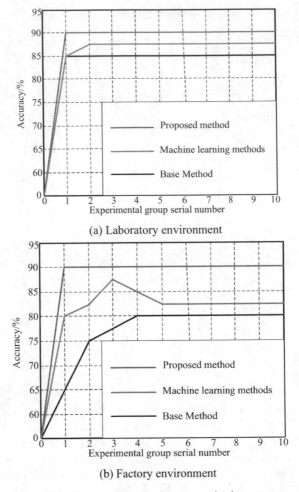

(a) Laboratory environment

(b) Factory environment

Fig. 4. Test results of abnormal state monitoring accuracy

By analyzing the experimental results in Fig. 4, it can be determined that in two different experimental environments, the method proposed in this paper has high accuracy in monitoring abnormal states of electrical equipment and will not change due to changes in the experimental environment. Compared with the methods in this article, the other two methods cannot obtain accurate monitoring results after application. In summary, the application effect of this method is better.

4.4   Time Consuming Test of Electrical Equipment Condition Monitoring

The test results of electrical equipment status monitoring time consumption are shown in Fig. 5.

Analyzing the experimental results in Fig. 5, it can be seen that under different experimental environments, the method proposed in this paper can obtain the operating

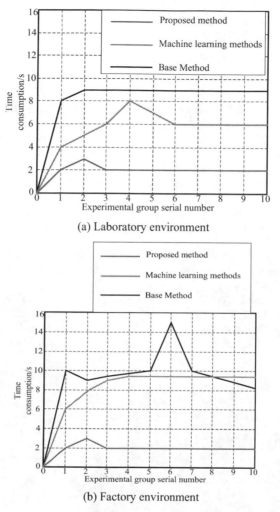

(a) Laboratory environment

(b) Factory environment

Fig. 5. Test results of electrical equipment status monitoring time consumption

data of electrical equipment in the shortest possible time, while the other two methods take relatively longer time. And when the experimental environment is a factory environment, the other two methods significantly increase the time consumption. From the above experimental results, it can be determined that the application effect of this method is better.

5   Conclusion

Aiming at the shortcomings of current electrical equipment condition monitoring methods in application, this study proposes a method of electrical equipment condition monitoring based on wireless communication technology, and verifies it through experimental demonstration. The experimental results show that this method has high application value, and it needs to be measured on a larger scale in the future research to ensure that this method has a wide range of application.

References

1. Wang, Y.: A new type of FBG sensor for condition monitoring of electrical equipment. Laser & Infrared **52**(1), 51–56 (2022)
2. Liu, J., Zhang, L.: State detection of electrical equipment based on semiconductor laser interference technology. Laser J. **42**(5), 52–56 (2021)
3. Deng, K., Hou, X., Lin, X., et al.: Online evaluation scheme for the performance of an electrical equipment monitoring terminal based on 5G communication. Power Syst. Protect. Control **49**(7), 39–47 (2021)
4. Yang, F., Li, Y., Lu, W., et al.: Ultra-high frequency signal monitoring method for partial discharge process of high-voltage electrical equipment. Adv. Power Syst. Hydroelectric Eng. **38**(11), 55–60+70 (2022)
5. Gao, C., Zhao, Y., Wang, D., et al.: Research status and prospect of condition based maintenance technology for offshore wind turbine electrical equipment. Trans. China Electrotech. Soc. **37**(z1), 30–42 (2022)
6. Lai, G., Gao, J., Luo, L.: Switching device status on-line monitoring of intelligent distribution power network. Power Electr. **55**(12), 60–64 (2021)
7. Xu, Z., Zheng, S., Kang, B., et al.: Overheat fault identification method for electrical equipment based on three-phase self-searching comparison method. Infrared Technol. **43**(11), 1112–1118 (2021)
8. Lai, C., Zhong, Q., Zhao, Y., et al.: Risk Assessment method for electric shock of electrical equipment in low-voltage DC power supply system. Autom. Electric Power Syst. **45**(19), 141–147 (2021)
9. Mu, J., Mu, J., Zou, J., et al.: Design of low-power self-powered sensor system to monitor the state of mechanical equipment. J. North Univ. China (Nat. Sci. Edn) **42**(4), 355–360 (2021)
10. Jiang, Y., Liu, Z., Wang, W., et al.: Research on multi-state monitoring system of substation equipment based on edge-cloud collaboration. Power System Protection and Control **49**(6), 138–144 (2021)

The Application and Research of Intelligent Mobile Terminal in Mixed Listening and Speaking Teaching of College English

Bo Jiang(✉)

Navigation Technology Department, Tianjin Maritime College, Tianjin 300000, China
LX190625@126.com

Abstract. With the rapid development of mobile technology, the coverage of Wlan, 3G and 4G networks is expanding day by day, and intelligent mobile terminal assisted English teaching and learning has become a hot research field. This study explores the application of intelligent mobile terminals in mixed listening and speaking teaching of college English from three aspects. The first aspect analyzes the application of mobile terminals in the collection of listening and speaking teaching resources. The second aspect analyzes the application of mobile terminals in the recommendation of listening and speaking teaching resources. The third aspect analyzes the application of intelligent mobile terminals in listening and speaking teaching scoring: intelligent mobile terminals extract the relevant features of students' input voice, and use SVR to give students' listening and speaking practice scores, which are presented on the mobile terminal learning page. The results show that the average absolute error is less than 1, indicating that the application of intelligent mobile terminals in the recommendation of college English mixed listening and speaking teaching resources is better. The correlation degree is more than 0.5, which indicates that the accuracy of the evaluation results is high, and the resource recommendation time is always below 80ms, proves the application effect of intelligent mobile terminals in college English mixed listening and speaking teaching.

Keywords: Intelligent Mobile Terminal · College English · Mixed Listening and Speaking Teaching · Application Analysis

1 Introduction

In the "College English Curriculum Requirements" revised by the Ministry of Education in July 2007, the teaching goal of college English is set to "cultivate students' comprehensive English application ability, especially listening and speaking ability". At the same time, it also points out that "colleges and universities should be supported by modern information technology, especially network technology, so that English teaching and learning can be developed towards personalized and self-help learning without being limited by time and place to a certain extent." College English teaching should pay more

L. Yun et al. (Eds.): ADHIP 2023, LNICST 548, pp. 277–294, 2024.
https://doi.org/10.1007/978-3-031-50546-1_19

attention to developing students' listening and speaking skills, At the same time, the ability and habit of autonomous learning in listening and speaking should be cultivated. At present, college English listening and speaking teaching in domestic colleges and universities generally faces the following problems: large class teaching (40–50 class), part of the content in listening and speaking textbooks is disconnected from social development, multimedia language classrooms and other teaching facilities are insufficient, and hardware and software equipment are aging and not updated in time. Influenced by these factors, there are few opportunities for students to really exercise their listening and speaking ability in the listening and speaking class and they can't get feedback from teachers in time, which leads to the decline of students' learning enthusiasm, let alone the cultivation of their habit and ability of autonomous learning. The emergence and popularization of mobile technology has provided favorable conditions for English listening and speaking teaching: mobile devices such as smart phones can make students no longer limited to a fixed learning time, space and mode, and can choose appropriate ways to learn languages at any time and place.

Under the above background, this research will integrate the application of mobile technology after class with the teaching of English listening and speaking in class, and try to build a mixed teaching model of college English listening and speaking courses based on mobile technology, supplemented by a teaching record model that combines the whole process assessment and electronic teaching files, in order to further cultivate students' autonomous learning ability and improve their listening and speaking ability.

2   The Application of Mobile Terminal in the Collection of Listening and Speaking Teaching Resources

For a long time, the classroom teaching of college English for non English majors in China has adopted the lecturing intensive reading mode, ignoring the cultivation of college students' English listening and speaking ability. College students often fail to achieve good results when they really need to communicate in English in their future work and life. Today, with the rapid development of information technology, the deep integration of information technology and English courses has become the core of the current college English teaching reform. How to build an English listening and speaking teaching model that conforms to the characteristics of the subject and the learning rules, and how to cultivate students' English communication ability, has become a major issue facing the current.

The rapid development of mobile terminal technology provides a new opportunity for the reform of college English listening and speaking teaching mode. Mobile terminal devices include smart phones, laptops, tablets, on-board smart terminals, wearable devices and other specific forms. In terms of technology and function realization, mobile terminals have multimedia functions such as audio and video, and intelligent tools supporting data transmission and processing capabilities. It can access the Internet to browse and download information, as well as submit data and interact with roles. At the same time, the mobile terminal is a good helper for learning. It can be equipped with a visual operating system, and can install customized learning software and intelligent companion for various applications. Mobile terminal technology supports learners to use mobile devices for anytime and anywhere learning. Mobile technology assisted language learning has incomparable advantages in expanding learning time and space, enriching learning interaction, improving learning efficiency, etc. Blending Learning has been studied at home and abroad for a long time. Margret believes that blended learning is the combination or mixing of network technology based schools to achieve a certain teaching goal. It is the combination of multiple teaching methods and teaching technologies to achieve the best teaching results together. It is the combination of teaching technology and specific teaching classes. This paper studies how to effectively use intelligent mobile terminal applications to carry out English listening and speaking teaching. The research is divided into three parts, namely, the application of mobile terminals in the collection of listening and speaking teaching resources, the application of mobile terminals in the recommendation of listening and speaking teaching resources, and the application of intelligent mobile terminals in listening and speaking scoring.

With the rapid development of information technology, computer network has been used more and more in various teaching processes. Among them, the construction of online teaching resources has attracted more and more attention. For example, the gradual development of teaching resource database, teaching website and online course construction has become one of the core contents of education informatization. Therefore, a large number of relevant college English mixed listening and speaking teaching resources can be obtained on the network by using the general search engine of mobile terminals (see Fig. 1).First, we should determine the target teaching content, and then use the general search engine of intelligent mobile terminal to obtain the listening and speaking teaching resources that have a certain degree of relevance to the target teaching content on the network; On this basis, the teaching resources obtained on the network are optimized and sorted, high-quality teaching resources with high relevance and quality to the target teaching content are selected, and a listening and speaking teaching resource database is constructed to facilitate users to query and access at any time[1].

The core of resource database design is to solve the problem of database classification management. According to the classification method of teaching resources of the Ministry of Education, listening and speaking teaching resources are divided into five types according to the types of documents. It covers text, multimedia and file resources. The specific classification is shown in Fig. 2.

Due to the variety of resources, the commonness of resources can be extracted from many resources, and the data structure of resources can be analyzed for resource storage. Therefore, according to the above five types of resources, we abstract two types of data for management, one is text information, the other is file information, and store them in the cloud storage space of intelligent mobile terminals. The connotation of cloud storage is storage virtualization and storage automation. Through cluster application, grid technology or distributed file system and other functions, a large number of different types of storage devices in the network are gathered together to work together through application software to jointly provide external data storage and business access functions. In other words, cloud storage is no longer storage but a service. Its core is to combine application software and storage devices, and realize the transformation of storage devices to storage services through application software. In general, cloud storage is a new concept extended and developed from the concept of cloud computing. Cloud

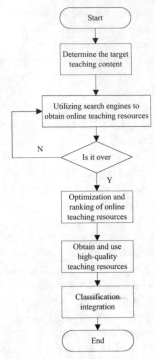

Fig. 1. Collection process of listening and speaking teaching resources based on intelligent mobile terminal general search engine

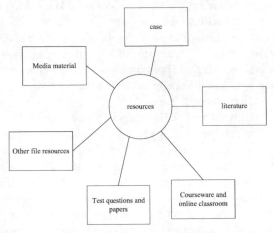

Fig. 2. Classification of College English Mixed Listening and Speaking Teaching Resources

storage is to store resources in a secure large-scale storage server through the network. Wherever you go and use any intelligent mobile terminal, you can access the listening, speaking and teaching resources[2]as long as you can connect to the storage server. The data structure of college English hybrid listening and speaking resources stored in the cloud of intelligent mobile terminal is shown in Fig. 3 below.

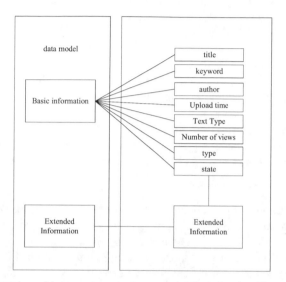

Fig. 3. Listening and speaking resource storage structure based on intelligent mobile terminal

The node consists of two parts. The first part is the basic information part. Each resource has its own basic information to facilitate the administrator's query and statistics. The basic information includes title, keyword, author, upload time, text type, number of views, type, and status. The second part is the extension information. The secondary extension information is divided into a variety of defined types, including text and file class structures. At the same time, the user-defined node type is reserved to prepare for future resource expansion. For text resources, information is directly stored in the defined data structure as an entity, sealed in data objects, and for file type resources, we use the form of storing file data objects, including the original name, current name, size, and physical address of the file.

3   Application of Mobile Terminal in Recommendation of Listening and Speaking Teaching Resources

Personalized recommendation is an important branch of the application of intelligent mobile terminals in college English mixed listening and speaking teaching. It analyzes the user's behavior characteristics, the context of mobile devices, social network relationships and other information through the recommendation program in mobile terminals, predicts applications that users may be interested in, helps users filter resources in advance, and tries to find applications that match user preferences [3]. The recommendation program in mobile terminal is developed based on the collaborative filtering algorithm. Collaborative filtering algorithm is a commonly used recommendation algorithm. It first calculates the similarity between users, and then recommends items to target users according to the preferences of similar users, so as to complete personalized recommendation, which has certain practicability and effect. Collaborative filtering recommendation algorithms are divided into BaseItemCF (project-based system filtering algorithm) and BaseUserCF (user based collaborative filtering algorithm). These two algorithms calculate user similarity and item similarity, respectively. BaseItemCF is to recommend similar items of interest to users BaseUserCF is to recommend similar items of interest to users. In the collaborative filtering technology, the following assumption is true: if user A and user B have similar interests, then the items that are of interest to Party A may also be of interest to Party B.

The collaborative filtering algorithm is to apply this idea in life to the recommendation system [4]. The application process of mobile terminal recommendation program in listening and speaking teaching resource recommendation is as follows:

(1) Student interest expression

The common interests of students are the basis for recommendation. Therefore, in collaborative filtering algorithms, data processing is mainly based on students' scores of teaching resources rather than content.

Assume that the number of students is m, the number of teaching resources is n, Scoring recommendation matrix A is a $m \times n$ the matrix of i line No j elements of columns a_{ij} indicates a student i for teaching resources j the scoring result of. The design of student teaching resource scoring matrix is shown in Table 1.

Table 1. Scoring matrix of student teaching resources

Students/Resources	1	2	...	n
1	a11	a12	...	a1n
2	a21	a22	...	a2n
...	...	...	...	...
m	am1	am2	...	amn

In the student teaching resource scoring matrix as shown in Formula (1), a_{ij} it is a score of 1–5, representing students i for teaching resources j the scoring result of. a_{ij} the higher the value of, the higher the student's rating of the teaching resource.

$$A = \begin{bmatrix} a_{11} & a_{12} & \cdots & a_{1n} \\ a_{21} & a_{22} & \cdots & a_{2n} \\ \vdots & \vdots & \ddots & \vdots \\ a_{m1} & a_{m2} & \cdots & a_{mm} \end{bmatrix} \tag{1}$$

(2) Nearest neighbor set selection

The generation of "nearest neighbor set" is to calculate the similarity between students' scores of teaching resources, and establish the nearest neighbor set of students' interests, which is the core of implementing collaborative filtering based recommendation algorithm.The Person formula is mainly used to calculate the relevant similarity of two students i and k similarity between $S_{(i,k)}$ the calculation formula is shown in Formula (2) [5].

$$S_{(i,k)} = \frac{\sum\limits_{i \in C_{(i,k)}} (a_{ij} - \overline{a}_i)(a_{kj} - \overline{a}_k)}{\sqrt{\sum\limits_{i \in C_{(i,k)}} (a_{ij} - \overline{a}_i)^2} \sqrt{\sum\limits_{k \in C_{(i,k)}} (a_{kj} - \overline{a}_k)^2}} \tag{2}$$

where, $C_{(i,k)}$ on behalf of students i and k common scoring item set of; a_{ij}、a_{kj} on behalf of students i and k listening and speaking teaching resources j scores; $\overline{a}_i$、$\overline{a}_k$ on behalf of students i and k the average score of.

(3) Generate recommendation results

According to the set of students' "favorite neighbors" realized in the previous step, recommend teaching resources to students from the following two aspects:

① Predict students' preference for all teaching resources;

② According to the students' preference for teaching resources, the teaching resources are sorted and N teaching resources that students most like are recommended, namely Top-N recommendation set.

Hypothetical students i the collection of assessed listening and speaking teaching resources is B_i for any teaching resources b does not belong to B_i the formula of the predictive value of the preference degree of is shown in Formula (3).

$$D = \frac{\vec{i} \sum\limits_{k=1}^{K} \sin k + \sum\limits_{k=1}^{K} \sin k (courk - \vec{u})}{\sum\limits_{k=1}^{K} \sin k} \tag{3}$$

Among them, $\vec{i}$ indicates a student i the average score of the assessed set of listening and speaking teaching resources, k for students i students of the nearest neighbor set, $\sin k$ indicates a student k and i similarity, $\vec{u}$ for students i for listening and speaking teaching resources b the average recommended score of.

Based on the preference prediction formula as shown in Formula (3), grade all teaching resources' preference, rank the scoring results, and select the highest recommendation set for recommendation[6].

4 Application of Intelligent Mobile Terminal in Listening and Speaking Teaching Scoring

The task of this chapter is to effectively apply the recommended high-quality teaching resources to the process of English listening and speaking teaching. The purpose of English listening and speaking teaching is to train students' listening and speaking ability. Listening and speaking teaching scoring is a technology that students listen to the specified text, then follow or translate into the target language simultaneously, and finally the intelligent mobile terminal feeds back scores according to students' practice results. Its goal is to endow the intelligent mobile terminal with the ability to act as a virtual teacher, conduct a fair, objective and efficient evaluation of students' listening and speaking practice, and alleviate the serious shortage of professional listening and speaking teachers. In learning, it can help students better understand the pronunciation level, improve the efficiency of listening and speaking learning and promote self-study; In examinations, it can assist or replace manual marking of listening and speaking tests, greatly improving the efficiency and quality of marking. The core of the application of intelligent mobile terminal in listening and speaking teaching scoring is to use the relevant algorithms of the voice automatic recognition module in the terminal. The specific application process is as follows:

4.1 Resource Retrieval Module

The intelligent mobile terminal has the functions of course classification, course resource navigation and course index, which provides students with recommended courses of different levels and popularity. Students can learn specific courses according to their own application level and learning needs to improve their sense of interaction with the system; You can also enter the corresponding text through the search box to conduct fuzzy query, presenting the most appropriate and personalized course search results. See Chapter 2 for the specific process.

4.2 Speech Processing

The generation and perception of speech depend on human voice system and auditory system. The speaker first generates voice information in his mind, and then packages the information in the form of rhythm, loudness, pitch cycle rise and fall, that is, language coding operation. After coding, the speaker sends out sound through the cooperation of the vocal organs, and then transmits the voice signal to the listener's ear through the sound wave as the medium. Its auditory system transmits the processed signal to the brain center and converts it into language coding, thus generating semantic information [7]. Speech processing mainly includes two aspects: one is processing speech signals, such as preprocessing speech signals to eliminate most of the useless information; On the other hand, the speech signal is analyzed and the feature parameters are extracted for subsequent learning. It mainly includes the following three aspects:

(1) Preemphasis

Since the energy loss caused by lip radiation is concentrated in the high frequency part, it is necessary to emphasize the high frequency part of speech to make the spectrum smooth. The "pre emphasis technology" is to pass the sampled signal through a FIR high pass filter, and its transfer function is as follows:

$$f(d) = 1 - \beta d^{-1} \tag{4}$$

Among them, β is the pre weighting coefficient, usually $0.9 < \beta < 1.0$. if t the input signal at the moment is $s(t)$, the output signal after pre emphasis $y(t)$ for:

$$y(t) = s(t) - \beta s(t-1) \tag{5}$$

(2) Framing windowing

Speech signal is a typical time-varying signal. It is difficult to study a long segment of speech signal, but in reality, when people speak, the movement of the mouth and throat is a continuous action, and the speed is not fast. According to this characteristic, a long speech signal is usually divided into several short segments using differential thinking for research, and these short segments are called "analysis frames" [8].Although there can be no overlap between frames when framing, this may cause the calculated pitch to jump. In order to prevent jumping and make the voice signal after framing more stable, it is necessary to overlap a part between the two analysis frames. This part is called frame shift. The frame shift should not be too long, generally less than 1/2 of the frame length. The current frame length is 256 and the frame shift is 80.Some window functions are needed for framing. The commonly used window functions are rectangular window and Hamming window. The voice signal can be segmented by moving the window for weighting. The window function selected in this section is Hamming window. The following is the introduction of rectangular window and Hamming window.

Window function of rectangular window $p(l)$ for:

$$p(l) = \begin{cases} 1, 0 \leq l \leq L-1 \\ 0, \quad otherwise \end{cases} \tag{6}$$

where, L indicates the window length; l represents a voice signal frame.

The window function of Hamming window is:

$$p(l) = \begin{cases} 0.54 - 0.46 \cos \frac{2\pi l}{L-1}, 0 \leq l \leq L-1 \\ 0, \qquad\qquad\qquad otherwise \end{cases} \tag{7}$$

(3) Endpoint detection

Generally, in order to ensure the integrity of voice information, an intelligent mobile terminal will leave a blank voice segment when recording voice signals. Therefore, the process of endpoint detection is to detect the information of voice segments and eliminate noise segments, so as to determine the starting and ending points of effective voice information, so as to improve the accuracy of subsequent operations [9]. In this section, the double threshold comparison method is used for endpoint detection. Two performance indicators are needed: short-time energy and short-time zero crossing rate. They are introduced below.

In reality, people's vocal organs are inertial, so the state of voice signals will not change abruptly, and the energy contained in voice signals is different. For example, the energy contained in voiceless and voiced sounds is obviously different, so short-term energy can be used to express the personality characteristics of voice signals. The extraction formula of short-term energy is as follows:

$$E(l) = \sum_{t=1}^{\infty} \left[s(t)p(l-t) \right]^2 \tag{8}$$

Among them, $E(l)$ it represents the No l frame voice signal $s(t)$ short-term energy.

In the time domain diagram of the voice signal, if the voice signal is continuous, when the waveform crosses the time axis, it indicates that zero crossing has occurred; If it is a discrete voice signal, it is necessary to find adjacent sampling points. If one voice signal is positive and the other is negative, it is also considered that zero crossing has occurred. Short time zero crossing rate is defined as:

$$e(l) = \sum_{-\infty}^{\infty} |sgns(t) - sgns(t-1)|p(l-t)$$
$$= |sgns(t) - sgns(l-1)|p(l) \tag{9}$$

Among them, sgn[] is a pseudo symbolic function whose expression is:

$$sgn[s(l)] = \begin{cases} 1, s(l) \geq 0 \\ -1, otherwise \end{cases} \tag{10}$$

Among them, $p(l)$ is a window function, l is the window length. $p(l)$ the expression is:

$$p(l) = \begin{cases} \dfrac{1}{2L}, 0 \leq l \leq L-1 \\ 0, otherwise \end{cases} \tag{11}$$

The double threshold method first sets two thresholds for short-term energy and zero crossing rate, so that endpoint detection can be divided into four stages:

(1) Mute stage: if one of the energy and zero crossing rate is below the low threshold, it is the mute stage;
(2) Transition section: if both parameters exceed the low threshold but none of them enter the high threshold, the transition section will be entered;
(3) Voice segment: if either of the two parameters exceeds the high threshold, the voice segment will be entered;
(4) End segment: if both the energy and zero crossing rate parameters are reduced below the low threshold, and the minimum time threshold is greater than the total time length, it will be marked as a noise signal, and then continue scanning. If the minimum time threshold is less than the total time length, it will be marked as the end point.

4.3 Voice Scoring

When students complete the English listening or voice following test, the intelligent mobile terminal will immediately extract the relevant features of the students' input voice and the corresponding features of the standard voice, and give the score of the students' input voice according to certain scoring rules, which will be displayed on the learning page. This module allows students to understand their own pronunciation level, and constantly improve their English listening and speaking level according to this scoring feedback.

(1) Intelligent mobile terminal extracts voice features

Feature extraction is a very important step before machine learning, which determines the credibility and accuracy of the scoring model. As long as the feature selection is accurate enough, even if the SVR model is not optimal, a scoring result with small error can still be obtained. SVR is a Nonlinear regression algorithm, which can handle the feature extraction task of nonlinear relationship. The model established by SVR has good anti noise performance, can accurately extract features from noisy data, and improves the accuracy of feature extraction. In the scoring model based on SVR studied in this paper, two types of features are mainly extracted: voice features and text features. Speech features are extracted directly from speech signals, and text features are extracted from the output of speech recognition engine of intelligent mobile terminal. Before using relevant technologies to extract features, feature screening[10]is often required. This article refers to the following guidelines when screening features:

① The importance of each feature can be measured by calculating its Pearson correlation coefficient with the manual score. Generally, features with a correlation coefficient lower than 0.2 should not be selected;

② The intelligent scoring system should describe the examinee's spoken language from multiple dimensions. Features with high similarity should not be included at the same time, so we calculate the correlation between the selected features. For each pair of features with a correlation coefficient greater than 0.9, delete one of them;

③ Since this paper is aimed at the mixed listening and speaking teaching score of college English, there is usually no fixed reference answer to this kind of question, so when making feature selection, this paper mainly selects universal features rather than features strongly related to the answers of the reference text, in addition to semantic similarity features. Table 2 shows a brief description of the features finally selected for use in this article.

In this paper, four phonetic features are extracted to evaluate the pronunciation quality, fluency and content richness of candidates' spoken English. Speech speed is mainly used to describe oral fluency, which can be calculated by the following formula:

$$V = \frac{N}{T - T'} \tag{12}$$

where, V represents speaking speed; N it represents the total number of words in students' spoken language, T it indicates the total duration of oral recording, T' indicates the mute duration in the recording. In addition to the speed characteristics, the number of silences in the recording can also reflect the oral fluency of the tester to some extent.

Table 2. Speech Features

Feature category	Feature Name
Phonetic features	articulationRate
	numSilence
	posteriorScore
	speakingRatio
Text-based features	eassyLength
	uniqueWords
	parseTreeDepth
	semanticSimilarity
	goodGrammerRatio

In terms of pronunciation quality evaluation [11], the posterior probability feature of pronunciation is used by many oral scoring systems. This paper also uses this feature to describe the accuracy of examinees' pronunciation. In addition, the time ratio of extracting pronunciation can also reflect the richness of oral content to a certain extent.

$$H = \frac{T - T'}{T} \tag{13}$$

where, H stands for pronunciation time ratio.

In the traditional oral evaluation for reading aloud questions, the standard oral sequence corresponding to the reference text is usually used as the label to force alignment the test speech, and then the average posterior probability of each phoneme is calculated through the classic GOP (Goodness of Pronunciation) algorithm. However, there is no reference text in the open oral scoring, so it is necessary to combine the speech recognition engine and an acoustic model trained with standard English pronunciation to calculate the average posterior probability as the pronunciation quality feature.

At the text level, grammar is the most basic criterion to distinguish the language proficiency of examinees. We use part of speech tags to determine whether there are grammatical problems in the use of words in spoken content. The famous English original novels generally do not have grammatical errors. This paper uses the method in the open source composition scoring system (EASE) Enhanced AI Scanning Engine) for reference to extract ternary tags and quaternary tag combinations from Sherlock Holmes' novel collections after tagging sentence, word and part of speech tags, and store the extracted results locally as a tag combination query database. In feature extraction, we will extract three tag combinations and four tag combinations of each sentence from speech recognition text. For each combination, we query in the tag combination library. If we cannot find it, it is considered that there is a syntax error. We use the following formula to calculate the text syntax accuracy μ (ς is the total number of ternary label combinations and quaternary label combinations contained in the text, ϖ indicates the total number of correct label combinations).

$$\mu = \frac{\varsigma}{\varpi} \tag{14}$$

SVR cannot directly recognize text data and audio data, so it cannot directly use speech recognition text as the input data of scoring model. We need to convert the above data into the numerical tensor form that SVR can handle. Text vectorization refers to the process of transforming text into numerical tensor. This paper will use word embedding technology to vectorize speech recognition text. This paper uses the pre trained word embedding model G1oVe to transform speech recognition text into vector representation. The whole text vectorization process: first, through data cleaning of speech recognition text, onomatopoeia and repeated words in the text due to recognition errors are removed. Then the cleaned text is segmented. Then, the embedded text file of G1oVe words is parsed. The file is a. txt file. Each line word string in the file and its corresponding vector representation. Finally, we build a word embedding matrix that can be loaded into SVR.

(2) Intelligent mobile terminal evaluates listening and speaking quality

Building an appropriate and efficient voice acoustic model is the last stage of the speech recognition system, which has a significant impact on the performance of the system. An ideal speech recognition network should have strong generalization ability and learning ability of sample features. It can learn a large number of training samples, so as to mine the corresponding relationship between various speech feature parameters and speech semantic information, and achieve accurate classification of test samples.

Support Vector Regression (SVR) algorithm is a machine learning algorithm based on structural risk minimization criteria. It makes full use of the advantages of machine learning, and can learn complex data patterns with only limited training samples, thus mapping feature scores to target scores. Therefore, this paper uses SVR algorithm as regression model to achieve effective fusion of multidimensional evaluation features. It is introduced below. Given training data set $\{(x_1, y_1), (x_2, y_2), \ldots, (x_m, y_m)\}$, where, $x_i \in X^n$ indicates that from the i extracted from segment reading voice n dimensional eigenvector, $y_i \in Y$ yes x_i corresponding manual scoring, m is the total number of samples in the training data set. Our goal is to train all sample pairs in the dataset (x_i, y_i), find a regression function that is as flat as possible $y = F(x)$ to approach the relationship between them and minimize the prediction error.

For non-linear data x because it is difficult to be linearly separable in the original space, the SVR algorithm uses a nonlinear function to solve this problem $R(x)$, will x map to high-dimensional feature space for processing. The regression function is defined as:

$$F(x) = \langle v, R(x) \rangle + r \tag{15}$$

Among them, v is the weight vector, r is offset $\langle , \rangle$ it is an inner product operation.

The most widely used SVR algorithm is ε Type of insensitive loss function, defined as:

$$q[F(x) - y] = \begin{cases} 0, |F(x) - y| < \varepsilon \\ |F(x) - y| - \varepsilon, ^{otherwiswe} \end{cases} \tag{16}$$

where, positive number ε is the preset error, when the regression function $F(x)$ for the actual target value: ε within the range, i.e. ε in the insensitive zone, the loss is recorded as 0.To measure ε the deviation on both sides of the insensitive zone defines two relaxation variables, namely g_i, h_i, the objective function is:

$$\min \frac{\|v\|^2}{2} + J \sum_{i=1}^{m} (g_i + h_i) \tag{17}$$

The following constraints are met:

$$\begin{cases} y_i - F(x_i) \leq \varepsilon + g_i \\ F(x_i) - y_i \leq \varepsilon + h_i \\ g_i, h_i \geq 0 \end{cases} \tag{18}$$

In the formula, constant J is the penalty coefficient for prediction errors. It can be seen from Formula (16) and Formula (17) that this is an optimization problem, so Lagrange function is constructed. Main variables of Lagrange function $v, r \sim g_i, h_i$ calculate the partial derivative in turn, and make its value 0.The result of partial derivation is substituted into Lagrange function and transformed into dual optimization problem. Finally, the SVR regression function obtained by solving is:

$$F(x) = \sum_{i=1}^{m} (\phi_i + \phi_i') K(x_i, x) + r \tag{19}$$

Among them, $K(x_i, x)$ is a kernel function; $\phi_i \sim \phi_i'$ are two Lagrange multipliers.

Finally, the pronunciation quality evaluation process based on support vector regression algorithm is as follows:

1) Based on students' spoken English pronunciation data, two types of features are extracted respectively, and feature scores are calculated;
2) The cubic polynomial function is used to normalize each feature score calculated to make it consistent with the range of manual scoring;
3) Construct SVR training sample set with multi-dimensional evaluation feature score as input and manual score as output;
4) Training parameters in SVR scoring model;
5) Use the same method to extract the evaluation features of each dimension of the pronunciation to be tested, and then use the trained SVR scoring model to fuse the features, so as to achieve an effective evaluation of the overall pronunciation quality of students.

5 Application Test

5.1 Experimental Configuration

(1) Front end pretreatment configuration

The pre emphasis coefficient used in pre emphasis is 0.97. The Hamming window with a window length of 25 ms is used to smooth the voice frame signal. The duration of each voice signal frame is 25 ms, and the overlap between adjacent voice frames is 15 ms.

(2) Knowledge base configuration

The acoustic model uses a context independent monophone model, and the HMM model corresponding to each phoneme consists of three emission states from left to right. The probability distribution on the 1V} 'CC acoustic eigenvector associated with the HMM state is simulated using GMM containing eight Gaussian components. The language model uses the ternary language model, and the pronunciation dictionary uses the CMU pronunciation dictionary of Carnegie Mellon University (CMU).

(3) Speech recognition engine

This paper is based on the open source intelligent mobile terminal of CML University to develop an assessment model of English pronunciation quality suitable for Chinese students.

5.2 Recommended Test of Listening and Speaking Teaching Resources

Two data sets are used to verify the effectiveness of smart mobile terminals in recommending college English mixed listening and speaking teaching resources. The quality of recommendation is measured by its prediction results, mainly by measuring the accuracy between the system's recommendation results and users' real scores. There are many existing evaluation strategies, among which the average absolute error MAE is easy to understand and easy to calculate, which is the most widely used measurement standard. Therefore, the average absolute error is recommended as the application effect. The results are shown in Table 3 below,

Table 3. Average Absolute Error

Data set	1	2
mean absolute error	0.54	0.27

It can be seen from Table 3 that the average absolute error is less than 1, which indicates that the application effect of intelligent mobile terminals in the recommendation of college English mixed listening and speaking teaching resources is good.

5.3 Listening and Speaking Quality Evaluation Test

The sentences with balanced phoneme coverage in the CMU ARCTIC corpus are selected as the reading text corpus. Every five sentences form a reading passage with a length of about 50 words. Ten students from Guilin University of Electronic Science and Technology are invited to read these passages at a normal speed, and the pronunciation is as clear as possible. Finally, 10 pieces of reading speech data are recorded, Save as 16 kHz sampling rate and 16bit mono WAV format. Invite English teachers from Foreign Languages Institute to give a full score of 100 points to the overall pronunciation quality of these voice data in terms of two types of characteristics. When evaluating pronunciation quality, the manual scoring is usually taken as the reference standard, and the system performance is evaluated by measuring the correlation between machine scoring and manual scoring. The results are shown in Fig. 4 below.

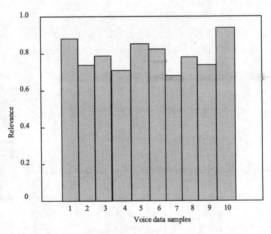

Fig. 4. Correlation

As can be seen from Fig. 4, the correlation degree exceeds 0.5, indicating that the evaluation results are highly accurate, which proves the application effect of intelligent mobile terminals in college English mixed listening and speaking teaching.

5.4 Recommended Time Test for Listening and Speaking Teaching Resources

Based on the above experimental configuration, the recommended time for listening and speaking teaching resources in this method was determined in five groups of experiments, and the specific results are shown in Fig. 5.

From Fig. 5, it can be seen that the resource recommendation time of the method in this article is always below 80ms, and the curve variation is relatively stable, proving that the recommendation efficiency of intelligent mobile terminals in mixed listening and speaking teaching of college English is relatively high.

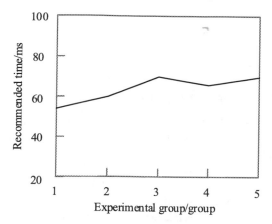

Fig. 5. Recommended time test

6 Conclusion

With the improvement of China's international status and the increasing frequency of international exchanges, the development of the country and society has put forward higher requirements for college students' English ability. The College English Curriculum Requirements issued by the Ministry of Education in 2007 clearly states that the teaching goal of college English is to cultivate students' comprehensive English application ability, especially listening and speaking ability. At present, college English courses in China are in the process of actively exploring the transformation from the traditional "teacher centered" teaching model to the "student centered" teaching model. As the main course to cultivate students' listening and speaking ability, college English listening and speaking courses are no exception. But generally speaking, listening and speaking courses are subject to large class teaching, and the teaching model is single The influence of various factors, such as insufficient teaching facilities and technical means, has not achieved much in improving students' listening and speaking ability, and even led to students' learning enthusiasm for listening and speaking courses getting worse. At the same time, compared with reading and writing courses, many teachers are not so calm when facing listening and speaking courses. However, with the rapid development of information technology and mobile technology, the coverage of Wlan, 3G and 4G networks is expanding, and mobile electronic terminal devices such as smart phones, PDAs and tablets are also becoming more and more popular. Mobile technology assisted English teaching and learning has become a hot topic in English education. Therefore, How to use mobile technology to build a college English listening and speaking teaching model that conforms to the characteristics of English subjects and language learning laws has also become a major research topic. The main contents of this study are summarized as follows:

(1) In this paper, the recommendation algorithm in intelligent mobile terminal is applied to practice to achieve personalized recommendation of listening and speaking teaching resources.

(2) SVR algorithm is introduced to fuse the evaluation features of different dimensions, which significantly improves the overall performance of pronunciation quality evaluation.

There are still some areas that can be improved and expanded in this study. Here are the following as the development directions of future research:

(1) In this paper, the method of feature comparison is used in intonation assessment, that is, each paragraph of students' reading voice needs to have a corresponding reference standard voice. In the future, we can explore the intonation assessment method using statistical modeling.
(2) The input parameters of the scoring model are improved.For the scoring model that needs feature engineering, the dimension of features can be appropriately increased, while removing some features that are less relevant to manual scoring.For the "end-to-end" scoring model, voice data and text data can be converted into better vector representations.

References

1. Gerd, K.: Writing virtual reality teaching resources. Phys. Teacher **61**(2), 107–109 (2023)
2. New Resources in TRAILS: The teaching resources and innovations library for sociology. Teach. Sociol. **51**(1), 108–111 (2023)
3. Yuan, X.: A balanced allocation method of English MOOC teaching resources based on QoS constraints. Inter. J. Continuing Eng. Educ. Life-Long Learn. **33**(1), 84–98 (2023)
4. Lin, Y.: A neural network-based approach to personalized recommendation of digital resources. Comput. Inform. Mech. Syst. **5**(4), 97–101 (2022)
5. Zou, F., Chen, D., Xu, Q., et al.: A two-stage personalized recommendation based on multi-objective teaching–learning-based optimization with decomposition. Neurocomputing **452**(6), 716–727 (2021)
6. Jiang, S., Ding, J., Zhang, L.: A Personalized recommendation algorithm based on weighted information entropy and particle swarm optimization. Mob. Inf. Syst. **2021**(4), 1–9 (2021)
7. Kang, Z., Sadeghi, M., Horaud, R., et al.: Expression-preserving face frontalization improves visually assisted speech processing. Int. J. Comput. Vis. **131**(5), 1122–1140 (2023)
8. Yoo, H., Seo, S., Im, S.W., Yong, G.G.: The Performance evaluation of continuous speech recognition based on Korean phonological rules of cloud-based speech recognition open API Inter. J. Netw. Distrib. Comput. **9**(1), 10 (2021)
9. HyeongJu, N., JeongSik, P.: Accented speech recognition based on end-to-end domain adversarial training of neural networks. Appl. Sci. **11**(18), 8412 (2021)
10. Ahmed, A., et al.: Connecting Arabs: B ridging the gap in D ialectal speech recognition. Commun. ACM **64**(4), 124–129 (2021)
11. Lee, D., Kim, D., Yun, S., Kim, S.: Phonetic variation modeling and a language model adaptation for Korean English code-switching speech recognition. Appl. Sci. **11**(6), 2866 (2021)
12. Liu, S., He, T., Li, J., et al.: An effective learning evaluation method based on text data with real-time attribution - a case study for mathematical class with students of junior middle school in China. ACM Trans. Asian Low-Resour. Lang. Inform. Process. **22**(3), 1–22 (2023)
13. Tan, S., Sun, L., Song, Y.: Prescribed performance control of Euler-Lagrange systems tracking targets with unknown trajectory. Neurocomputing **480**, 212–219 (2022)

Research on Anti-interference Dynamic Allocation Algorithm of Channel Resources in Heterogeneous Cellular Networks for Social Communication

Hongbo Xiang[✉]

Heilongjiang University of Technology, Jixi 158100, China
professorxiang2023@163.com

Abstract. The current channel allocation method does not consider the user transmission power problem, which leads to the problems of high user power consumption and low average transmission capacity, so a new anti-interference dynamic allocation algorithm for social communication channel resources in heterogeneous cellular networks is proposed. Determining a reusable set of channel resources for social network users; Under the premise that a given social network user reuses an arbitrary set of resources, the transmission power of the user is adjusted to measure the throughput of each network user on different channel resource sets. At the same time, the undirected graph theory in graph theory and ant colony genetic algorithm are used to cluster social network users, and as heterogeneous network scenarios change, the undirected graph will dynamically change, forming a new clustering scheme. Using intra cluster orthogonal inter cluster multiplexing as a criterion, auction method is used to allocate channel resources for social network users to reduce inter user interference. According to the selfishness of user behavior, a non cooperative game model is established, which combines fixed point theory and iterative algorithms to allocate power to users who complete channel allocation, maximizing user energy efficiency. The experimental results show that the proposed algorithm can reduce the power consumption and greatly increase the average user data amount.

Keywords: Social Communication Heterogeneous · Cellular Network · Channel Resources · Anti-Interference · Dynamic Allocation

1 Introduction

Due to the development of mobile communication network and Internet technology, mobile communication has become an indispensable part of people's life. Currently, with the passage of time, the number of users of mobile communication networks and the amount of data generated in the network are showing an exponential growth trend at an alarming speed. To deal with the above situation and more complex application scenarios, it is urgent to study the fifth generation mobile communication system to adapt to the

L. Yun et al. (Eds.): ADHIP 2023, LNICST 548, pp. 295–313, 2024.
https://doi.org/10.1007/978-3-031-50546-1_20

current development [1]. Presently, although the data transmission speed of 4G mobile communication system is fast, the spectrum efficiency is high, and the compatibility is good, there are still some defects, such as waste of resources and limited communication capacity. Compared with 4G communication system, 5G system can effectively make up for the shortcomings of 4G communication system, and can greatly improve the data transmission capacity and system capacity.

The use of social communication technology can directly realize the direct communication between user equipment without base station relay. Therefore, social communication can not only alleviate the load of the base station, but also bring high-speed data services to users [2]. Research shows that social communication technology has many advantages, such as reducing base station load and mobile terminal battery consumption, improving spectrum utilization and robustness of network infrastructure failure. For example, Document [3] the file first identifies the Small Cell Base Station (SBS) with the strongest interference to User Equipment (MUE), and dynamically allocates subchannels for SBS through cross layer interference information to serve its Small Cell User Equipment (SUE) and achieve network resource allocation. Literature [4] studied a joint optimization method for mode selection and channel allocation. Firstly, the distributed multi-agent deep Q-network algorithm is utilized to redefine the reward function based on the goal of maximizing system and rate. Then, a local information sharing strategy is adopted to reduce signaling overhead, allowing the agent to learn and select the optimal channel allocation in a low-cost mode. Reference [5] proposes a resource allocation algorithm with robust security. This method constrains the minimum security rate and cross layer interference power of each cell user, as well as the maximum transmission power of each base station, a resource allocation model based on energy efficiency maximization is established under bounded channel uncertainty. Using the Dinkelbach method, the worst case method, and the successive convex approximation method, the optimization problem is transformed reasonably, and the Lagrangian dual function method is used to solve the problem, thereby achieving reasonable allocation of channel resources.

Although the above methods have achieved significant research results at this stage, due to the lack of consideration of the user's transmission power issues, resulting in large power consumption and a small average amount of data sent by the user. Therefore, an anti interference dynamic allocation algorithm for channel resources in heterogeneous cellular networks for social communication is proposed. After verification, the proposed method can reduce power consumption and increase the average data transmission volume of users.

2   Algorithm

2.1   Adjustment of User Transmit Power

In social communication system, adjacent users can establish the communication links directly and thus to complete information transmission. In the process of link establishment, auxiliary control is needed to ensure the stability and reliability of the link [6]. In terms of network control, the architecture of social network communication can be divided into two parts: network central control and network auxiliary control (Fig. 1).

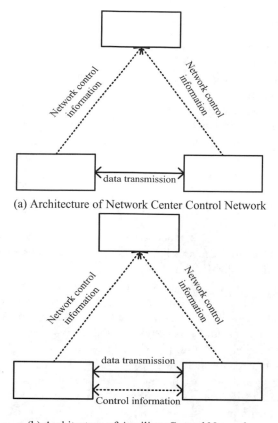

(a) Architecture of Network Center Control Network

(b) Architecture of Auxiliary Control Network

Fig. 1. Architecture of social communication Network

(1) Network center control architecture

Network Central Control Network architecture is a centralized control architecture in which there is a central control node to manage and coordinate the operation of the entire network. This control node is responsible for processing and allocating network resources, scheduling and directing traffic, managing connections and routing, and other functions. By collecting and analyzing network information, the central control node realizes intelligent decision-making and optimizes resource allocation to improve network efficiency and performance. This architecture is usually suitable for small-scale or domain-specific networks, such as sensor networks or factory automation networks. Such a system can realize the management of inter-user interference [7], and can determine whether neighboring users have established direct communication links in time by the environment where users are located. However, using this control structure, the signaling overhead is large.

(2) Architecture of auxiliary control network

Network-assisted control architecture is a distributed control architecture in which there is no single central control node, but the control functions are distributed to each node of the network. Each node has certain control capabilities to achieve network management and coordination based on local information and local decisions [8]. The nodes in the network auxiliary control architecture communicate and work together to complete the tasks of network resource allocation, route selection, data transmission control and so on. This architecture is more suitable for large-scale and complex networks, such as the Internet or mobile communication networks. Using this control structure, devices can be discovered independently, transmission power can be dynamically adjusted, and communication links can be directly established between users. The difference between this structure and the above structure is that the signaling overhead is smaller, but the efficiency is lower.

On the basis of the above communication architecture of social networks, the modes of social networks mainly include cellular mode, dedicated mode and sharing mode, which are described as follows:

(1) Cellular mode

Cellular communication is a common wireless communication mode, in this mode, there is no direct communication link between users, can only rely on the base station to achieve data transmission, each user is assigned to a specific cell. However, compared with the social communication mode, the advantage of this mode is that it can avoid interference between users and provide more stable and reliable communication[9]. However, there is the problem of low interest rate of spectrum resources, because the spectrum resources of the same frequency band are idle and wasted between different cells, and the allocation of spectrum resources requires higher management and coordination costs. The schema structure is shown in Fig. 2.

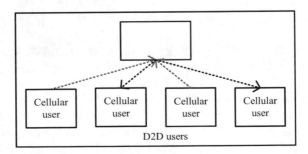

Fig. 2. Cellular Mode

(2) Dedicated mode

By using the dedicated mode, direct communication between adjacent users can be realized, avoiding the interference and delay of intermediate nodes, thus improving the efficiency and responsiveness of communication. Compared with cellular mode, dedicated mode only needs one downlink spectrum or upstream spectrum to complete social communication, so it has a higher utilization rate of spectrum resources [10]. However, in dedicated mode, the user device needs to have the ability to establish a communication link directly. This involves the support of device hardware and communication protocols to ensure that devices can communicate in the appropriate frequency band and achieve efficient data transmission. At the same time, interoperability and compatibility between devices also need to be considered, so that devices from different manufacturers can seamlessly connect and communicate to achieve the interconnection of social networks.

(3) Sharing mode

In shared mode, social network users can reuse licensed frequency bands in the cellular network to transmit data. This practice of reusing frequency bands enables more users to share spectrum resources, thus improving the efficiency of spectrum utilization. Compared with the above two modes, the shared mode has higher spectral efficiency. However, the reuse of cellular users' spectrum resources may cause interference problems among users. Therefore, it is necessary to prevent interference problems through certain resource allocation methods and interference management schemes [11, 12]. These algorithms and schemes can be intelligently adjusted based on information such as users' communication needs, network topology and spectrum conditions to optimize resource allocation and reduce interference.

A complete social communication process mainly includes three steps: device discovery, communication establishment and data transmission. In the cellular network, according to the needs of the service and the relevant information among the users, the cellular user can obtain the users who communicate with themselves independently or through the base station. Therefore, communication links are built through base stations to complete data transmission between users [13]. This process needs to ensure interoperability and compatibility between devices while safeguarding the security of communication and signal quality. The implementation process of this pattern is shown in Fig. 3.

The analysis and research of the process of social communication, taking two adjacent users as an example, describes the realization process as follows.

(1) User A needs to send a request to the base station in its area to establish a communication link with User B. The request will contain user B's basic information.
(2) When the base station receives the request from user A, the base station will determine the specific location of user B according to the basic information of user B sent by User A. After this process is completed, the base station will send social measurement signaling to the two users respectively to measure their social distance.
(3) After the two users receive the measurement signaling, the two users will respectively measure the channel state between them to ensure the accuracy of the judgment results. The channel status message measured by the two users is then sent to the base station.

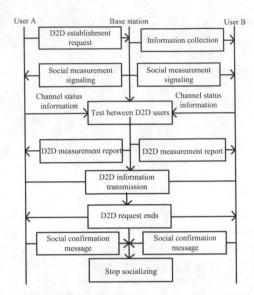

Fig. 3. Realization Flow of social communication

(4) After receiving the measurement information of the two users, the base station will match the received measurement information with the preset establishment conditions to complete the final judgment and determine whether the two meet the conditions for the establishment of the social link. If the conditions of establishment are met, the base station will send the judgment result to the two users.

(5) When the two users receive the construction results, the communication link will be directly established to complete the data transmission.

(6) At the end of the transmission, both users will send a request to stop social communication to the base station.

(7) After receiving the request, the base station will modify the network parameter configuration to a certain extent, and extract and confirm the termination information request sent by the two users again to ensure the accuracy of the operation. After confirming the error, the social communication between the two users will be terminated.

In the social communication system, according to the number of users participating in the social communication and the different functions, the social communication has three modes.

(1) Unicast communication mode:

In fact, the unicast communication mode is a one-to-one data transmission for users. According to the coverage range of social network users the and different relay functions, the unicast communication mode can be divided into:

1) User-to-user direct communication

2) Relay and forward communication mode

(2) Multicast communication mode

In social communication system, the multicast communication mode is to combine with multiple social network users, and make them share the same content and realize the information transmission of multiple social network users at one time[14]. Thanks to the addition of multicast technology, social network users can combine with multiple users to form the multicast communication, and transmit the same data resources to other users, so as to reduce the number of communications between social network users and improve the utilization efficiency of resources[15]. In a multicast communication group, each user plays the role of sender and receiver. A user can choose to send data to one of the users in the communication group or to multiple users at the same time. This flexible data transmission method can meet the individual needs of users and diversified social relationships. The communication mode is shown in Fig. 4.

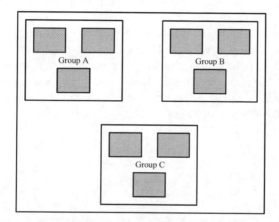

Fig. 4. Social Network Multicast (Multicast) Communication Method

(3) Broadcast communication mode

In social network broadcast communication, social network users can transmit information so that other users within a certain distance can receive information. However, the increase in the number of users leads to a corresponding increase in the base station load, and it is difficult to ensure that all users can provide better communication services.

To solve this problem, social communication has become one of the key technologies in communication systems such as LTE and 5G. It allows data routing between adjacent users to communicate directly under the control of the cellular system without passing through the base station. The technology reduces the load of the base station significantly, and also reduces the delay of the end-to-end information communication between users. In addition, it also allows social network users to reuse the channel resources of cellular users, which greatly improves the utilization of high-frequency spectrum and system capacity. However, this causes the problem of common frequency interference, which leads to the interference between users in the process of communication, and reduces the communication effect. Therefore, in order to solve this problem, it is necessary to develop interference management and resource allocation schemes. These schemes can maximize the throughput of social network links in the system and reduce interference levels through spectrum allocation, power control, and channel scheduling.

In social communication, the channel model takes many factors into account. In addition to path loss, fast fading due to multipath effect and slow fading due to shadow effect are also considered. This more accurately reflects the channel characteristics between social network users and cellular users. The channel gain can be modeled and analyzed according to these characteristics to optimize the performance of the system. Therefore, the channel gain between a cellular user and a social network user receiver is expressed as:

$$h_{cd} = k \cdot \delta_{cld} \cdot \xi_{cld} \cdot d_{cld}^{-\alpha} \tag{1}$$

In Formula (1), k represents the coefficient of path loss; α represents the index of path loss; δ_{cld} represents the fast fading factor following exponential distribution; ξ_{cld} represents the slow fading factor following logarithmic positive attitude distribution; $d_{cld}^{-\alpha}$ represents the distance between cellular user and social network user receiver.

If there are D pairs of social network users in the system, and C cellular users communicate with each other in orthogonal channels. When social network user d reuses the channel resource of cellular user i, the formulas of calculating SINR γ_i and rate r^i of cellular user i are shown as formula (2)–(5):

$$\gamma_i = \frac{P^i h_{cd}}{P_d^i h_{cd} + N_0} \tag{2}$$

$$r^i = \lg(1 + \gamma_i) \tag{3}$$

In the formula, P^i represents the signal power received by the cellular user i; P_d^i represents indicates the signal power of communication between cellular users i and social network users d; N_0 represents indicates interference noise power.

The SINR (signal to interference plus noise ratio) and speed of social network user d are shown as follows:

$$\gamma_d^i = \frac{P_d^i h_{cd}}{P^i \lg(1 + N_0)} \tag{4}$$

$$r_d^i = \lg\left(1 + \gamma_d^i\right) \tag{5}$$

Co-channel interference occurs when social network users and cellular users use the same channel at the same time. There is a certain connection between the interference intensity and the distance, in general, the smaller the distance, the stronger the interference intensity. In general, the number of users who simultaneously conduct social communication in a residential district is less than that of cellular users, so multiple cellular channel resources may become potential multiplexing objects of social network users. When the interference of cellular users to social network users is less than a certain threshold, the base station will allow social network users to reuse channel resources, and through such constraints, ensure the QoS requirements of social network users. The areas that restrict the reuse of cellular users are shown as follows:

$$P_{c,\max} h_{cd} \leq I_{c,d} \tag{6}$$

In the formula, $P_{c,\max}$ represents the maximum power received by cellular users; $I_{c,d}$ represents the power limit for signal reception in this area.

After the set of channel resources reused by social network users is given, the transmit power of social network users is optimized by power control, and maximize the throughput of social network communication links under resource constraints. Based on the previous step, any social network user D and the reusable potential cellular channel can be determined. On this basis, the reusable resource set S_d of social network user D is obtained. Therefore, the transmission power target of optimized social network users can be expressed as:

$$\begin{cases} \max u_d^k = \sum_{i \in S_d} \log_2 \left(1 + \frac{P_d^i h_{cd}}{P^i \log_2 (1+\gamma_d^i)} \right) \\ 0 \leq P_d^i \leq P_{c,\max} \\ \sum_{i \in s_d} \leq P_{c,\max} \end{cases} \tag{7}$$

In the formula, u_d^k represents the transmission power target of the optimized social network user.

Next, under the optimization objectives and constraints, the water filling algorithm is applied to achieve the rational distribution of power for social network users. According to Karush-Kuhn-Tucker (KKT) condition, we can get:

$$L\left(P_d^i, \alpha, \beta\right) = \sum_{i \in S_d} \log_2 \left(1 + \frac{P_d^i h_{cd}}{P^i \log_2 (1+\gamma_d^i)} \right) - \alpha \left(P_d^i - P_{c,\max} \right) - \beta \left(\sum_{i \in S_d} P_d^i - P_{c,\max} \right) \tag{8}$$

In the equation, α and β are both distribution condition coefficients.

After calculating the partial derivatives, we can get:

$$P_d^i = \left[\varphi - \frac{1}{H} \right]^+ \tag{9}$$

where,

$$[\varphi]^+ = \max\{0, \varphi\} \tag{10}$$

$$H = \frac{h_{cd}}{P^i \log_2 (1+\gamma_d^i)} \tag{11}$$

Based on the power control, we can see that the maximum throughput and the corresponding transmission power of reusing any resource set by social network users in the set time. In the process of channel allocation, the scheme of power control is used to complete the power adjustment of user on different channels.

2.2 An Anti Interference Dynamic Allocation Algorithm for Channel Resources in Heterogeneous Cellular Networks in Social Communications

With the development of wireless services, users need to consume more energy. However, the battery capacity of mobile terminals is finite, and the existing battery technology has not been further developed. How to effectively improve the energy efficiency, spectrum efficiency and user experience is an urgent problem to be solved. In order to improve the spectrum efficiency, a cellular channel resource is allowed to be reused by multiple social network users at the same time. Because many pairs of social network users reuse the same channel resource, the interference environment in the system becomes more complex. It not only includes the cross-layer interference between the original cellular users and social network users, but also the same layer interference between social network users who reuse the same channel resource.

The uplink transmission system of heterogeneous network with dense social network users includes cellular users and social network users randomly distributed in the residential area. The eNB is located in the center of the residential area. During the uplink cycle of cellular network, the cellular user sends data to eNB, and eNB suffers interference from social communication transmitters. In the communication process, social network user receivers will be interfered by cellular users and other social network user transmitters sharing the same channel. It is assumed that the channel condition follows Rayleigh distribution. In order to completely simulate the real environment, different channel selection will lead to different channel selective fading (frequency selective fading).

Let's suppose that there are D pairs of social network users in the system, and C cellular users use the orthogonal channel for communication, but there is no residual channel resource in the system. If q denotes the set of social network users that share the resources of the k th cellular channel, the Signal to Interference plus Noise Ratio (SINR) of the cellular user on the k th channel is:

$$\gamma_c^k = \frac{P_d^k h_{cd}}{\sum_{k \in C} p_d^k h_{cd} + N_0} \tag{12}$$

In the formula, p_d^k represents the signal power of the k-th channel in the orthogonal channel used by social network users d.

The SINR of the D th social network user on the k th channel is:

$$\gamma_d^k = \frac{P_d^k h_{cd}}{\sum_{k \in C} p_d^k h_{cd} + P_d^k h_{cd} + N_0} \tag{13}$$

Next, r_c^k and r_d^k represent the rates of the k th cellular user and the d th social network user, respectively:

$$r_c^k = \log_2\left(1 + \gamma_c^k\right) \tag{14}$$

$$r_d^k = \log_2\left(1 + \gamma_d^k\right) \tag{15}$$

The battery capacity of mobile devices is limited, lacking of new technological breakthroughs, so the energy consumption becomes particularly important. Therefore, new optimization criteria should be adopted to fully balance the energy consumption and transmission quality. The battery life will be an important optimization parameter. The energy consumption of each user in the system includes two parts: transmitting energy and circuit energy. The circuit energy refers to the energy consumed by all circuit blocks along the signal path. It has an important impact on battery life, so it can't be ignored. Without loss of generality, it is assumed that all users have the same constant circuit power consumption P_0. In order to capture nonlinear effects, Peukert law is used to model the battery life.

$$T = \frac{P_0 B_c}{I^\alpha} \tag{16}$$

In the formula, B_c represents the circuit energy; I^α represents a constant current in the circuit.

Therefore, for the user whose transmitting power is p_i and working voltage is V_0, the battery life is:

$$T_i = \frac{B_c V_0^a}{(p_i + p_0)^a} \tag{17}$$

In the equation, a represents the loss coefficient.

Next, measure the user's energy efficiency by setting the expected data volume of the battery. That is to say, the maximum amount of data can be transmitted within a limited battery capacity.

$$u_i = \left(r_c^k + r_d^k\right) T_i \tag{18}$$

In order to ensure the normal communication between users, this section mainly considers the QoS requirements of users, and the utility function is shown in formula (19).

$$\begin{cases} \max u_i\left(p^i, p_{c,\max}\right) = \left(r_c^k + r_d^k\right) T_i \\ s.t \left(r_c^k + r_d^k\right) \geq r_{i,\min} \\ \sum_{i=0}^{N} \frac{p_i^k}{V_{i,c}^k} \leq I_{k,\max}^c \\ 0 \leq p_k^i \leq p_{c,\max} \end{cases} \tag{19}$$

In the formula, $r_{i,\min}$ represents the minimum transmission rate of cellular network users i; p_k^i represents the signal power of the k-th channel in the orthogonal channel used by cellular network users i; $V_{i,c}^k$ represents the working voltage of the k-th channel in the orthogonal channel C used by cellular network users i; N represents the number of users; $I_{k,\max}^c$ is defined as the maximum tolerable interference of the k th cellular user.

If B-bits content must be sent within time $T_{i,\max}$, $T_{i,\max}$ is the maximum delay. Let's assume that the channel is static during the optimization, the utility function can be expressed by Formula (20) under the QoS constraint of maximum delay.

$$\begin{cases} \max u_i\left(p^i, p_{c,\max}\right) = \left(r_c^k + r_d^k\right)T_i \\ s.t \ \dfrac{B_{i,\min}^k}{\left(r_c^k + r_d^k\right)} \leq T_{i,\min} \\ 0 \leq p_k^i \leq p_{c,\max} \end{cases} \tag{20}$$

In the equation, $B_{i,\min}^k$ represents the minimum circuit energy of the cellular network user i in the k-th channel.

The interference threshold is also an important factor affecting QoS, which is particularly important to ensure the normal communication of cellular user. If $I_{k,\max}^c$ is defined as the maximum tolerable interference of the k th cellular user, the utility function under the QoS constraint of maximum interference threshold can be shown in formula (21).

$$\begin{cases} \max u_i\left(p^i, p_{c,\max}\right) = \left(r_c^k + r_d^k\right)T_i \\ s.t \ \displaystyle\sum_{i=0}^{N} \dfrac{p_i^k}{v_{i,c}^k} \leq I_{k,\max}^c \\ 0 \leq p_k^i \leq p_{c,\max} \end{cases} \tag{21}$$

The undirected graph is defined as $G = (W, V, E)$. V is the set of vertices, which represents social network users in the system. E represents the set of edges connecting each point. W represents the set of weights of edges. And the greater the weight W, the greater the interference between social network users. The maximum weight between social network users is represented by w_{ij}.

Based on the graph coloring principle, the social network users with large interference are divided into the same cluster, and the social network users with small interference are divided into different clusters to maximize the total interference in the same cluster. D_k denotes social network users in the same cluster. The optimization target is expressed in formula (22).

$$\begin{cases} \max \displaystyle\sum_{i=1}^{D}\sum_{j=1}^{D}\sum_{m=1}^{M} e_{im}e_{jm}w_{ij} \\ s.t \ \displaystyle\bigcup_{k=1}^{M} D_k = D \\ e_{im} = \{0, 1\} \\ e_{jm} = \{0, 1\} \end{cases} \tag{22}$$

In the formula, e_{im}, e_{jm} represents the clustering coefficients of different social network user divisions.

Clustering is still a NP difficult problem. In this section, the ant colony genetic algorithm (ACGA) is used to solve the social network user clustering problem. The ACGA is based on genetic algorithm. Firstly, a solution set of excellent initial population can be obtained through one iteration of ant colony algorithm, and then the solution set is taken as the initial population of genetic algorithm. Finally, the genetic algorithm is used to find the clustering result and thus to find out the optimal solution of the original problem. The operation flow of ACGA is described as follows:

(1) Initialization

Set the parameters of ant colony algorithm and the amount of information on the corresponding path;

(2) m ants are randomly placed on n vertexes;
(3) Ant k selects the next colored vertex according to the state transition probability and adds it to the tabu list;
(4) Repeat Step (3) to find the first traversal path of m ants respectively, and $S(D \times P)$ denotes the initial colored matrix solved by ant colony algorithm;
(5) The vertices with 1 element in each column of $S(D \times P)$ are combined to form a substring population. And then, the genetic algorithm parameters are initialized.
(6) The fitness function of individual in the population is calculated;
(7) The selection, crossover and mutation are performed to generate new clustering results;
(8) $t < NG, t = t+1.$. Return to Step (6), otherwise the genetic algorithm is terminated.

The channel allocation is to improve the energy efficiency of users as much as possible. According to the clustering results, the interference between users in the same cluster is larger than that in different clusters. In order to reduce the interference between users, the channel allocation follows the principle of intra-cluster orthogonality and inter-cluster reuse. Next, the auction algorithm will be used to complete channel allocation. Social network users are regarded as the bidders, and the cellular channels are regarded as lots. If the social network user d wins the channel resource of the i th cellular user, the private valuation of the social network user d can be expressed as $v(d, i)$.

In the auction, social network users improve their energy efficiency by sharing channel resources, but they will also pay some costs. In order to fully reflect the fairness of auction, the linear anonymous price is adopted, and the formula of calculating the payment price is shown as formula (23).

$$P_d(i) = \beta_\tau p_d^i \tag{23}$$

In the formula, β_τ represents the elasticity coefficient of resource prices; p_d^i represents the sharing channel resource price between social network user d and cellular user i.

Therefore, the utility function of the bidder (the satisfaction with the channel resource of the i th cellular user) can be defined as formula (24).

$$U_d(i) = v(d, i) - P_d(i) \tag{24}$$

After the channel allocation, the problem of resource allocation is transformed into the power allocation of user. Because users are only interested in maximizing their energy efficiency, the problem of user power allocation is modeled as a model of non-cooperative game theory, and the participants are the users in the same channel. The fixed-point theory and iterative algorithm are combined to allocate the power to the users who complete the channel allocation, so as to maximize the energy efficiency.

3 Simulation Experiment

In order to verify the comprehensive effectiveness of the proposed algorithm, simulation experiments in heterogeneous cellular networks are required. The specific experimental parameters are shown in Table 1.

Table 1. The parameter of simulation experiment

Parameter	Setting value
Cellular network radius	1500 m
Number of cellular users	25
Maximum power of cellular users	20 dBm
Rate requirements for cellular users	1000 bps
Number of social network user pairs	1–50
Maximum power of social network communication sender	20 dBm
Channel bandwidth	1.8e5 Hz
Noise power	−100 dBm

In the MATLAB platform, 1000 distributed scenes are randomly generated each time by Monte Carlo method. The experimental system configuration is as follows: the CPU is Intel dual core e8400, the main frequency is 3.0 GHz, the memory is 4.0 gb, and the operating system is windows 2008, 64-bit system. Taking power consumption and average data transmission as factors, the algorithm proposed in this paper is compared with the dynamic channel allocation (DCA) algorithm (reference [3] algorithm), the joint optimization algorithm for mode selection and channel allocation (reference [4]), and the robust security resource allocation algorithm (reference [5] algorithm).

(1) Power consumption/(W)

In order to verify the effectiveness of the algorithm in this paper, 100 simulation experiments are carried out under different number of users. The results are shown in Table 1.

As shown in Table 2, The power consumption of each algorithm will increase with the increase of the number of experiments. However, by adjusting the user's generating power, the interference generated by the transmitter is reduced, thus effectively reducing the power consumption of the whole algorithm.

(2) Average amount of data sent for users / (bit /Hz)

Figure 5 shows the comparison results of average data transmission volume of different algorithms.

Table 2. The comparison for power consumption of different algorithms

Number of experiments/(times)	Power consumption/(W)			
	The method	The dynamic channel allocation (DCA) algorithm	The joint optimization algorithm for mode selection and channel allocation	The robust security resource allocation algorithm
10	1.25	1.30	1.28	1.40
20	1.30	1.38	1.34	1.47
30	1.42	1.47	1.46	1.52
40	1.47	1.58	1.52	1.60
50	1.54	1.67	1.59	1.67
60	1.60	1.74	1.63	1.74
70	1.66	1.80	1.69	1.78
80	1.71	1.88	1.74	1.83
90	1.80	1.94	1.83	1.90
100	1.86	2.00	1.87	2.01
Average value / (W)	1.561	1.676	1.595	1.692

Analysis of the experimental data in Fig. 5 shows that, the average amount of data sent by users increases with the number of users. Although the average user data transmission volume of this algorithm is higher than the other three methods, the algorithm effectively solves the user's transmission power, avoids the interference generated by the transmitter, and improves the average transmission data amount. Therefore, the algorithm in this paper is superior to the other three algorithms.

To further verify the superiority of the proposed algorithm, the transmission time was used as an evaluation indicator to compare the data transmission rates of the four algorithms. The results are shown in Fig. 6.

According to the results obtained in Fig. 6, there is a certain gap in the results obtained by the four algorithms as the amount of data increases. The transmission time shows an increasing trend. When the transmission data volume reaches 1000, the transmission time consumption of the algorithm proposed in this paper, along with the dynamic channel allocation (DCA) algorithm (reference [3] algorithm), the joint optimization algorithm for mode selection and channel allocation (reference [4] algorithm), and the robust security resource allocation algorithm (reference [5] algorithm), is 1s, 3.4s, 4s, and 3.3s. Comparing the results obtained from the four algorithms, it can be seen that the algorithm proposed in this paper can effectively shorten the time required for data transmission, improve transmission efficiency, and have strong practicality.

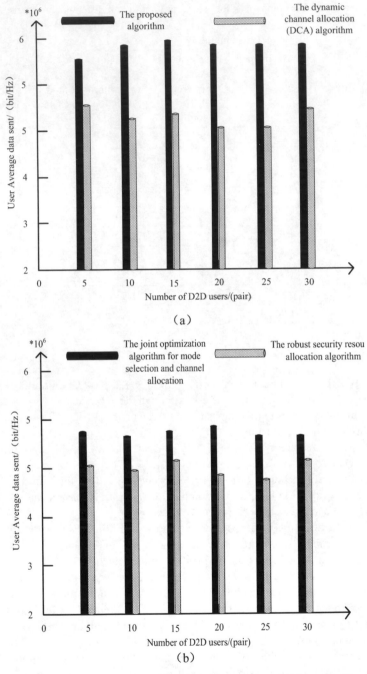

Fig. 5. The comparison results for average data sent of different algorithms

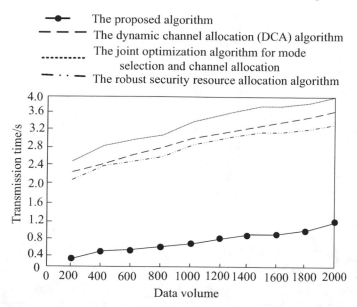

Fig. 6. Data Transmission Rate

4 Conclusions

With the rapid growth of terminal devices such as smartphones and wearable devices, as well as the continuous development of wireless services, people's demand for bandwidth is increasing, and wireless spectrum resources are becoming increasingly scarce. Therefore, how to improve the utilization of spectrum resources has become a research hotspot in recent years. At the same time, a dynamic anti-interference channel resource allocation algorithm for heterogeneous cellular networks in social communication was designed and proposed to address a series of issues with traditional algorithms. This algorithm addresses the common frequency interference problem in heterogeneous networks composed of social network users and cellular users, and combines a resource allocation mechanism based on combinatorial auctions to maximize the throughput of social traffic information links in the system. At the same time, using undirected graph theory and ant colony genetic algorithm in graph theory to cluster social network users, as heterogeneous network scenarios change, the undirected graph will dynamically change, forming a new clustering scheme. Based on the principle of reuse between orthogonal clusters within a cluster, auction method is used to allocate channel resources for social network users to reduce inter user interference. Based on the selfishness of user behavior, a non cooperative game model is established, which combines fixed point theory with iterative algorithms to allocate electricity to users who complete channel allocation, maximizing user energy efficiency. The results show that the average power consumption of this algorithm is 1.561W, which can effectively reduce power consumption. The data transmission volume is maintained above 5bit/Hz, and the transmission time is less than 1s, effectively improving the user data transmission volume and having high transmission efficiency. However, due to limitations in research time and personal academic abilities, there are still shortcomings in research work. The future research focus will be on the following aspects:

(1) Research resource allocation algorithms in multi to multi scenarios to further improve spectrum utilization and system performance.
(2) The algorithm in this article only considers slow motion within the specified range. But in actual network environments, there are high-speed mobile users and cross community users. Therefore, we will improve our algorithms in this area in the future.

References

1. Chen, Y., Ai, B., Niu, Y., et al.: Sub-channel allocation for full-duplex access and device-to-device links underlaying heterogeneous cellular networks using coalition formation games. IEEE Trans. Veh. Technol. **69**(9), 9736–9749 (2020)
2. Li, J., Lei, G., Manogaran, G., et al.: D2D communication mode selection and resource optimization algorithm with optimal throughput in 5G network. IEEE Access **7**, 25263–25273 (2019)
3. Ban, I., Kim, S.J.: Interference-aware dynamic channel allocation for small-cells in heterogeneous networks with FFR. IEICE Trans. Fundam. Elect. Commun. Comput. Sci. E102.A(10), 1443–1446 (2019)
4. Zhi, Y., Tian, J., Deng, X., et al.: Deep reinforcement learning-based resource allocation for D2D communications in heterogeneous cellular networks. Digital Commun. Netw. English version **8**(5), 834–842 (2022)
5. Yongjun, X.U., Cao, Q., Wan, Y., et al.: Robust secure resource allocation algorithm for heterogeneous networks with hardware impairments. J. Electron. Inf. Technol. **45**(1), 243–253 (2022)
6. Lee, J., Lee, J.H.: Performance analysis and resource allocation for cooperative D2D communication in cellular networks with multiple D2D pairs. IEEE Commun. Lett.Commun. Lett. **23**(5), 909–912 (2019)
7. Asuhaimi, F.A., Bu, S., Klaine, P.V., et al.: Channel access and power control for energy-efficient delay-aware heterogeneous cellular networks for smart grid communications using deep reinforcement learning. IEEE Access PP(99) (2019)
8. Xue, Y., Xu, B., Xia, W., et al.: Backhaul-aware resource allocation and optimum placement for uav-assisted wireless communication network. Electronics **9**(9), 1397 (2020)
9. Rehman, W.U., Salam, T., Almogren, A., et al.: Improved resource allocation in 5G MTC networks. IEEE Access **8**, 49187–49197 (2020)
10. Alemaishat, S., Saraereh, O.A., Khan, I., et al.: An efficient resource allocation algorithm for D2D communications based on NOMA. IEEE Access **7**, 120238–120247 (2019)
11. Jan, A., Parah, S.A., Malik, B.A., et al.: Secure data transmission in IoTs based on CLoG edge detection. Future Generation Comput. Syst. **121**(3), 20–35 (2021)
12. Nie, Z.X., Long, Y.Z., Zhang, S.L., et al.: A controllable privacy data transmission mechanism for Internet of things system based on blockchain. Inter. J. Distrib. Sensor Netw. **18**(3), 303–315 (2022)

13. Rahimian, A., Hosseini, M.R., Martek, I., et al.: Predicting communication quality in construction projects: A fully-connected deep neural network approach. Autom. Const. **139**(7), 1–15 (2022)
14. Fan, N., Shen, S., Wu, C.Q., et al.: A hybrid trust model based on communication and social trust for vehicular social networks. Inter. J. Distrib. Sensor Netw. **18**(5), 161–166 (2022)
15. Gong, W., Pang, L., Wang, J., et al.: A Social-aware K means clustering algorithm for D2D multicast communication under SDN architecture. AEU - Inter. J. Electr. Commun. **132**(2), 10–26 (2021)

Numerical Simulation of Dual Laterolog Response Based on Wireless Communication Technology

Hongbo Xiang[✉]

Heilongjiang University of Technology, Jixi 158100, China
professorxiang2023@163.com

Abstract. The current numerical simulation of dual laterolog response is mostly unidirectional, and the efficiency of numerical simulation is low, resulting in high resistivity and potential safety hazards when conducting potential processing. Therefore, the design and verification research of the numerical simulation of dual laterolog response based on wireless communication technology are proposed. Firstly, the feature extraction of numerical simulation is carried out, and the multi-level method is adopted to improve the efficiency of numerical simulation, realize multi-level grid generation of dual laterolog, establish a wireless communication numerical simulation model, and implement numerical simulation by boundary constraint processing. The final test results show that the measured resistivity is well controlled below 7 in combination with the electrical coefficient of each set point through numerical simulation processing at the depths of 1.1 m, 1.3 m, 1.5 m, 1.8 m, 2 m, 2.3 m, 2.5 m and 3 m, which indicates that this numerical simulation has relatively large coverage and strong pertinence, it has practical application value to optimize the unit simulation structure, improve the efficiency and quality of the overall numerical simulation, and minimize the difference in the simulation process.

Keywords: Wireless Communication Technology · Dual Laterolog · Response Numerical Simulation

1 Introduction

As a commonly used multi-dimensional exploration and development technology, dual laterolog processing has achieved relatively good results in current social engineering activities, greatly improving the quality and efficiency of project construction [1]. Generally, the reservoir of dual laterolog is relatively thick, and the dual laterolog response law is also very different from that of vertical wells. The existing vertical well logging interpretation methods can no longer be used in dual laterolog. Therefore, it is necessary to carry out numerical simulation processing [2] in combination with high and new technologies, equipment and devices. In fact, the research on the response characteristics of dual laterolog is relatively simple at present, which is mainly due to the influence

L. Yun et al. (Eds.): ADHIP 2023, LNICST 548, pp. 314–329, 2024.
https://doi.org/10.1007/978-3-031-50546-1_21

of the sedimentary characteristics of the formation. In addition, the dual laterolog is generally shale formation, which has strong isotropic vertical weighting characteristics. The general rule is that components are evenly distributed along the direction of the layer interface, which is isotropic. In the direction perpendicular to the layer interface, the degree of lateral opportunism will also change correspondingly, which is characterized by high resistance. Therefore, under such a background, it is relatively difficult to conduct numerical simulation processing for dual laterolog [3]. Moreover, the formation structure of dual laterolog is relatively complex, and the reservoirs of various rocks have many fractures. The fracture dip angle will also have a great impact on the permeability of shale gas reservoirs, which is very likely to directly lead to problems such as cracking or subsidence of gas reservoirs, which will reduce the actual production efficiency [4] to a certain extent. In addition, the angular rock layer or the built-in structure of the dual laterolog often occurs the phenomenon of diameter expansion, which causes the local size change of the borehole to produce pressure on the dual laterolog response, which will also cause errors in the results of the numerical simulation. Therefore, the analysis and verification of the numerical simulation of dual laterolog response based on wireless communication technology are proposed. The so-called wireless communication technology is actually a multi-dimensional communication mode, which mainly refers to a communication form [5] that uses the characteristics of electromagnetic wave signals that can spread in free space to exchange information. The integration and practical application of this technology with the numerical simulation of dual laterolog response can further expand the coverage of actual numerical simulation and improve the accuracy and integrity of simulation [6] to a certain extent. Then, based on this, the finite element method can also be used to optimize the results of numerical simulation, and the corresponding numerical simulation scheme [7] can be formulated for the changes of anisotropy and fracture dip angle of the dual laterolog and the underlying rock layer. In addition, the current dual laterolog is widely distributed and usually has a large scale, and the reservoir space beneath the rock layer is highly heterogeneous. It is necessary to make a comprehensive analysis of the response characteristics, environmental correction and sensitive factors of dual laterolog, and describe and process the details of model construction, grid division and parameter setting in numerical simulation, it provides reference basis and theoretical reference for subsequent relevant research and practical verification [8].

2 Design a Numerical Simulation Method for Wireless Communication of Dual Laterolog Response

2.1 Feature Extraction of Numerical Simulation

In general, in the process of numerical simulation processing, the basic features are extracted first according to the actual simulation requirements based on the conditions of the stratum itself, a core point of numerical simulation is selected, and extended simulation and numerical establishment are carried out to lay a foundation for subsequent numerical simulation [9]. This time, based on the actual situation of the dual laterolog project, the response characteristics are analyzed in many aspects, which can be divided into three aspects. One is the horizontal fracture. In the process of feature extraction, the depth of the cave center is the same as the midpoint of the fracture, and the fixed distance between the cave and the borehole wall is 20 cm. Assuming that the drilling is very deep, and the properties of cave fluid and fracture fluid are the same, the diameter of the borehole is 18.2 cm, the electrical resistivity is 0.1 $\Omega \cdot$ m, and the matrix fracture opening is 56.8 μm. The resistivity is 2350 $\Omega \cdot$ m, and the resistivity of borehole drilling fluid is 0.65 $\Omega \cdot$ m [10–12]. Next, we will collect and analyze other numerical indicators according to the actual extraction demand, as shown in Table 1 below:

Table 1. Numerical simulation characteristic index setting table

Numerical simulation characteristic indicators	Initial standard value	Measured standard value
Surrounding rock resistivity/$\Omega \cdot$ m	5.5	6
Borehole diameter/m	3.5	3.6
Anisotropic coefficient	16.35	18.55
Lateral resistivity/$\Omega \cdot$ m	8.5	9
Resin simulation directional crack angle/°	50	65
Electrode coefficient	14.25	16.34
True resistivity/$\Omega \cdot$ m	6.5	7.5
Reservoir Porosity	65	75

According to Table 1, set the characteristic indicators of numerical simulation. Then, on this basis, through the response analysis of horizontal fracture cave dual laterolog, it can be found that under the above circumstances, the existence of fractures will lead to a significant reduction in resistivity, and the results of deep logging are negatively correlated in the difference of characteristics; In the case of a single fracture, the influence of caves on dual laterolog is relatively small. In the analysis, the logging tool, fracture and cave center are all placed at the same depth, and the corresponding impact is analyzed. Through the analysis, it can be found that if the cave is relatively close to the borehole, the value of bilateral resistivity will be smaller. However, it should be noted that the lateral wells of horizontal fracture cave combination have been negative amplitude difference for a long time [13–15].

The second is the dip fracture characteristics of dual laterolog. This part is mainly in the case of inclined fractures, the response of lateral wells will be controlled by the fracture occurrence. When low angle fractures are developed, the difference in resistivity of lateral wells is positive. In this case, compared with the response of lateral wells with a single fracture, the response of lateral wells is controlled by the presence of caves to a certain extent, and in the specific analysis process, it can be found that the inclined fracture will not affect the response of lateral wells due to the change of inclination angle. Under different environments, the data of critical angle of positive and negative amplitude difference change of lateral wells are basically the same.

The third is the vertical fracture characteristics of dual laterolog. Different from inclined fractures, vertical fractures have a wide range of characteristics. When analyzing vertical fractures, it can be found from the analysis results that caves will cause the response value of adjacent dual laterolog to gradually decrease. In the case of vertical fractures, the positive difference of resistivity is very obvious. In the case of vertical fractures, the specific response identification range of lateral wells is very similar to that of a single cave. So far, the extraction of numerical simulation characteristics has been completed, and the collection of basic data and information has been realized, which can be used to measure and verify the subsequent numerical simulation in the project area.

2.2 Multilevel Grid Generation of Dual Laterolog

After the feature extraction of numerical simulation is completed, then, combined with wireless communication technology, the grid of dual laterolog is divided. In mesh generation, the more the number of mesh nodes and the finer the elements, the higher the accuracy of the solution, but the calculation time will also increase. The mesh division is too sparse. Although the required calculation memory is small and the calculation time is short, the calculation accuracy is correspondingly low. Combined with the wireless communication technology of knight errant, the grid is divided scientifically and reasonably, and different density grids are selected in different solution areas to ensure the calculation accuracy of the model while reducing the time consumed in calculation. Triangle grid is used for division. Due to the large size contrast of geometric model, the wellbore, fractures and strata are divided into sub domains. The mesh is dense at the instrument and fracture, and sparse at the infinite distance of the stratum. The running time of the program can be greatly reduced by using different meshes to divide the domain. The mesh generation in the solution area shall meet the following requirements:

(1) At the junction of different dual laterolog domains, the elements dissected cannot cross the interface;
(2) Elements of dual laterolog do not overlap each other;
(3) The grid of the generated dual laterolog covers the whole solution area;
(4) The closer the triangulation mesh is to the equilateral triangle, the better the quality of the triangulation mesh and the more stable the construction state;
(5) Combined with acquisition characteristics, ensure proper grid density of dual laterolog.

Then, based on this, directional analysis of dual laterolog electrode, borehole, fracture and formation grid generation is conducted in a two-dimensional axisymmetric manner. The mark subdivision processing is carried out through triangular extremely refined grid, and the subdivision is densified near the electrode of the dual laterolog tool, with the minimum unit size of 0.016 m. Due to the large potential change at the contact surface between the dual laterolog tool and mud, the contact surface between the borehole and the fracture, the contact surface between the borehole and the formation, and the contact surface between the fracture and the formation, the refined grid subdivision is selected, combined with wireless communication technology, to ensure the accuracy of the simulation. An auxiliary point is added at the center of all electrodes to facilitate the reading of the overall potential of the electrodes and to improve the quality of grid generation. From the electrode to the infinite distance, the number of grids gradually decreases, the grid becomes more and more sparse, and the side length of the triangle gradually increases. This reduces the time of model calculation, which is conducive to the establishment of the later numerical simulation environment and the control of errors. The numerical simulation will be balanced and symmetrical. The specific principle is shown in Fig. 1 below:

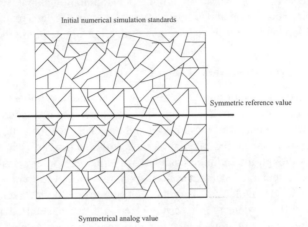

Fig. 1. Symmetry diagram of dual laterolog response numerical simulation

According to Fig. 1, complete the analysis and research on the symmetry of numerical simulation of dual laterolog response. Next, because the solution area of the upper and lower axisymmetric model is 1/2 of the axisymmetric model, the reduction of the solution area is convenient for calculating the dual laterolog response value of the micro fracture. Triangular extremely fine mesh is also used for mesh generation. The mesh generation is relatively dense near the electrode system, and sparse in the remote stratum. Based on the two-dimensional model, the three-dimensional model is meshed. A reasonable mesh generation can not only ensure the correctness of the solution, but also effectively shorten the time of model calculation, which affects the calculation accuracy of the model. In the 3D model, the mesh is generated from inside to outside, from dense to sparse. Dense dissection is carried out near the electrode system. However, in this part, it should be noted that because the electrode occupies a small part of the whole stratum, the infinite region and part of the stratum boundary can be hidden first when processing the grid and setting the materials. Recovery calculation shall be performed after application. The finer the regional grid near the well is, the thinner the regional grid far away from the well is. If the potential is zero at infinity, the coarser quadrilateral grid is selected. In the 3D model, the contrast between the fracture and the electrode and the stratum area is large. Combined with wireless communication technology, the difficulty of mesh generation is increased, and the calculation time of mesh generation is increased. Subdivision subdivision can avoid the large amount of computer memory occupied by mesh subdivision, shorten the calculation time, and achieve accurate simulation results. Through the establishment of three-dimensional model, it is considered that the two-dimensional axisymmetric and two-dimensional up and down symmetric models cannot meet the simulation of fractures with different dip angles. Even though the borehole, instruments and formation have axial symmetry, the fractures with different dip angles in the formation do not have axial symmetry. Due to the large amount of 3D model calculation, each mobile instrument needs to be re divided and solved. Combined with wireless communication technology, multi-level local division is carried out to lay a foundation for subsequent numerical simulation.

2.3 Establishment of Wireless Communication Numerical Simulation Model

After completing the multi-level grid generation of dual laterolog, next, combined with wireless communication technology, a wireless communication numerical simulation model is established. Corresponding conductivity devices shall be set in the numerical simulation area and marked measuring points. The conductivity value is the reciprocal of resistivity. The set conductivity and dielectric constant are convenient for subsequent analysis of dual laterolog response characteristics of the same filled mineral fractures. According to different materials, the default response numerical simulation boundary of the system does not need to be set separately, and the boundary conditions are automatically met during the solution process. The potential at infinity is 0, the electrode surface potential is equal, and the normal derivative of the potential of the insulating material is 0, preventing the current from directly passing through the insulation boundary. Combined with wireless communication technology, the electric field distribution in the research area is calculated by potential superposition in the bilateral numerical simulation. In the deep detection mode of the two-dimensional model, the reference ratio of the power supply electrode is 2.3 and the shielding electrode ratio is 5.4. Combined with the principle of electric field superposition, the three electrodes send out currents to form three electric fields, forming the basic numerical simulation environment of the model.

Next, combined with wireless communication technology, the response state of dual laterolog is visualized, so that the reservoir can be visually described and closely combined with geological evaluation; However, at present, imaging logging is expensive, and its detection depth is shallow, so it has limited effect on fractures and holes that extend farther. As a conventional logging technology, dual laterolog has the advantages of low cost and large detection range, so it is of great significance to study the response characteristics of dual laterolog of fractures and vugs for logging interpretation of fractured vuggy reservoirs such as carbonate rocks. Firstly, a basic numerical simulation model is designed based on wireless communication technology. One is stratum model. The key point of this model design is to carry out forward modeling of the dual laterolog response of the fractured vuggy formation, grasp the characteristics of the dual laterolog response of the hole, preliminarily discuss the problem of hole identification, and further study the impact of the hole on the fractured vuggy formation. Therefore, the dual laterolog numerical simulation model can be divided into two cases, namely, dual laterolog hole model and dual laterolog fracture hole model, as shown in Fig. 2 and 3 below.

According to Fig. 2 and Fig. 3, the setup and analysis of dual laterolog hole model numerical simulation and dual laterolog fracture hole model numerical simulation are completed. Next, the model is reasonably processed and adjusted in combination with wireless communication technology to form a complete numerical simulation process

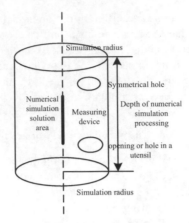

Fig. 2. Numerical simulation of dual laterolog hole model

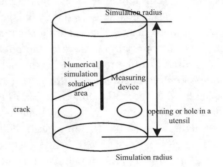

Fig. 3. Numerical simulation of fracture cavity model of dual laterolog

and a targeted processing structure, so as to further strengthen the application capability of the model.

2.4 Numerical Simulation by Boundary Constraint Processing

After the establishment of the numerical simulation model of wireless communication, the final numerical simulation is realized by boundary constraint processing. When collecting and analyzing the response state of dual laterolog, due to the convenience of finite element mesh generation and the advantages of complex boundary conditions processing, specific boundary constraint conditions or standards can be set in combination with wireless communication technology. The wireless communication technology and finite element method are used to set the basic numerical simulation solution interval for discrete calculation, so as to transform the calculation of partial differential equation with continuous solution area into the linear measurement of finite nodes (degrees of freedom). The specific numerical simulation setting process is as follows:

(1) The solution area of dual laterolog is discretized according to the range of wireless communication coverage unit and node identification range

(2) Determine cell range difference function or numerical simulation shape function
(3) Combine the stiffness matrix of each element
(4) Assemble the element stiffness equations to form the total stiffness equations for numerical simulation
(5) Directly impose the given boundary constraints on the numerical simulation conditions
(6) Load and solve the total stiffness equations, and calculate the potential function of dual laterolog
(7) Processing of dual laterolog electrode system and calculation of dual laterolog response

Based on the above process, the numerical simulation processing of dual laterolog response can be realized. Then, the next stage of discretization processing is carried out. Because the finite element mesh is relatively free, the shape and size of the formed finite element elements are also more irregular. Therefore, the finite element method is more applicable and the calculation accuracy is relatively higher when solving the complex boundary conditions in the region. The size and distribution of the field field quantity in the internal response process of dual laterolog are properly measured and numerically collected. Under the condition of using fewer nodes, the ideal calculation accuracy is achieved through the reasonable distribution of nodes. In general, the discretization of the finite element solution area has the following discrete principles, as shown in Table 2:

Table 2. Setting of Discrete Indexes for Numerical Simulation of Dual laterolog

Discrete index for numerical simulation of dual lateral logging	Directional parameter values	Controllable parameter values
Medium comprehensive ratio	3.2	4.1
Coincidence ratio in numerical simulation of dual lateral logging	6.5	6.8
Boundary basic values	16.35	18.44
Infinite boundary simulation element values	2.1	2.6
Grid encryption difference	0.21	0.16
Concave four times/time	10	18

According to Table 2, complete the setting and analysis of discrete indicators for numerical simulation of dual laterolog. The accuracy and stability of numerical simulation are improved by using boundary constraints, which lays the foundation for subsequent engineering construction.

In summary, the flow chart of numerical simulation research on dual lateral logging response based on wireless communication technology is shown in Fig. 4.

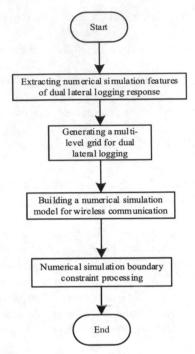

Fig. 4. Flow chart of numerical simulation research on dual lateral logging response based on wireless communication technology

3 Method Test

The main purpose of this study is to analyze and validate the practical application effect of the bidirectional side nail response digital simulation method based on wireless communication technology. To ensure the authenticity and reliability of the test results, comparative analysis was conducted and the G bidirectional side nail construction project was selected as the main test object. Collect basic data and information of the project using professional equipment and facilities, summarize and integrate them, and classify them according to different categories for future use.

Based on the actual needs of numerical simulation processing and changes in standards, comparative research was conducted, and basic testing environments were established and set up by integrating wireless communication technology. Dedicated to analyzing and verifying the effectiveness of this digital simulation method in practical applications, in order to evaluate its feasibility and effectiveness. By comparing and analyzing the actual situation, conclusions can be drawn and suggestions for improvement and optimization of this method can be provided. This research is of great significance for improving the practical application value of the digital simulation method for bidirectional nail response.

3.1 Test Preparation

Combined with wireless communication technology, set and overlap the numerical simulation processing environment of G dual laterolog construction project. First, the basic numerical simulation boundary conditions are set in combination with the actual numerical simulation processing direction and the directional conversion standard, and the higher-order basis function is calculated first, as shown in Formula 1 below:

$$D = \alpha^2 - \sum_{i=1} \beta i + (1 - \alpha)^2 \tag{1}$$

In Formula 1: D represents a higher-order basis function, α represents the conversion coefficient, β indicates the directional numerical coverage, i Indicates the number of conversions. According to the above settings, the calculation of higher-order basis functions is completed. Set it as the most basic processing dimension value in the numerical simulation structure. Combined with wireless communication technology and three-dimensional finite element analysis principle, reasonable mesh generation method is used to estimate the initial interpolation ratio of higher-order basis function. Combining with the selected two-way logging response of Project G, the value of sparse mesh generation simulation unit is scientifically adjusted, as shown in Formula 2 below:

$$L = (1 - v \times \frac{\beta v}{m + n})^2 + mv^2 \tag{2}$$

Equation 2: L represents the value of the subdivision simulation unit of the sparse grid, v indicates coverage, m represents the analog accuracy value, n indicates the precision deviation, β represents year-on-year linear interpolation. According to the above settings and analysis, the calculation of the simulation unit value of the subdivision of the sparse grid is completed. At this time, judge whether the numerical simulation environment is reasonable according to the change of the simulation unit value of sparse grid subdivision, and calibrate the change position of the dual laterolog response, connect with the initial simulation structure to form a complete processing system, and clarify the processing standards of each link for future use.

In the G project, 6 points are selected as the actual numerical simulation measurement area, and the electrode coefficient of the logging response at this time is measured in combination with the high-order basis function and the variation of the simulation unit value of the sparse grid division.

In general, the response of different logging locations will be different, and the traditional numerical simulation will only simulate one point. Although this method can achieve the expected simulation tasks and objectives, it is not targeted, and will also form uncontrollable problems in the face of complex processing and measurement environments. In addition, due to the changes in the external environment and the impact of specific factors, the final processing results of Dow numerical simulation will also have errors, affecting the implementation and improvement of subsequent work. Therefore, the range of numerical simulation and basic index parameters are set to improve the simulation structure and ensure the authenticity and reliability of the final numerical simulation, as shown in Table 3 below:

Table 3. Basic indicators and parameter settings of numerical simulation

Basic indicators for numerical simulation	Initial controllable parameter standards	Standard for measured conversion parameters
Year on year linear interpolation basis function	+11.35	+15.24
Same unit simulation difference	3.05	4.21
conversion rate	1.2	1.6
Deep lateral truncation error	0.21	0.15
Coefficient of dual laterolog electrode system	+10.15 −3.2	+13.18 −2.5
Number of local simulations/time	12	18
Vertical response ratio/%	89.34	90.12

According to Table 3, the basic indicators and parameters of numerical simulation are set and verified. Next, according to the changes of the collected values, combined with the selected six point locations, the information is summarized and classified, and combined with wireless communication technology and finite element analysis principles, the grid division and boundary processing are carried out to ensure the stability and reliability of the initial numerical simulation measurement environment. Set the setting process of numerical simulation conditions in combination with boundary standards, as shown in Fig. 5 below:

According to Fig. 5, complete the design and analysis of the flow structure of the setting of numerical simulation conditions. At this time, a directional numerical simulation structure is designed according to the terrain characteristics of the area, the preset values and reference values are imported into the structure, the response state at this time is measured, the wireless communication technology and finite element design method are integrated, the set numerical simulation boundary conditions are fused, and the basic test environment is set. Next, the wireless communication technology is used to, conduct specific test and verification analysis.

3.2 Test Process and Result Analysis

In the above built test environment, combined with wireless communication technology, set the actual requirements and standards for numerical simulation of G project, and conduct specific testing and analysis. Analyze the response of the dual laterolog built in the selected area, and extract the actual response characteristics. This part mainly carries out analysis and research from different angles, namely different widths, different inclinations and different filling degrees. Combining with the actual measurement requirements, the potential function under multiple environments is calculated, as shown in Formula 3 below:

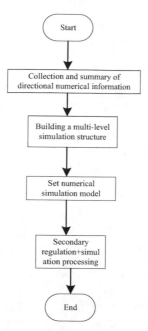

Fig. 5. Flow structure diagram of numerical simulation condition setting

$$H = h - \gamma \times \frac{\Im_2(\gamma + 1)}{\kappa \Im_1} \tag{3}$$

In Formula 3: H represents the potential function, h represents the constant value of conversion potential, γ indicates the recognition range of directional features, κ represents the cell value, $\Im_1$ and $\Im_2$ represents the controllable response range and the measured response range respectively. The specific value of the potential function can be obtained based on the calculation requirements of the. Then, after the basic setting is completed, the surrounding environment needs to be adjusted and integrated. In dual laterolog, the logging tool axis is vertical or nearly vertical to the formation level, and whether the formation, borehole or mud invasion shape is considered to be rotationally symmetric around the tool axis. For the change of construction requirements and standards of the project, the dual laterolog does not actually have strong symmetry, and the symmetry of borehole and mud invasion shape no longer exists.

Therefore, in order to increase the authenticity and reliability of numerical simulation, in the process of formation exploration, analyze the data and information related to the response of logging tools, such as formation attributes, borehole shape, mud invasion status, and instrument measurement location. The formation properties of dual laterolog include dip angle, strike, lithology, porosity and fluid properties in the porosity. When the logging tool is oblique or parallel to the formation plane, the measurement result of the tool is related to the measurement orientation. If the formation is evaluated in the vertical well mode, it will produce a large error. When the well passes through the reservoir, the upper and lower dense surrounding rocks also affect the instrument measurement response, the unequal thickness of the reservoir relative to the borehole also affects the measurement results of some instruments, the pore type of the formation and the fluid property in the pore affect the mud invasion and the fluid distribution in the space around the borehole, these will affect the measurement results.

After the above settings and adjustments are completed, the next step is to integrate the wireless communication technology and finite element simulation program for the next stage of settings. Set the wireless communication unit area within the coverage range, and set a certain number of nodes to obtain real-time data and information. The program is used to analyze the influence of the target stratum, intrusion zone, surrounding rock, instrument eccentricity and tilt angle and other stratum structures on the instrument measurement results in detail, and to set the basic measurement simulation indicators and related parameters, as shown in Table 4 below:

Table 4. Basic Numerical Simulation Indexes and Parameter Setting Table

Basic numerical simulation indicators	Directional indicator parameter standards	Measured index parameter standards
Constant ratio	2.01	2.16
Analog conversion deviation limit	16.35	17.55
Shallow lateral apparent resistivity	6.5	8.5
Deep lateral apparent resistivity	10.5	16.5
Stratum thickness/m	3.5	4
Virgin zone resistivity	8.5	9
Motor coefficient	3.16	4.22
Electrode reflux value	21.38	26.34

Set and adjust the basic numerical simulation indicators and parameters according to Table 4. Next, based on the actual situation of the current engineering stratum and wireless communication technology, a stable numerical simulation environment is built. Four positions are selected as the marker points for numerical simulation, and the output responses of different electrodes are measured, as shown in Fig. 6 below:

According to Fig. 6, the collection and display of output responses of different electrodes are completed. Then, based on this, the collection and collection of values and information are carried out for the four set points. Wireless communication technology is used to associate the set communication monitoring nodes for conversion and replacement of real-time data and information. Perform numerical simulation on the ground at different levels and get basic simulation results. Set multiple depths and determine the final numerical simulation index parameters to determine the final resistivity, as shown in Formula 4 below:

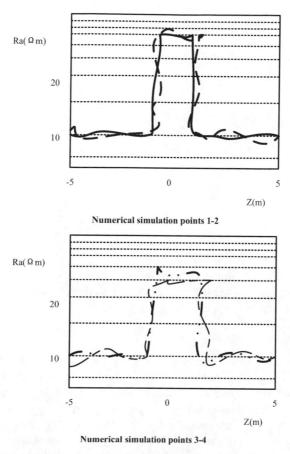

Ra(Ω m)

Numerical simulation points 1-2

Ra(Ω m)

Numerical simulation points 3-4

Fig. 6. Diagram of output response of different electrodes

$$U = \eta^2 \times \sum_{e=1} \Re e - \gamma\eta + \Im e^2 \qquad (4)$$

In Formula 4: U is the resistivity of numerical simulation, η represents the conversion ratio, η represents the value of lateral analog unit, e Indicates the number of one-way simulations, γ represents the conversion deviation, $\Im$ represents the inverse mean. According to the above determination, complete the comparative analysis of the test results. At this time, set the depth of 1.1 m, 1.3 m, 1.5 m, 1.8 m, 2 m, 2.3 m, 2.5 m and 3 m for numerical simulation measurement, and measure the resistivity of each marking point, as shown in Table 5 below:

According to Table 5, complete the analysis of the test results: conduct numerical simulation processing for the depths of 1.1 m, 1.3 m, 1.5 m, 1.8 m, 2 m, 2.3 m, 2.5 m and 3 m, and combine the motor coefficient of each set point, the measured resistivity is well controlled below 7, indicating that this kind of numerical simulation has relatively large coverage and strong pertinence, it has practical application value to optimize the

Table 5. Comparison and Analysis of Test Results

Numerical simulation to determine depth	Motor coefficient	Numerical simulation of resistivity
1.1 m	1.03	3.5
1.3 m	1.16	4.1
1.5 m	1.23	4.5
1.8 m	1.29	4.9
2 m	1.31	5.3
2.3 m	1.36	5.6
2.5 m	1.42	5.8
3 m	1.52	6.4

unit simulation structure, improve the efficiency and quality of the overall numerical simulation, and minimize the difference in the simulation process.

4   Conclusion

This article studies a numerical simulation method for bidirectional nail response based on wireless communication technology. Adopting multi-level methods to improve the efficiency of numerical simulation, achieving multi-level grid generation of bidirectional side nails, establishing a wireless communication numerical simulation model, and achieving numerical simulation through boundary constraint processing. In a word, the above is the design verification study of the numerical simulation of dual laterolog response based on wireless communication technology. With the assistance and support of wireless communication technology, the numerical simulation effect of dual laterolog is further guaranteed. Although the acquisition and acquisition of response values will be affected by caves, geological layers and other conditions, the daily numerical simulation error is greatly reduced, and the simulation accuracy is further improved. Moreover, under the effect of wireless communication technology, the numerical simulation range of dual laterolog has gradually become more scientific and reasonable, showing a normal linear correlation, which also provides reliable data for the determination of the location of dual laterolog and the marking of monitoring points, thus providing technical support for subsequent logging work.

References

1. Myshakin, E., Garapati, N., Seol, Y., et al.: Numerical simulations of depressurization-Induced gas hydrate reservoir (B1 sand) response at the Prudhoe Bay Unit Kuparuk 7-11-12 pad on the Alaska North Slope. Energy Fuels **5**, 36 (2022)
2. Shishkina, O.A., Indrupskiy, I.M., Kovalenko, K.V., et al.: Simulation of time-lapse resistivity logging during two-phase well testing in petroleum reservoirs. J. Phys: Conf. Ser. **1730**(1), 012101 (2021)

3. Xie, G.: Establishment of logging evaluation criteria for the cementing quality of low-density cement slurries. Petrol. Drill. Tech. **50**(1), 119–126 (2022)

4. Song, Y., Chen, H., Li, C., et al.: Symmetrically partitioned isotropic model for quantitative equivalent simulation of circumferential anisotropy. J. Geophys. Eng. **1**, 1 (2022)

5. Zhang, Y., Wei, Y., Houquan, Y.U., et al.: Simulation and experimental studies on the influencing factors of a thermal flowmeter with constant temperature difference. Petrol. Drill. Tech. **49**(2), 121–126 (2021)

6. Yi, Q., Zhang, G., Amon, B., et al.: Modelling air change rate of naturally ventilated dairy buildings using response surface methodology and numerical simulation. Build. Simul. **14**(3), 827–839 (2021)

7. Zhou, X., Zhang, Z., Zhang, C.: Bi-LSTM deep neural network reservoir classification model based on the innovative input of logging curve response sequences. IEEE Access 1 (2021)

8. Kang, Z., Li, X., Ni, W., et al.: Using logging while drilling resistivity imaging data to quantitatively evaluate fracture aperture based on numerical simulation. J. Geophys. Eng. **3**, 3 (2021)

9. Yu, L., Wang, H., Wang, H., et al.: 3D FV simulation of the orthogonal azimuth electromagnetic tool response logging while drilling with multi annular grooves using potentials. IOP Conf. Ser. Earth Environ. Sci. **660**(1), 012043 (2021)

10. Fan, Xiao, L.: Parameter simulation of permeable brick in concrete pavement based on particle accumulation. Comput. Simul. (005), 039 (2022)

11. Wang, S.: Artificial intelligence applications in the new model of logistics development based on wireless communication technology. Sci. Program. (9), 1–5 (2021)

12. Wu, Z., Chen, R., Pan, S., et al.: A ratiometric fluorescence strategy based on dual-signal response of carbon dots and o-phenylenediamine for ATP detection. Microchem. J. **164**, 105976 (2021)

13. Wu, H.T., Chen, M.Y.: A multi-function wearable radio transceiver device based on radio communication technology. Comput. Electr. Eng. **91**(4), 107062 (2021)

14. Li, X., Zhou, Y., Wong, Y.D., et al.: What influences panic buying behaviour? A model based on dual-system theory and stimulus-organism-response framework. Int. J. Disaster Risk Reduct. (5), 102484 (2021)

15. Li, Z., Wang, J., Xiao, L., et al.: A dual-response fluorescent probe for Al3+ and Zn2+ in aqueous medium based on benzothiazole and its application in living cells. Inorganica Chimica Acta **516**, 120147 (2021)

Sharing Method of Online Physical Education Teaching Resources in Higher Vocational Colleges Based on Soa Architecture and Wireless Network

Zhipeng Chen[✉]

Chongqing Vocational Institute of Engineering, Jiangjin 402260, China
`chenzhipeng1993@126.com`

Abstract. Due to the large amount of data and complicated information processing of online physical education teaching resources, the reliability of traditional teaching resources sharing method is difficult to guarantee, and it is easy to collapse when multiple users run together. In order to improve the transmission efficiency of online educational resources sharing, SOA architecture and wireless network are introduced, and an online physical education teaching resource sharing method is designed. Design an overall architecture of data sharing based on SOA architecture, and establish an online physical education teaching resource base according to this framework; On this basis, the entity E-R diagrams of different modular teaching resources are established, and the data information is converted into DBMS records. By dividing the attributes of physical education teaching resources and classifying teaching resources, the effective transmission of resource sharing can be realized. Complete the sharing of online physical education teaching resources in higher vocational colleges in the data exchange center. The experimental results show that the proposed method has a good effect in practical application, which can improve the network load rate, ensure the transmission speed of resource sharing at a higher level, and have higher reliability and better resource sharing performance.

Keywords: SOA Architecture · Wireless Network · Higher Vocational Colleges · Online Physical Education Teaching Resources · Sharing Method

1 Introduction

With the continuous advancement of education, teaching and information construction, the widespread application of network multimedia and the promotion of higher vocational education, the education and teaching in colleges and universities have presented a variety of learning methods, and modern higher vocational education has been inseparable from the support of information technology [1]. Among them, online learning is the most convenient and popular learning method in colleges and universities, and online physical education teaching is a novel teaching method. At this stage, most colleges

L. Yun et al. (Eds.): ADHIP 2023, LNICST 548, pp. 330–342, 2024.
https://doi.org/10.1007/978-3-031-50546-1_22

and universities have established campus networks, and in the process of construction, teaching resources such as textbooks and teaching courseware have been electronically processed and placed in the websites of various schools to facilitate the learning of students and teachers [2, 3]. However, in different campus networks, students' learning resources are relatively independent. Without correct identity authentication, it is difficult for students and teachers to share the teaching resources in the campus network, and they can't log in to the school website and teaching website [4]. This way of resource circulation can not meet the needs of users for learning resources, but also lead to the phenomenon of repeated use of learning resources within the school. In order to realize the circulation of high-quality teaching resources and optimize students' learning resources, technical units put forward special guiding measures for building campus resource sharing methods.

Reference [5] puts forward the application of Internet-based mobile information system in college physical education classroom teaching. Firstly, the survey data is analyzed by induction. Secondly, the characteristics of college courses are studied by case analysis. Finally, after discovering the existing problems, the author uses the method of comparative study to compare the influence of students, teachers and leaders in charge on the existing educational problems. Physical education in colleges and universities lacks theoretical guidance and concept sorting, which makes students not form a good sense of lifelong physical education; The positive influence of physical exercise on a person greatly promotes the development of personal lifelong sports. Reference [6] puts forward the application of massive open online course in physical education teaching mode under the background of big data, and applies big data technology to online teaching platform. According to the analysis results of big data, learners can flexibly choose the time and place of class according to their own needs and characteristics, and study personalized and customized teaching resources. Through the comparative experiment of two classes of students majoring in physical education in a university, the results of big data analysis show that the application of massive open online course in physical education teaching mode can improve students' learning enthusiasm to a certain extent, and students' satisfaction with massive open online course is as high as 92%. It can also help students improve their academic performance, and the average grade of classes using massive open online course is 10 points higher than that under the traditional education mode.

Although the above research has made some progress, due to the lack of modern technology as a support for the development and construction of the shared system, after the system is put into use, there are some problems such as high maintenance cost and failure to play the expected effect. Therefore, this paper puts forward a method of online physical education teaching resources sharing in higher vocational colleges based on SOA architecture and wireless network. The innovation of this method is to share resources by using the service combination technology of SOA architecture, and at the same time, to break through the academic and cultural barriers between campus and off-campus by using wireless network technology, so as to make the sharing of physical education teaching resources more efficient, convenient and popular. Establish entity E-R diagrams of different modular teaching resources, and convert data information into DBMS record form; Divide the attributes of physical education teaching resources, classify teaching resources and realize the effective transmission of resource sharing. The research shows that the proposed method has a good effect in practical application, which can improve the network load rate, ensure the transmission speed of resource sharing at a higher level, and have higher reliability and better resource sharing performance.

2 Teaching Resource Sharing Based on SOA Architecture

2.1 Based on SOA Architecture to Establish the Overall Architecture of Data Sharing

SOA architecture is a service-oriented architecture, which can complete the interoperability of different program functions through good interfaces and component architecture, and it is a technical system to solve the development and maintenance problems of application systems in complex distributed environments [7, 8]. Its core idea is to decompose the application into multiple reusable, decentralized and autonomous service units, and build a complex application system through the combination of these service units. Through SOA architecture, we can realize standardized management of different resources, reuse resources and improve development efficiency.

In order to ensure the real-time sharing of online physical education teaching resources information in higher vocational colleges, this paper designs an interactive model of information sharing, and the overall structure is shown in Fig. 1.

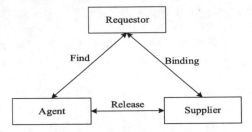

Fig. 1. SOA physical education teaching resources information sharing model architecture

The SOA sports teaching resources information sharing model shown in Fig. 1 includes three parts: the agent, the supplier and the requester of sports teaching resources information sharing. The supplier means that the platform provides various services, including providing information to the database and publishing existing data. The main task of the supplier is to provide available data resources to the agent platform to meet the positioning and query needs of the requester [9]. Requester refers to the platform that needs to download information. People can get the registered service description by searching the directory information and bind it with the corresponding service provider network. The main task of the requester is to obtain the required teaching resource information through the agent platform. Agent means that the platform completes the agent task by providing various services for the requester and the supplier. The agent acts as an intermediary, which connects the supplier and the requester, so that the requester can obtain the required teaching resource information through the agent without directly interacting with the supplier. The agent can also provide other services to meet the needs of the requester.

Generally speaking, the framework of SOA sports teaching resources information sharing model aims to connect the supplier and the requester through the agent as the intermediary to realize the sharing and exchange of teaching resources information. In this way, the acquisition and utilization of teaching resources become more convenient and efficient.

2.2 Establishing Online Physical Education Teaching Resource Database Under Wireless Network

Wireless network technology is a high-tech technology to realize wireless communication. At present, wireless network covers many places, has a wide range of applications, and has also been widely used in the field of education. The use of wireless network technology can break through the academic and cultural barriers between higher vocational colleges and make the sharing of teaching resources more convenient and popular [10]. In order to meet the demand of resource sharing, after building the information sharing model of SOA physical education teaching resources, the online physical education teaching resource database is established to fill the data layer inside the system. According to the requirements of system development and design, the database in this system includes teaching resource database, user information database and management database, in which the teaching resource database is mainly used to store the professional knowledge of ideological and political courses, which belongs to the foundation of resource sharing system development, and the resource information database includes learning resources and announcement information [11, 12]. Teachers can upload their personal information in a regular format to the database, including knowledge points, cases, etc., and also upload it to other resources that other students think are beneficial to learning. Students can browse the resources on the web page and download the required materials according to their personal learning needs. Managers can organize, store and manage the contents in this database. In this process, students of different majors can choose their own learning methods, so as to better cultivate students' awareness of autonomous learning. The public information in the database includes information posted on the website and linked resources. Administrators need to update resources in time according to students' learning progress and teaching needs.

The establishment of online physical education teaching resource database under wireless network is shown in Fig. 2.

According to Fig. 2, the information base is mainly used to store system login records and personal privacy information of users. The student information database contains personal basic information, study attendance and network status [13, 14]. The personal information database contains the registered user's name, email address, personal profile and other information, and the administrator needs to manage the data after logging into the system and completing his personal identity authentication. The learning sign-in status database can record students' online learning data, including each learning time and learning content [15, 16]. The online learning exchange meeting records the problems encountered by students in the interaction and discussion between groups.

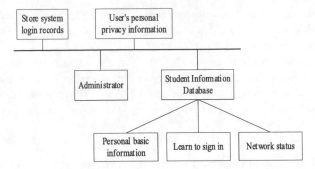

Fig. 2. The structure of online physical education teaching resource pool is established under wireless network.

Take the online physical education teaching resource database as an example, design the information input format in the database, as shown in Table 1.

Table 1. Information input format in online physical education teaching resource database

Data type	Field	Constraint condition	Meaning
Grade	String	Without	Grade
Subject	String	Without	Subject
Coding	Int	Primary key	Encode
Creator	String	Without	Creator
Content	String	Without	Content
Difficulty Level	String	Without	Difficulty Level
Source	String	Primary key	Source
Type	String	Without	Type
Access Control	String	Without	Access Control
Comments	String	Without	Comment Information

According to Table 1, the format of information input in the online physical education teaching resource database is provided, including the data types, constraints and meanings of each field. This information is very important for entering and managing physical education teaching resources in the database, which can help ensure the accuracy and consistency of the data. According to the above way, the establishment of online physical education teaching resource database under wireless network is completed.

2.3   Attribute Division and Resource Sharing Transmission of Online Physical Education Teaching Resources in Higher Vocational Colleges

On the basis of the above-mentioned online physical education teaching resource database, SOA architecture and wireless network technology are introduced to divide the attributes of online physical education teaching resources, and different types of resources are provided to effectively share transmission channels through the classification of teaching resources. In this process, the entity E-R diagrams of different modular teaching resources are established, and the data information is converted into the form of DBMS records. In this way, the internal sharing between information resources is constructed, and this mode is used as the interactive interface between system programs and databases. Integrate the data at the communication transmission interface into a data set and upload it to the system cloud. Call the data transmission function in SOA architecture and wireless network to upload resource data. The expression of the transfer function is as follows.

$$A_\alpha = \frac{B}{C \times D} \tag{1}$$

In formula (1), A_α stands for data transfer function; B indicates the time required for uploading resources to the cloud; C represents the system operation buffer coefficient; D represents the amount of data uploaded at a time. It is assumed that each SOA architecture and wireless network module have the same perception of data, and different types of data have different attributes. Therefore, it is necessary to set an attribute parameter when dividing resource attributes, and express this parameter as E, so the relationship between E and cloud data can be expressed by the following calculation formula.

$$E = \frac{E_{11}(K_{11} + E_{21})}{A_\alpha} \times F_G \tag{2}$$

In Formula (2), E_{11} and E_{21} represent the amount of data that can be carried by the cloud at different times; F_G stands for data attribute perception coefficient. According to the above formula, a delay parameter condition of the shared signal in transmission is set. In this way, an attribute pattern for online physical education teaching resources is generated, and the attribute pattern is used as the basis for the division of resource attributes. This process is shown in the following calculation formula.

$$G_G = F \begin{bmatrix} u_k \\ v_k \end{bmatrix} + e \begin{bmatrix} 1, j \\ 0, E \end{bmatrix} \tag{3}$$

In formula (3), u_k represents the stability coefficient of spatial data; v_k represents the sharing efficiency of spatial data transmission; F represents the attribute parameters of online physical education teaching resources; e represents the resource transmission timing; j indicates the sensitivity of the processor in the system to data perception. Through the above methods, we can master the basis of attribute division of resources in the system. On this basis, in order to share resources, we should unify the data format and improve the timeliness of cloud sharing transmission through pre-processing. This process is shown in the following calculation formula.

$$H_J = \frac{K_L \times T}{E \times j} \times G_G \tag{4}$$

In Formula (4), G_G represents the unification of data formats; K_L represents the working frequency of the data processor; T stands for the reliable period of single shared transmission. In the process of sharing, teachers need to set the multicast address of resources in advance. On this basis, they use multimedia technology to receive data from the teaching resource server, set different multicast addresses in different shared files, and form a teaching resource flow communication channel according to the structure of address data. Through the above methods, the attribute division and resource sharing transmission of online physical education teaching resources based on SOA architecture and wireless network are realized.

2.4   Realizing the Sharing of Online Physical Education Teaching Resources in Higher Vocational Colleges by Data Exchange Center

After completing the above-mentioned attribute division and resource sharing transmission design of online physical education teaching resources in higher vocational colleges based on SOA architecture and wireless network, in order to ensure the absolute security of teaching resources in the sharing process, online physical education teaching resources in higher vocational colleges can be used as metadata and a data exchange center can be constructed to realize the comprehensive sharing of online physical education teaching resources in higher vocational colleges. This structure can be expressed as a process as shown in Fig. 3.

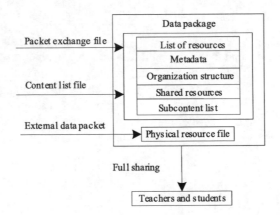

Fig. 3. Metadata interactive sharing structure

According to the data interactive sharing structure of 3 yuan, the storage format of online physical education teaching resource A can be adjusted to format-A, and the storage format of teaching resource B can be adjusted to format-B, and the process of resource sharing can be regarded as the interactive process of A and B. According to the following steps, metadata exchange is carried out in the data center where format resources are stored. The specific steps are as follows:

Step 1: Organize the remote online physical education teaching resources to be transmitted and shared, and determine the data transmission standard. Store the corresponding data in the template to complete the relevant preparation work.

Step 2: Set the shared terminal as the receiving format, and explain, remark, register and record the data information to be transmitted at the receiving terminal. At the same time, a clear file is established to store the entity information, so as to ensure that the teaching resource sharing agreement in the data exchange center has been generated.

Step 3: Dock the user server application interface with the shared transmission communication interface, as shown in Fig. 4.

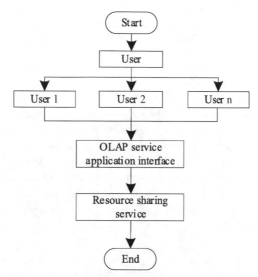

Fig. 4. The user server application interface interfaces with the shared transmission communication interface.

According to Fig. 4, the user server application interface refers to the interface where the application running on the user server interacts with other systems or services. It provides a standardized way, which enables the user server to perform data transmission and information query with other systems. Through the docking of user server application interface and shared transmission communication interface, data sharing and communication between different systems can be realized, and the interoperability and integration between systems can be improved. Applications on the user server can interact with other systems more flexibly to meet various business needs.

Step 4: Check the identity information of the sharing end to ensure the security and privacy of personal information and resource information during transmission. The authentication process of identity information can be expressed by the following calculation formula.

$$\theta = \frac{P_d \times G_d \times H_J}{N_0 + \sum_m y_c^d \times P \times G_{dc}} \tag{5}$$

In Formula (5), P_d is represented as an information transmitter; G_d denotes controllable or uncontrollable interference of the transmission channel; N_0 stands for fixed communication base station; y_c^d is expressed as shared channel bandwidth; P is expressed as the effective gain value of the communication channel; G_{dc} stands for user identity information category. Usually, there are three types of m, corresponding to m_1, m_2 and m_3, which are respectively represented as receiving end users, sending end users and intermediate users. The specific value of m is set according to the category of shared transmission peer users.

After the user's identity information is checked, remote online physical education teaching resources are shared and transmitted according to the resource sharing agreement, so as to realize the research on online physical education teaching resources sharing method in higher vocational colleges based on SOA architecture and wireless network.

3 Experimental Analysis

In order to verify the effect and feasibility of online physical education teaching resources sharing method in higher vocational colleges based on SOA architecture and wireless network, an experiment was designed. Before the experiment, a pilot higher vocational college in a certain area was selected as the test unit, and the existing online physical education teaching resources in colleges and universities were called, and the teaching resource sharing system in schools was developed by using this method. In order to meet the experimental requirements, a comparative experimental test environment is built according to Fig. 5.

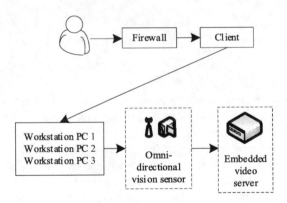

Fig. 5. Structure diagram of experimental environment

On the basis of the above experimental environment structure of Fig. 5, in order to ensure that the test environment meets the actual needs, the technical parameters of the equipment are designed after the selection of the equipment in the test environment is completed. In order to ensure that the distance education terminal can effectively carry online physical education teaching resources in higher vocational colleges, relevant arrangements are made in the experimental operation terminal. The specific contents are shown in Table 2.

Table 2. Table of shared information of experimental operating equipment terminals

Serial number	Information list	ID terminal address information
1	Terminal server 1	@S1 Domain
2	Terminal server 2	@S2 Domain
3	Intermediate agency	@Agent
4	Shared service user 1	@Com-1
5	Shared service user 2	@Com-2
6	Shared service user 3	@Com-3
7	Security audit log server	@Log Server
8	Data backup server	@Backup Server
9	Emergency response server	@Emergency Server
10	System monitoring server	@Monitor Server

After the comprehensive deployment of the test environment is completed, the proposed method is integrated into the computer terminal according to the standard. According to the channel of online teaching resources in higher vocational colleges, the professional course education information is obtained, the database is introduced, and the online physical education teaching resources-driven transmission mode in higher vocational colleges is designed in combination with the WEB terminal. On this basis, a data exchange center is designed in terminal server 1 and terminal server 2, and teaching resources are shared in the center.

In order to ensure a certain contrast of the experiment, the reference [5] mobile information system based on the Internet in college physical education classroom teaching method and reference [6] massive open online course's application method in physical education teaching mode were selected as the control group, and the sharing performance of the methods was tested by calling the terminal server to run data in the background. In order to test the feasibility of the proposed method in real educational environment, the proposed method was used as an experimental group to test it.

Establish the communication connection between multiple terminals in the school, ensure that after different terminals are in good communication state, drive the program and share the online physical education teaching resources among multiple terminals. Take the network speed of resource sharing transmission under different concurrent access numbers as the evaluation index, call the background data, record the running network speed under different concurrent access numbers, and count the experimental results, as shown in Fig. 6.

As can be seen from Fig. 6, with the increase of the number of concurrent visitors, the network speed of physical education teaching resources sharing transmission decreases under the method of reference [5], and the network interruption occurs when the number of concurrent visitors reaches 240, that is, the number of simultaneous online users has exceeded its own load. Only the resource sharing network speed of the proposed method does not increase with the increase of the number of concurrent visitors. Therefore,

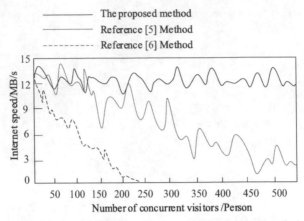

Fig. 6. The results of network speed of PE teaching resources sharing transmission under different methods.

the proposed method has a good effect in practical application, which can improve the network load rate and ensure the transmission speed of resource sharing at a higher level.

In order to verify the resource sharing efficiency of the proposed method, we compared the method in reference [5] with the method in reference [6], conducted multi-user data search, and recorded the running situation and resource matching results of the three methods. The specific experimental results are shown in Table 3.

Table 3. Comparison Table of Experimental Results

Number of users	Number of requests	Average response time/s			Matching degree of shared resources/%		
		The proposed method	Reference [5] method	Reference [6] method	The proposed method	Reference [5] method	Reference [6] method
50	30	0.17	0.28	0.34	98.9	84.5	84.2
100	150	0.19	0.32	0.36	98.6	72.8	83.2
150	300	0.22	1.02	0.39	97.7	71.1	83.1
200	400	0.23	1.98	1.02	97.4	69.3	82.5
250	450	0.31	2.35	1.09	97.0	69.0	82.0
300	550	0.35	2.69	1.15	96.5	68.2	81.6
350	680	0.46	2.74	1.26	96.1	67.6	81.2
400	770	0.56	2.78	1.34	95.6	67.2	80.9
450	890	0.69	3.54	1.45	95.4	67.0	80.1
500	990	0.85	4.56	1.56	95.2	66.6	79.2

By comparing Table 3, from 50 to 500 users, the number of requests per user gradually increased from 30 to 990. With the increase of the number of users and requests, the average response time also increases gradually; The data of matching degree of shared resources fluctuates in increase and decrease. And it can be seen that the proposed method is superior to the methods in reference [5] and reference [6] in all aspects. With the increase of the number of users, the reliability of the proposed method is still guaranteed, and it can maintain a stable and fast response speed, which can meet the smooth operation when there are many users at the same time. Compared with the method of reference [5] and the method of reference [6], the response speed is normal in the early stage, but it is not optimistic when multiple users run at the same time. And in terms of matching effect of resource sharing, because the proposed method increases resource category information, the search efficiency is obviously faster, while the reference [5] method and the reference [6] method have no relevant information processing, and can only rely on similar names to search, so the search matching situation is poor or even mismatched. Therefore, the online physical education teaching resource sharing method based on SOA architecture and wireless network in higher vocational colleges has higher reliability and better resource sharing performance.

4   Conclusion

This paper puts forward a method of sharing online physical education teaching resources in higher vocational colleges based on SOA architecture and wireless network, and draws the following conclusions through research:

(1) The proposed method has a good effect in practical application, which can improve the network load rate and ensure the transmission speed of resource sharing at a higher level.
(2) With the increase of the number of users, the reliability of the proposed method is still guaranteed, and it can maintain a stable and fast response speed. At the same time, it runs smoothly when there are many users, and the resource sharing method has higher reliability and better resource sharing performance.

The research method is a research field with great potential and development prospect. Future research work can be prospected in the following aspects:

(1) Improving system performance and stability: Future research can be devoted to improving system performance and stability. By optimizing the system architecture and improving the data transmission and processing mechanism, the response speed, concurrent processing ability and stability of the system are further improved, and the user experience is improved.
(2) Strengthen security and privacy protection: Online physical education teaching resources sharing involves users' personal information and sensitive data, so security and privacy protection is an important research direction. Future research can explore more secure and reliable identity authentication and access control mechanisms, strengthen data encryption and transmission security, and design privacy protection strategies and mechanisms to ensure the security and privacy of user data are not infringed.

In a word, the future research can continue to explore and innovate in depth, and achieve more outstanding results in system performance, security and privacy protection, intelligent and personalized service, cross-platform adaptation and quality evaluation of teaching resources, so as to provide better support and services for online physical education teaching resources sharing in higher vocational colleges.

References

1. Abirami, K.R., Ajaye, K.P., Amrrish, R., et al.: Efficient method for storing health record in cloud using integrity auditing and data sharing. J. Phys: Conf. Ser. **1916**(1), 012191–12121 (2021)
2. Fangzhou, L., Yang Yang, D., et al.: Study on the building of network sharing system for the database of genuineness of Chinese medicinal materials. J. Med. Inform. **42**(7), 62–67 (2021)
3. Merz, K.M., Amaro, R., Cournia, Z., et al.: Editorial: method and data sharing and reproducibility of scientific results. J. Chem. Inf. Model. **60**(12), 5868–5869 (2020)
4. Cooper, K.A.: Data sharing attitudes and practices in the plant sciences: results from a mixed method study. J. Agric. Food Inf. **22**(1), 37–58 (2021)
5. Li, J.: Application of mobile information system based on internet in college physical education classroom teaching. Mob. Inf. Syst. **2021**, 1–10 (2021)
6. Wang, L., Wang, M.: Application of MOOC in physical education teaching mode under the background of Big Data. J. Phys. Conf. Ser. **1744**(4), 042233 (7p.) (2021)
7. Yao, S., Li, D., Yohannes, A., et al.: Exploration for network distance teaching and resource sharing system for higher education in epidemic situation of COVID-19 - ScienceDirect. Procedia Comput. Sci. **183**, 807–813 (2021)
8. Quane, E.C., Quane, S.L., et al.: Assessment of technological setup for teaching real-time and recorded laboratories for online learning: implications for the return to in-person learning. J. Chem. Educ. **98**(7), 2221–2227 (2021)
9. Nes, A., Hybakk, J., Zlamal, J., et al.: Mixed teaching methods focused on flipped classroom and digital unfolding case to enhance undergraduate nursing students' knowledge in nursing process. Int. J. Educ. Res. **109**(3), 101859 (2021)
10. Hu, S., Liu, Y., Wang, S.: Teaching exploration of case-based data modeling optimization for database system. Open J. Soc. Sci. **08**(3), 514–521 (2020)
11. Sato, E., Chen, J.C.: Rise to the occasion: the trajectory of a novice Japanese teacher's first online teaching through action research. Lang. Teach. Res. **25**(2), 306–329 (2021)
12. Talaván, N., Lertola, J., Moreno, A.I.: Audio description and subtitling for the deaf and hard of hearing: media accessibility in foreign language learning. Transl. Translang. Multiling. Contexts **8**(1), 1–29 (2022)
13. Liu, S.X., Zhang, D.P., Wu, Y.N., et al.: Wireless sensor network positioning algorithm based on RSSI model. Comput. Simul. **39**(1), 427–431 (2022)
14. Tucker, B.V., Kelley, M.C., Redmon, C.: A place to share teaching resources: speech and language resource bank. J. Acoust. Soc. Am. **149**(4), A147–A147 (2021)
15. Liu, X., Chen, S., Song, L., et al.: Self-attention negative feedback network for real-time image super-resolution. J. King Saud Univ. – Comput. Inf. Sci. **34**(8), 6179–6186 (2022)
16. Liang, X., Yin, J.: Recommendation algorithm for equilibrium of teaching resources in physical education network based on trust relationship. J. Internet Technol. **23**(1), 133–141 (2022)

Wireless Networks for Social Information Processing, Image Information Processing

Application of Intelligent Mobile Terminal in Virtual Building Construction Training Teaching

Shida Chen[1(✉)], Xiaodan Liang[1], and Pan Zhao[2]

[1] Shanghai Urban Construction Vocational College, Shanghai 200438, China
csd_succ@163.com
[2] Chizhou University, Chizhou 247000, China

Abstract. The conventional teaching method of building construction training is mainly desktop real-time data interaction. Although the investment cost is low, its immersion is poor, which affects the effectiveness of practical teaching courses. Therefore, the application of intelligent mobile terminal in virtual building construction training teaching is studied. Establish the function module of virtual building construction training teaching, and improve the immersion function of the training teaching course. Simplify the teaching task of virtual building construction training based on intelligent mobile terminals, reduce other costs to the minimum, increase the investment in interactive equipment, and meet the immersive experience of students in practical teaching. Manage the virtual building construction training report, and conduct the whole life cycle training management for students in the whole training teaching process, so as to achieve the immersion and effectiveness of the training teaching. The simulation experiment proves that the average benefit function of this teaching method is 24.6, and its intelligent mobile terminal has low latency and energy consumption, which has high terminal efficiency.

Keywords: Intelligent Mobile Terminal · Virtual Building Construction Training Teaching · Teaching Task

1 Introduction

The construction industry is one of the three pillar industries of China's national economy, and its sustainable development can simultaneously drive the continuous development of multiple related industries. It plays a very important role in promoting employment, accelerating the transfer of rural surplus labor, and driving the development of related industries. Virtual reality technology is a new field of computer research. In recent years, it has gradually attracted the attention of all circles, and has been further developed in the application field. As a new technology in the computer field that integrates a variety of science and technology, it has involved many research and application fields, and is considered to be an important development discipline in the 21st

© ICST Institute for Computer Sciences, Social Informatics and Telecommunications Engineering 2024
Published by Springer Nature Switzerland AG 2024. All Rights Reserved
L. Yun et al. (Eds.): ADHIP 2023, LNICST 548, pp. 345–360, 2024.
https://doi.org/10.1007/978-3-031-50546-1_23

century and one of the important technologies that affect people's lives. Virtual reality technology is an advanced computer user interface, which integrates computer graphics, computer simulation technology, multimedia technology, artificial intelligence technology, human-computer interface technology, sensor technology and other cross technologies. It achieves a special purpose by providing users with various intuitive and natural real-time perception interactive means such as viewing, listening, and touching. "Immersion", "interactivity" and "conceptualization" reflect the key characteristics of the VR terminal, and emphasize the leading role of human in the VR environment. Modern architecture has high technical content, diverse structural forms, complex process, and great difficulty in construction organization and management. It is necessary for higher architectural vocational education to cultivate construction technicians and management talents who have both college level professional knowledge and advanced skills and are good at transforming engineering drawings into physical entities. The practical teaching of construction engineering technology is an important link in cultivating students' hands-on and innovative abilities. The implementation of virtual reality training room technology will have a profound impact on exploring and developing modern educational ideas, improving educational technology levels, improving experimental and training environments, optimizing teaching processes, and cultivating talents with innovative awareness and ability.

Reference [1] proposes an exploration of a blended virtual training teaching model for vocational colleges based on deep learning. Combining deep learning with vocational training teaching, analyzing the theoretical and practical foundation of integrating deep learning into vocational training teaching. Supported by theories such as blended learning and virtual simulation technology, analyzing problems and other five stages to form a blended virtual training teaching mode for deep learning in vocational education. By conducting experiments to verify the promoting effect of this model in cultivating students' innovative thinking, we aim to assist in the cultivation of innovative and highly skilled talents. Reference [2] proposes the construction of a logistics training and teaching center based on Internet of Things technology. Starting from the development of the logistics industry and the demand for logistics talents, this paper elaborates on the goal, significance, and feasibility of constructing a logistics teaching and training center based on Internet of Things technology. The overall planning and specific design of the construction plan of the training and teaching center are also carried out, and the expected teaching results of the Internet of Things intelligent logistics laboratory are discussed.

Due to the limitations of teaching conditions and the impact of construction season, economy, site, traffic and safety, students have fewer opportunities to go to the construction site to carry out hands-on construction. It is difficult to carry out practical training in the real environment for each link of the construction process and quality inspection. There are great limitations in understanding and mastering the construction technology, which not only involves a lot of funding issues, it also involves personnel training cycle, construction safety, resources, environment and other issues. Virtual construction technology can track every link in the construction process, and carry out training, verification, optimization of construction technology and construction organization in the whole process of construction and production. At the same time, because virtual reality technology has the advantages of low cost, good versatility, easy modification, high security and other traditional technologies in application, it makes the application of virtual technology in building construction urgent and possible. Intelligent mobile terminal is the mainstream of the development of mobile Internet industry today [3–5]. It has the ability of high-speed access to the network, can carry various open operating systems, has independent operating space, and supports users to install apps independently. Common mobile intelligent terminals in life mainly include smart phones, vehicle mounted intelligent terminals, laptops, wearable devices, etc. Driven by these emerging technologies, enterprise planning and national policies, mobile intelligent terminals have developed rapidly and become the main tool for people to store and obtain information in their daily work, study and life, especially the intelligent mobile terminals that dominate the market. Therefore, this paper studies the application of intelligent mobile terminal in virtual building construction training teaching, and makes full use of virtual reality technology, computer technology, network technology, multimedia technology, etc. To realize the true reproduction of construction technology and quality inspection operation process on the computer, so that students can interact with the terminal realistically and immersively. Achieve the same learning effect as the on-site training, promote the teaching modernization reform of the course, improve the teaching quality, reduce the training cost and risk, and improve the students' professional ability and quality.

2 Design of Virtual Building Construction Practice Teaching Method Based on Intelligent Mobile Terminal

2.1 Establish Virtual Building Construction Training Teaching Function Module

This paper integrates "teaching, learning and doing" as a whole to strengthen the cultivation of students' ability, and adopts the "action oriented" teaching mode based on the project teaching method. The realistic three-dimensional audio-visual effect and natural interactive operation function of the virtual terminal of the construction practice training strongly attract students to participate in the scene mode, explore freely and learn independently, actively build their knowledge structure, cultivate innovative ability and exploration spirit, and truly reflect the teaching method of teaching according to materials. From this, a virtual teaching framework for building construction training is constructed, as shown in Fig. 1 below.

As shown in Fig. 1, this paper divides the virtual building construction training teaching into five functions: students use virtual training terminals to make the teaching mode more vivid; Students can operate through virtual training and complete practical training that is difficult to carry out under real conditions; Students can repeatedly operate key practical training without limitation of time and space, so as to achieve the effect of repeated review and consolidation of learning, and further promote the reform of practical training teaching methods and contents; The interaction between students and virtual training scenes is completed through visual training objects, and the detailed description of relevant scenes makes the design of training objects more effective to highlight the training objectives; Use the mouse to interact with visual training objects. Virtual training objects and students can interact by clicking or dragging the mouse. When developing

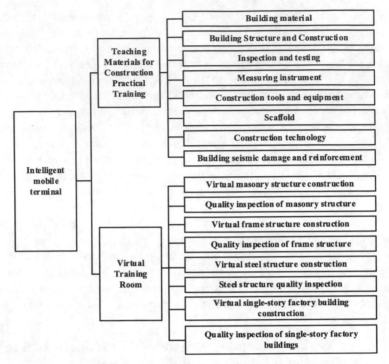

Fig. 1. Virtual Teaching Framework of Building Construction Training

an intelligent mobile terminal [6, 7] website, the terminal realizes dynamic interaction through HTML technology, and the generated pages are HTML pages. Users can directly double-click the homepage of the training website to enter the virtual training, and can set pages on the server for sharing. Intelligent mobile terminal websites can be optimized based on the characteristics and screen size of mobile devices, enabling users to easily browse and use website content on their phones or tablets, providing a better user experience. At the same time, developing intelligent mobile terminal websites can expand the user coverage of the website and attract more mobile users to visit the website. Considering students' habitual thinking and qualitative thinking, the operation design in virtual training conforms to people's logical thinking, and has strong operability and interactivity. Based on the characteristics of virtual reality technology and building construction courses, this paper divides the teaching route of virtual building construction training, as shown in Fig. 2 below.

As shown in Fig. 2, the building construction training virtual intelligent mobile terminal is built by self-development. The training teaching team is composed of vocational education experts, professional leaders, backbone teachers, industry and enterprise experts, and computer professionals. Based on students' learning, it is divided into practical teaching curriculum system, curriculum standards, evaluation standards, productive training standards on post training standards, etc. Guided by constructivism and humanism learning theories and referring to the virtual building construction training materials, this paper improves the immersive function of the practical teaching course from the

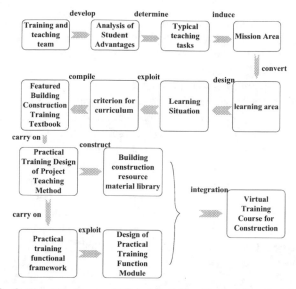

Fig. 2. Teaching Route of Virtual Building Construction Training

aspects of practical teaching objectives, practical teaching contents, practical learning environment analysis, practical teaching method design, practical learning evaluation, etc., in combination with the characteristics of virtual reality technology and courses. To really improve the learning quality of students majoring in construction engineering.

2.2 Simplify Virtual Building Construction Training Teaching Tasks Based on Intelligent Mobile Terminals

In this paper, the intelligent mobile terminal is applied to the virtual building construction training teaching process, utilizing the high-performance processors, operating systems, and various applications of intelligent mobile terminals to perform various functions and tasks, as well as supporting the installation of various applications. By downloading and using various software through the application store, the advantages of meeting users' personalized needs and reducing other training costs are met, increase the investment in interactive equipment, and meet the immersive experience of students in practical teaching.

The main task of virtual building construction training teaching is to study the general laws of construction technology of building engineering, the construction technology and process principle of each major type of work of building construction, and the development of new technology and process of building construction. Through learning and training, students can understand and master the construction technology and process principle of each major type of work in construction engineering, highlight the cultivation of professional post ability of construction workers, and cultivate students' basic ability to independently analyze and solve construction technology problems in construction engineering. Use "immersion", "interaction" and "conceptualization" to create the authenticity of architectural construction training courses.

Immersion, also called telepresence, refers to the emotional reflection of the observer on the virtual world. Users are fully integrated into this virtual world, just like people interact with nature in the real objective world, giving people a sense of immersive. Interactivity. Virtual reality is an open environment that can respond to user input. People can use some sensor devices such as keyboard, mouse, data glove, etc. to operate with objects in the virtual environment to get feedback information. Conceptual, virtual reality technology has a broad space for imagination, which can broaden the scope of human cognition. It can not only reproduce the real environment, but also conceive any objective environment at will. Users immerse themselves in the virtual environment and acquire new knowledge to improve their perceptual and rational understanding, thus enabling users to deepen their concepts and sprout new associations. The characteristics of intelligent mobile terminals include portability, versatility, real-time interconnection, personalized customization, Multi-touch and interaction, information acquisition and sharing, and personal assistant functions. Based on the characteristics of intelligent mobile terminals [8–10], this paper analyzes the virtual building construction training teaching, as shown in Fig. 3 below.

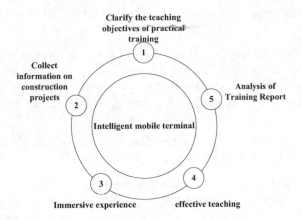

Fig. 3. Teaching analysis model of intelligent mobile terminal

As shown in Fig. 3 1 is to clarify the practical teaching objectives; 2 To collect information on construction projects; 3 is immersive experience; 4 is effective teaching; 5 is the analysis of training report. In order to ensure the orderliness of the whole teaching process, this paper assumes that there is a time slot structure in the operation of intelligent mobile terminals, and each terminal is connected to a server, then the communication established between the terminal and the designated practical teaching course is expressed as:

$$P = \sum_{m \in M} X_s^m(t), \forall s \in S, t \in T \tag{1}$$

In formula (1), P the communication range established for the terminal and the designated practical training teaching course; $X_s^m(t)$ communication request established for training course s and terminal m at time t; S is a practical teaching terminal; T for the whole training teaching time. When $P = $ At 1 h, $X_s^m(t)$ means that the communication between the training course s and the terminal m is successfully established at time t, and the next teaching operation can be carried out. In order to reduce the cost of practical teaching courses, this paper predicts the time when the terminal sets up communication for mobility, and simplifies the practical teaching tasks. The probability density function expression of the simplified model of practical teaching tasks is as follows:

$$f_{\tau_s}(t) = \begin{cases} \frac{1}{\tau_s}e^{-\frac{1}{\tau_s}}, t \geq 0 \\ 0, \qquad t < 0 \end{cases}, \quad s \in S, \quad t \in T \tag{2}$$

In formula (2), $f_{\tau_s}(t)$ is the probability density function expression of the simplified model of practical teaching tasks; τ_s refers to the time occupied by the training course in the training task. Order τ_s obeys the Gaussian distribution, and the task occupation probability follows t the training task can be simplified by increasing and decreasing. Assuming that the construction practice teaching task does not transfer, that is, under the condition of completing a teaching task, the constraints of the simplified model are expressed as:

$$p_s^0 = \int_{t_S^M}^{+\infty} f_{\tau_s}(t)dt \tag{3}$$

$$p_s^1 = \int_0^{t_S^M} f_{\tau_s}(t)dt \tag{4}$$

$$p_s^0 + p_s^1 = 1, \forall s \in S, t \in T \tag{5}$$

In formula (3–5), p_s^0 the probability of completing a practical teaching task for students; p_s^1 probability of completing two practical teaching tasks for students; t_S^M adjust the time of training course S on terminal M. The intelligent mobile terminal itself carries limited computing power, so the local computing delay of the terminal is:

$$T_s^L(t) = \frac{m_s \gamma_s}{f_{\tau_s}(t)}, \forall s \in S, \forall t \in T \tag{6}$$

In formula (6), $T_s^L(t)$ local computing delay for training course s; m_s data volume of virtual building construction training teaching task; γ_s is the calculation task of training course s. The energy consumption generated by the calculation training teaching task is mainly related to the terminal computing capacity. The energy consumption generated by each training task is expressed as:

$$E(t) = \kappa(f(t))^2, t \in T \tag{7}$$

In formula (7), $E(t)$ energy consumption for each training task; κ the conversion coefficient of periodic energy consumption; $f(t)$ calculate the energy consumption at the terminal for the training task. The data volume of known virtual building construction training teaching task is m_s the time delay of simplified practical training teaching task is:

$$T_s^U(t) = \frac{m_s}{R_s^U(t)}, \forall s \in S, \forall t \in T \tag{8}$$

In formula (8), $T_s^U(t)$ is the time delay of the simplified practical teaching task; $R_s^U(t)$ is a shortage of uplink scenarios for practical teaching tasks. $R_s^U(t)$ smaller the size, the higher the quality of the whole building construction training teaching, and the better the effect of teaching task simplification.

2.3   Management Virtual Building Construction Training Report

In this paper, virtual simulation technology is used to determine the entity model of the training together. The Java3D scene map is constructed from examples, integrating objects such as defining graphics, sound, light, position, direction, and the surface properties of visual objects and sound objects. A common graphic definition is a data structure consisting of nodes and arc edges. A node is a data element, and an edge is the relationship between data elements. The nodes in the scene graph are Java 3D instances, and the edge shows the relationship between these Java 3D instances. Java 3D scene graph is a tree structure constructed from a bunch of nodes with parent-child relationship. In the tree structure, there is only one node that is the root node, and other nodes can be accessed along the arc starting from the root. The nodes in the tree structure have no loops. A scene graph is formed by a tree rooted in a Locale object. Each scenario map has a unique Virtual Universe, which has a series of Locale objects. A Locale object that provides a reference point in the virtual world. The Locale object can be regarded as a sign to define the position of visual objects in the virtual world. Structured and modular programming methods are mainly used when planning multi scene navigation. Each solid model has its own unique functions and can be reused. The entity model can be either a static model or a dynamic model, and interaction can be achieved by adding action scripts. If there are multiple scenes, multi scene navigation is generally used to connect the scenes. Multi scene navigation needs to design a master control module to control the playing sequence of scenes. The virtual training part of this terminal is the multi scene navigation mode. This article uses the form of virtual animation to provide students with practical training experience, as shown in Fig. 4 below.

As shown in Fig. 4, (a) is the construction animation of prefabricated buildings; (b) Animation for building construction; (c) Animate the construction of steel plates; (d) Aerial shot scene animation for the building. In this paper, animation is added to the intelligent mobile terminal to enhance the effect of practical training and teaching. Most demonstration animations are two-dimensional animations made in Flash, but according to the special visual requirements of some animations, some three-dimensional animations are made in Java3D. There are three basic methods for Flash to produce animation: frame by frame animation: frame by frame animation is to edit the content of any number of keyframes, and then play the content of keyframes in order to form animation,

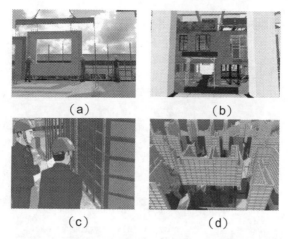

(a)   (b)

(c)   (d)

Fig. 4. Animation of Virtual Building Construction Training Teaching

which is tedious to produce; Compensation animation: pre define the content of the start point and end point of the animation. The process from the start point to the end point is compensation animation; Timeline special effect: it can provide users with "automatic" animation and visual effects that can be applied to objects, and can achieve effects such as copying to mesh, deformation, transformation, separation, expansion, projection, etc. The running effects of Java3D and VRML programs are basically the same. In Java3D, there are interpolator objects similar to VRML interpolator nodes, which can be combined with Alpha class to write various types of 3D animation programs. The Alpha class in Java3D has the same function as the key field in VRML. It is used to output the normalized time, but it can handle more complex scheduling problems. The functions of various interpolator objects in Java3D are similar to those of VRML interpolation nodes, which can be used to rotate the shape, move the coordinates of the shape, change the color of the shape, etc. In the animation, the TimeSensor continuously sends the time slice information. After receiving the time slice, the interpolator node will send it to the corresponding domain of the Transform node through interpolation calculation and path according to the set key points to realize the animation. When encountering complex animation design, the interpolator node is replaced by a code added to the Script node. These codes can generate and output animation data according to different animation requirements, which improves the flexibility of animation design. According to the virtual situation of building construction training, the teaching configuration is analyzed, as shown in Table 1 below.

As shown in Table 1, according to the characteristics of building construction, this paper collects a large number of construction cases all over the country, including building materials, structures and structures, detection and testing, measuring instruments, construction equipment, scaffolding, construction technology, engineering management cases, building seismic damage and reinforcement, current building regulations and technical standards, etc. The resource material library mainly includes text resources, picture resources, animation resources and video resources. The resource material database can search the building construction material resources by searching, which can realize the

Table 1. Teaching Configuration of Building Construction Training

Category	To configure
The server	CPU: main frequency 1.6G
	Memory: 512MB DDR
	Video card: 256M video memory
	Network card: 10/100M adaptive network card
	Hard disk: 160G
Client	CPU: main frequency 1.6G
	Memory: 512MB DDR
	Video card: shared video memory 128M
	Network card: 10/100M adaptive network card
	Hard disk: 80G
Network	Network transmission rate: 10Mbps
	Transmission medium: Class V twisted pair

retrieval of all materials, and can also realize the classification retrieval by category. Due to the new characteristics of virtual reality technology, the establishment of virtual training rooms only requires less funds, and the funds required for software upgrading and maintenance are less than those required for traditional instrument training rooms. Virtual reality technology has basically matured, has been widely used in the field of testing, and is also the development direction of future testing technology. The rapid development of network technology provides a strong technical guarantee for the realization of virtual training terminals. Computers are the hardware foundation of virtual instruments. General training rooms have a considerable number of computers. You can build virtual instrument terminals only by purchasing some relevant virtual instrument hardware, and the cost of purchasing these hardware is far lower than that of purchasing traditional instruments and equipment. It can be seen that after the application of intelligent mobile terminals to virtual building construction training teaching, it has economic feasibility, technical feasibility and resource feasibility, and can improve the quality of building construction training teaching to the greatest extent.

3 Simulation Experiment

In order to verify whether the practical teaching method designed in this paper has practical value, this paper has built a simulation experiment platform and carried out simulation analysis on the above methods. The practical performance of the conventional virtual building construction training teaching method based on deep learning, the conventional virtual building construction training teaching method based on the Internet of Things, and the virtual building construction training teaching method based on intelligent mobile terminals designed in this paper are analyzed to find the best teaching scheme. The specific experimental preparation process and the final experimental results are shown below.

3.1 Experiment Preparation

The simulation hardware condition is a small portable computer equipped with 3. 2 GHz AMD R7 processor and 16 G RAM, and the software condition is Intelli J IDEA software with Java-JDK1. 8 environment. The task is set as a radar emitter identification task with a data volume of (20005000) MB. The data volume of large tasks in this chapter is several times that of small and medium-sized tasks, and the computing resources of the cloud computing center are also several times that of the MEC server. The simulation parameters are shown in Table 2 below.

Table 2. Simulation Parameter Settings

Parameter description	Parameter value setting
Initial number of terminals	30
Energy consumption for terminal mode conversion	300 J
Maximum tolerance delay	13 min
Local computing resources of the terminal	Random (1, 2) GHz
Computing resources assigned to a single task by the MEC server	Random (10, 15) GHz
Computing resources assigned to a single task by the cloud computing center	Random (1000, 1100) GHz
Size of task upload data	Random (2000, 5000) MB
Task return fuzzy prediction coefficient	0.3
Cloud computing communication delay estimation coefficient	1.1
Energy consumption conversion coefficient	10–19
Weight of delay in revenue	0.5
Weight of energy consumption in income	0.5
Terminal standby mode power	12 dBm

As shown in Table 2, building construction training teaching tasks are arranged within 100 s in this experiment. The average power of uploaded teaching tasks is about 30 dBm, and the noise power is about -88 dBm, which can ensure the normal operation of mobile terminals. In this experiment, according to the basic needs of training teaching, the user roles are divided into the director of the training center, training administrator, training teacher, student, terminal administrator, etc. Among them, the director of the training center is the main participant in teaching, and is mainly responsible for the tasks of reviewing the training plan, viewing the training plan, reviewing the training application, putting forward and confirming reports for the inventory of construction consumables and tools, making statistics and reviewing the standing book, reviewing the application for modifying the training results, reviewing the application for make-up examination, and reviewing the application for reeducation.

The training administrator is the main participant in teaching and is mainly responsible for publishing the approved training plan, making training arrangements according to the training application form, publishing the training schedule, confirming the lending registration and return registration of construction consumables and tools, counting construction consumables and tools, establishing and registering account information for account management. Set the sign in and sign out time and form the management record of the training process during the training process, register and review the training results and then publish them, modify the training results, register the make-up examination for students' training, arrange the make-up examination, review the make-up examination results, publish the make-up examination results, assign classes for re education, review the re education results, publish the re education results and other tasks. As the main participants in teaching, the training teachers are mainly responsible for writing the training plan, viewing the training plan and review results, training use application, viewing the use application results and training schedule, borrowing and returning applications for construction consumables and tools, reviewing the training process, reviewing the training report, evaluating the training results, filling in the training results, putting forward the application for modifying the training results view the modified practical training results, assess make-up examination results, generate retake results and other tasks.

Students are the main participants in teaching. They mainly view training plans, training schedules, sign in and sign out during training, write training reports, modify applications for training results, view training results, apply for make-up examinations, participate in make-up examinations, view make-up examination results, apply for revision, view class information for revision, participate in revision, and view revision results. The terminal administrator is responsible for teacher information management, student information management, professional information management, department information management, training type information management, construction consumables and tools information maintenance, user type information management and other tasks. The use case diagram of construction practice teaching management is drawn, as shown in Fig. 5 below.

As shown in Fig. 5, the training teacher enters the terminal login interface, enters the administrator user name and password in the user name and password box, and click the "Login" button; Enter the main interface of the terminal, select and click the "Basic Information Management" button; Enter the "Basic Information Management" interface and select the "Teacher Information Management" menu button; Enter the "Teacher Information Management" interface and select the "Teacher Information Maintenance" menu button; Display the teacher information list, and select information maintenance operations: add, modify, or delete; The teacher information editing interface is displayed. If you select Add or Modify, fill in or edit the required items of teacher information; Save the new or deleted teacher information to the teacher information table in the terminal database. Students enter the terminal login interface, enter the administrator user name and password in the user name and password box, and click the "Login" button; Enter the main interface of the terminal, select and click the "Basic Information

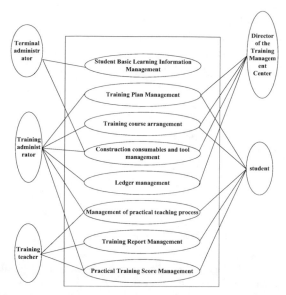

Fig. 5. Use case diagram of construction training teaching management

Management" button; Enter the "Basic Information Management" interface and select the "Student Information Management" menu button; Enter the "Student Information Management" interface and select the "Student Information Maintenance" menu button; Display the student information list and select the information maintenance operation: add, modify or delete; The student information editing interface is displayed. If you select Add or Modify, fill in or edit the required items of student information; Save the new or deleted student information to the terminal database student information table. The teacher enters the terminal login interface, enters the training teacher's user name and password in the user name and password box, and click the "Login" button; Enter the main interface of the terminal, select and click the "Training Plan Management" button; Enter the "Training Plan Management" interface and select the "Training Plan Writing" menu button; Enter the "Training Plan Writing" interface and click the "Add" button; Enter the training plan content editing interface and edit the content of the training plan, including: training name, training teacher, training project, training object, training semester, training purpose, training content, training requirements, training class hours, training textbooks, training reference books, training tools, training consumables, etc.

The training teacher can query the corresponding training report in the terminal by the name of the instructor and by the name of the student, directly review the training report, and give the evaluation results of the training report; Fill in the training results in the terminal and save them; The training administrator directly views the evaluated training results in the terminal, reviews them, and then publishes the training results. Students can directly query the training results through the terminal, and can view the training results by querying by student name and by training name respectively. By reviewing the data constraints of the business steps of the training report, filling in the data constraints of the business steps of the training results, saving the data constraints of the business steps of the training results, reviewing the data constraints of the business steps of the training results, publishing the data constraints of the business steps of the training results, querying the data constraints of the business steps of the training results, etc., Ensure the orderly implementation of the whole virtual building construction training teaching course.

3.2 Experimental Results

Under the above experimental conditions, this paper randomly selects 10–80 terminals, and used the methods of reference [1], the methods of reference [2], and the intelligent mobile terminal based virtual building construction training teaching method designed in this paper. The experimental results are shown in Table 3 below.

Table 3. Experimental Results

Number of terminals	The benefit function of reference [1] method	The benefit function of reference [2] method	The benefit function of the virtual building construction training teaching method based on intelligent mobile terminal designed in this paper
10	8.25	8.24	8.20
20	17.19	17.37	15.43
30	26.21	26.49	19.98
40	34.54	32.18	22.36
50	44.14	45.63	27.43
60	52.26	49.52	30.34
70	63.45	52.18	34.66
80	71.36	69.40	38.07

As shown in Table 3, in the process of virtual building construction training teaching, the benefit function is the key indicator to test the teaching quality. The smaller the benefit function, the lower the delay and energy consumption of the intelligent mobile terminal, and the higher the terminal benefit. From the experimental results, it can be seen that when the number of terminals is small, the benefit function gap of the three teaching methods is small. With the increasing number of terminals, the benefit function of the virtual building construction training teaching method based on intelligent mobile terminals designed in this paper is significantly lower than the benefit function of reference [1] method and the benefit function of reference [2] method, and has always been in the optimal position. In other words, the teaching performance of the three teaching methods is ranked as follows: the virtual building construction training teaching method based on intelligent mobile terminal designed in this paper > the methods of reference [1] >

the methods of reference [2], which plays an important role in improving the quality of virtual building construction training teaching.

4 Conclusion

Virtual training teaching is a combination of computer technology, software technology, network technology and traditional training instruments to change the construction mode of training terminals, improve the overall performance of training, and break through the space-time constraints of training operations. It is a change in traditional training methods and the third scientific research method after theoretical research and training research. This kind of training neither consumes equipment nor is restricted by the external conditions such as the site. It can be operated repeatedly and has strong interactivity. It can more truly reflect the training object and process. It is a mirror image of the real science and technology field.

In this paper, intelligent mobile terminals are applied to virtual building construction training teaching. By establishing a virtual building construction training teaching function module, the immersive function of the training course is enhanced, and the immersive nature of the training teaching is effectively achieved. The use of terminals can simplify virtual building construction training teaching tasks and reduce other costs. The use of terminals can maximize the potential of virtual training courses, provide a better learning place for students in the College of Architecture and Engineering, and improve the overall teaching quality.

Aknowledgement. "Chen Guang" project supported by Shanghai Municipal Education Commission and Shanghai Education Development Foundation (20CGB11).

References

1. Li, Y., Feng, R., Zhang, D.: Design and application of blended virtual training instruction mode for deep learning in vocational colleges. Vocation. Tech. Educ. Forum **38**(11), 58–65 (2022)
2. Yang, H.: Construction of logistics practical training center based on IoT technology. Logist. Technol. **39**(01), 131–134+160 (2020)
3. Zhou, H., Hu, X.B., Zhou, J., et al.: A new city air terminal service mode: urban mobile station for luggage check-in service and evolutionary approach. IEEE Trans. Intell. Transp. Syst. 1–17 (2021)
4. Zhao, Z., Yang, J., Xi, H., et al.: Research on mobile terminal technology supporting intelligent maintenance of substation. J. Phys. Conf. Ser. **1802**(3), 032139 (6p.) (2021)
5. Li, X., Pan, H., Zhu, R.S.: Simulation of vulnerability identification of mobile terminal software cache side channel. Comput. Simul. **002**, 039 (2022)
6. Terzioglu, T., Turkoglu, H., Polat, G.: Formwork systems selection criteria for building construction projects: a critical review of the literature. Can. J. Civ. Eng. **4**, 49 (2022)
7. Ismail, Z.A.B.: Thermal comfort practices for precast concrete building construction projects: towards BIM and IOT integration. Eng. Constr. Archit. Manag. **3**, 29 (2022)
8. Mishra, A.K., Korala, N., Khadka, S., et al.: Perception based study of variation order in building construction projects. Soc. Sci. Electron. Publish. (2) (2021)

9. Dixit, S.: Impact of management practices on construction productivity in Indian building construction projects: an empirical study. Org. Technol. Manag. Constr. Int. J. **13**(1), 2383–2390 (2021)
10. Xfw, A., Cyy, B., Wha, C., et al.: Integrated design of solar photovoltaic power generation technology and building construction based on the Internet of Things. Alex. Eng. J. **61**(4), 2775–2786 (2022)

Numerical Simulation Model Construction of Swept Frequency Dielectric Logging Response Based on Wireless Communication

Liang Pang[✉]

Public Basic Course Department, Wuhan Institute of Design and Sciences, Wuhan 430025, China
pangliang0611@163.com

Abstract. The current method is faced with the interference of logging signals caused by mud invasion, which cannot meet the real-time requirements. In order to effectively solve geological problems and improve the ability to solve complex formations, it is necessary to carry out numerical simulation of swept frequency dielectric logging response. Therefore, a numerical simulation model of swept frequency dielectric logging response based on wireless communication is proposed. By defining directional signals, the inclined transmitter coil structure has the ability of formation evaluation and geosteering azimuth detection, and a variety of formation models are used for numerical simulation of logging tools. The experimental results show that the numerical simulation model is effective.

Keywords: Wireless Communication · Sweep Frequency Dielectric Logging · Response Value · Simulation Model · Directional Signal

1 Introduction

With the continuous progress of communication technology, wireless communication technology has been widely used in the field of logging. With the rapid development of social economy, the demand for oil resources in the market is also gradually increasing. Because oil is a non renewable resource, the storage capacity is decreasing year by year with the continuous increase of market demand, and the oil storage environment to be exploited is becoming more and more difficult to develop. In actual engineering, more stratum information needs to be obtained before oil exploitation, which is conducive to the maximum benefit of oil exploitation [1, 2]. With the development of Shengli Oilfield, Zhongyuan Oilfield and other large oilfields, the oil and gas difficult to exploit in oil and gas reservoirs are gradually replaced by fresh water or sewage injection, and become a water flooded layer with small or complete water production [3, 4]. However, after water flooding treatment of oil and gas reservoirs, the resistivity changes slightly, and it is impossible to distinguish between water flooded layers and oil and gas reservoirs according to the resistivity change. In order to make the oilfield stable and high-yield for a long time, it is necessary to drill adjustment wells in the original developed well pattern. At this time, the effect of using resistivity parameters to distinguish oil, gas

L. Yun et al. (Eds.): ADHIP 2023, LNICST 548, pp. 361–374, 2024.
https://doi.org/10.1007/978-3-031-50546-1_24

and water layers is not good [5, 6]. It is gradually found that the propagation process of electromagnetic wave, a physical wave, is not only related to the conductivity of the medium. The higher the frequency of electromagnetic wave, the greater the impact on the dielectric constant. Thus, the dielectric logging for detecting the acoustic properties was developed [7]. The oil reserves are described by using the acoustic velocity and amplitude output from the acoustic logging tool. At the initial development stage of the acoustic logging technology, it is necessary to use a combination of various technologies to obtain formation information. The amount of formation parameters and rock information provided is small, and the reliability is low. With the rapid development of computer technology and measurement and control technology, the development of logging instrument technology has made great progress [8, 9].

At present, relevant scholars have made significant contributions to the development of logging technology. The earliest widely used logging method was to reflect formation information through low-frequency to high-frequency electromagnetic waves. Its basic principle is to use electrical methods to measure underground environmental data. When electromagnetic waves pass through underground formation minerals, different electrical parameters are generated. By analyzing the temporal and spatial distribution of electromagnetic waves at the receiving end, Obtain the distribution status of different minerals underground [10]. With the development of the times, the development of integrated circuits has promoted the logging technology to an unprecedented level, including induction logging, lateral logging, logging while drilling and other logging technologies. For different downhole detection purposes, different logging instruments are used in the actual detection process. Most of these detection mechanisms use logging circuits to form transmission currents or induced electromotive forces around the area to be measured downhole. The receiving end of the logging instrument changes the phase and amplitude by receiving the transmission characteristics of electromagnetic waves, and further calculates the resistivity or conductivity of the formation under logging. However, these methods have strong practicality in situations where there is a significant difference in conductivity between underground formations, and poor practicality in situations where there is a small difference in conductivity between local layers. Therefore, it is necessary to study the response characteristics of downhole resistivity logging directionally to provide Effective theory basis for the realization of oil and gas reservoir evaluation.

To address the above issues, the article proposes a simulation model for the numerical response of swept frequency dielectric logging based on wireless communication. Propose a combination of horizontal and vertical coils to construct inclined coils to detect underground formations, analyze the equivalence between the isotropic protective layer of electrical parameters and its macroscopic anisotropy, and finally comprehensively simulate and evaluate the swept frequency dielectric logging response of multi-layer inclined anisotropic formations. The experimental results show that this model is suitable for analyzing the response of sweep frequency dielectric logging tools with coils placed in any direction. It has the advantages of fast calculation speed and high accuracy, meets real-time requirements, and improves the ability to solve complex formations.

2 Sweep Frequency Dielectric Logging Model and Wireless Communication Method

This study first achieved the measurement of the dielectric constant of underground media by constructing a swept frequency dielectric logging model. In terms of constructing the swept frequency dielectric logging model, the complexity and nonlinear characteristics of the formation were considered, and an accurate and reliable dielectric logging model was established. An optimized inversion method is proposed for wireless communication problems in swept frequency dielectric logging. By optimizing the design of wireless communication between sensor nodes, the reliability and efficiency of data transmission are maximized. This method combines multi-objective optimization algorithms and signal processing techniques to achieve accurate inversion of the dielectric constant of underground media.

2.1 Construction of Swept Frequency Dielectric Logging Model

The swept frequency dielectric logging needs to measure the range of conductivity, measurement resolution and detection depth to select the operating frequency [11–13]. The general swept frequency dielectric logging model is a three coil system structure r_c is the coil radius of swept frequency dielectric logging model, r_w represents the wellbore radius, T represents the transmitting coil, R_1, R_2 represent the near receiving coil and far receiving coil of swept frequency dielectric logging model, $\varepsilon(\vec{r})$, $\sigma(\vec{r})$, $\mu(\vec{r})$ representative $\vec{r}$ for the dielectric constant, conductivity and permeability at the location [14, 15], after selecting the downhole background medium, divide and calculate the medium objects with different dielectric and electrical parameters from the downhole background medium, and select the formation with the same properties as the borehole parameters as the background medium, ε_0, σ_0, μ_0 respectively represents the dielectric constant, conductivity and permeability of the downhole background medium [16, 17]. Utilize the amplitude ratio of two receiving coils A_{R1}, A_{R2} and phase difference θ_{R1}, θ_{R2} formation with known sweep frequency dielectric logging conductivity is calibrated, and the amplitude ratio and phase difference are converted into corresponding conductivity, which is expressed by the following formula:

$$\begin{cases} A_R = A_{R1}/A_{R2} \\ P_D = \theta_{R1} - \theta_{R2} \end{cases} \tag{1}$$

Symmetrical compensation coil system is used to describe the boundary between formation and water layer during sweep frequency dielectric logging, measure the conductivity of underground formation at multiple depths, and calculate the voltage on any receiving coil of conductivity by calculating the electronic field generated by magnetic dipole [18, 19]. The finite element method is usually used to integrate the electric field on the coil, which can be obtained by using the magnetic field component in the normal direction of the coil H_{RH} Obtain the finite element wireless communication equation:

$$V_R = j_w \times \mu_0 \times S_R \times H_{RH} \times (A_R + P_D) \tag{2}$$

Where, j_w represents the electromagnetic conductivity of sweep frequency dielectric logging, S_R represents the area of receiving coil of swept frequency dielectric logging conductivity sensor.

Assumptions, ε_0 represents the relative permittivity of the swept frequency dielectric measurement of the anisotropic formation background medium, σ_0 represents the conductivity of swept frequency dielectric logging sensor, j represents the current parameter, and uses the following formula to give the Complex permittivity $\hat{\varepsilon}_0 = \varepsilon_0 - j\sigma_0/\omega$ induced current of swept frequency dielectric logging sensor is defined according to the volume equivalent mechanism $\vec{J}(\vec{r})$ since all induced currents are in uniform downhole formation background medium, the total field of sweep frequency dielectric logging sensor is located in all downhole formation areas $\vec{E}(\vec{r})$ I.e. incident field $\vec{E}^i(\vec{r})$ secondary induction field generated by scattering with downhole formation $\vec{E}^s(\vec{r})$ sum is expressed by the following formula:

$$\vec{E}(\vec{r}) = \left[\vec{E}^s(\vec{r}) + \vec{E}^i(\vec{r})\right] \times V_R \tag{3}$$

Where, the scattering field of regular and uniform formation $\vec{E}^s(\vec{r})$ and vector bit $\vec{A}(\vec{r})$ and scalar bit $\phi(\vec{r})$ relevant, $\vec{E}^s(\vec{r}) = -j_w\vec{A}(\vec{r}) - \Delta\phi(\vec{r})$, according to Green's function, downhole formation vector potential $\vec{A}(\vec{r})$ and swept frequency dielectric logging induced current $\vec{J}(\vec{r}')$ relevant:

$$\vec{A}(\vec{r}) = \frac{\mu_0}{4\pi} \int_V \vec{J}(\vec{r}') \frac{e^\sigma |\vec{r} - \vec{r}'|}{|\vec{r} - \vec{r}'|} dv' \times \vec{E}(\vec{r}) \tag{4}$$

Where, $\vec{r}'$ represent the direction of observation point P vector of point, $\vec{r}$ Represents the origin point P Vector of points, scalar bit $\phi(\vec{r})$ related to the volume charge of swept frequency dielectric logging receiving coil:

$$\phi(\dot{r}) = \frac{1}{4\pi\hat{\varepsilon}_0} \int_V \rho(\vec{r}') \frac{e^\sigma |\vec{r} - \vec{r}'|}{|\vec{r} - \vec{r}'|} k_0 dv' \times \vec{A}(\vec{r}) \tag{5}$$

Where, k_0 represents that in the process of sweep frequency dielectric logging, dv' Jacobi determinant representing regular transformation, $\rho(\vec{r}')$ represents the induced charge density, the wave number in the background medium, where, $k_0 = \omega\sqrt{\hat{\varepsilon}_0\mu_0}$ according to the current continuity equation, use the following formula to describe the relationship between the charge density and the induced current of the swept frequency dielectric logging receiving coil:

$$\nabla\vec{J}(\vec{r}) = -j_w\rho(\vec{r}) \times \phi(\vec{r}) \tag{6}$$

Because of the sweep frequency dielectric logging potential shift vector in the background medium $\vec{D}(\vec{r}')$ satisfy $\vec{D}(\vec{r}') = \bar{\varepsilon}(\vec{r})\vec{E}(\vec{r})$. Therefore, swept frequency dielectric logging receiving coil inductor current and electric displacement vector meet the following constraints:

$$\vec{J}_v(\vec{r}) = j_w\bar{\kappa}(\vec{r}) \times \vec{D}(\vec{r}') \times \nabla\vec{J}(\vec{r}) \tag{7}$$

Among them, $\tilde{\kappa}(\vec{r})$ stands for contrast tensor, i.e. $\tilde{\kappa}(\vec{r}) = (\bar{\varepsilon}(\vec{r}) - \vec{I}\hat{\varepsilon}_0) \cdot \bar{\varepsilon}^{-1}(\vec{r})$ expression based on the potential shift vector of swept frequency dielectric logging receiving coil can be obtained:

$$\vec{E}^i(\vec{r}) = \vec{D}(\vec{r}') \times \bar{\varepsilon}^{-1}(\vec{r}) + j_w\vec{A}(\vec{r}) + \nabla\phi(\vec{r}) \times \vec{J}_v(\vec{r}) \tag{8}$$

According to the analysis formula (8), under the condition that the frequency of the induced current of the swept frequency dielectric logging receiving coil is not high $\sigma/w \geq \varepsilon$, indicating that the complex permittivity in homogeneous background media is mainly related to frequency and downhole formation conductivity, and the relative permittivity ε has little effect on the response of sweep frequency dielectric logging [20, 21]. The inhomogeneous formation except the underground formation background medium is divided into tetrahedral grids, and the SWG basis function is used to discrete sweep frequency dielectric logging to receive coil displacement current and potential displacement vector. In each tetrahedron, the contrast tensor $\tilde{\kappa}$ are constants, we can obtain:

$$H_{JJ} = Tr(\tilde{\kappa}^{\pm}) \times \vec{E}^i(\vec{r}) \tag{9}$$

where, $Tr(\tilde{\kappa}^{\pm})$ trace representing the matrix, that is, the sum of diagonal elements, needs to obtain the induced charge density according to the continuity equation of formation logging current in Formula (9).

2.2 Optimization Inversion of Wireless Communication

Wireless communication refers to the communication mode of information transmission without the use of wired cables or other physical connections. It uses electromagnetic waves to propagate in the air, allowing information transmission between two communication parties [22, 23]. The advantage of this communication mode is that it can avoid the restrictions brought by the physical connection, so that the communication can be carried out more freely without worrying about the line being damaged or affected by other factors. In the inversion process of sweep frequency dielectric logging, the developed forward inversion method is used to simulate the response law of tilt coil with sweep frequency dielectric logging tool in highly inclined wells and its application in geosteering [24, 25]. In order to effectively ensure the accuracy of inversion results of downhole geosteering, adaptive damping factors are added and inequality constraints are imposed to transform the forward problem of sweep frequency dielectric logging propagation resistivity into a nonlinear equation $F = F(x), x = (x_1, \cdots, x_N)^T$ downhole formation parameters representing the location to be measured, $F = (F_1, \cdots, F_M)$ represents different function values, assuming f represent the function value of known measurement, then solve the unknown quantity x inversion problem of can be transformed into minimization f and $F(x)$ variance of is expressed by the following formula:

$$F' = \min\|f - F(x)\| \times H_{JJ} \tag{10}$$

In general, in order to make the whole iterative process stable, constraints need to be set when solving formula (20) $\|\delta x\| \leq \Delta$, Δ representing the maximum value of the given boundary, the inversion process of the entire tilt coil swept frequency dielectric logging model is transformed into solving the linear minimum equation problem with constraints:

$$D' = \min\{\|b - J\delta x\| : \|\delta x\| \leq \Delta\} \times F' \tag{11}$$

To solve formula (11), you need to $F(x)$ carry out forward calculation to obtain b On this basis, Jacobian matrix can be obtained through differentiation by forward modeling J, from which the satisfaction $\|\delta x\| \leq \Delta$ correction amount of.For Jacobian matrix J^T singular value decomposition is performed and expressed by the following formula:

$$J^T = U \times \Lambda V^T \times D' \tag{12}$$

Where, U and V On behalf of $M \times N$ and $N \times N$ matrix of order, Λ representative $M \times N$ diagonal matrix of order. In order to effectively reduce the multiplicity of the inversion problem, the change interval of the parameters of the swept frequency dielectric logging response numerical simulation model is taken as the prior information, which is expressed by the following inequality constraints:

$$\rho_{\min}^i \leq x_i \times J^T \leq \rho_{\max}^i, i = 1, 2, \cdots, m \tag{13}$$

Where, x_i represents the parameters to be inverted, i represents the parameter number, $\rho_{\min}^i$ and $\rho_{\max}^i$ for and on behalf of i upper and lower limits of the parameters to be inverted. It should be noted that during the inversion process of sweep frequency dielectric logging, the change range of each inversion parameter can be more accurate according to other methods such as logging interpretation.

3 Numerical Simulation of Sweep Frequency Dielectric Logging Response

3.1 Scanning Frequency Dielectric Logging Directional Signal and Geological Directional Vector Angle

A variety of formation models are used to carry out inversion numerical simulation for swept frequency dielectric logging tool, and a new directional signal is defined for the structure of swept frequency dielectric logging tool receiving tilt coil, so that the structure of swept frequency dielectric logging receiving coil has the ability of downhole formation evaluation and geosteering azimuth detection. According to the above theoretical derivation. The dyadic Green's function of the magnetic dipole source can be obtained in the rectangular coordinate system of the downhole formation $\tilde{G}^{HM}$, $\tilde{G}_{xz}^{HM}$ by z magnetic dipole in the direction of x For the magnetic field component generated in the direction, in order to effectively realize swept frequency dielectric azimuthal logging, it is necessary to extract the azimuthal signal, obtain new directional detection signal by using a new definition method, and subtract the phase difference and amplitude ratio signal on the receiving coil of the swept frequency dielectric logging tool from the phase difference and amplitude ratio signal on different receiving coils of the lower transmitting coil:

$$\Delta\varphi' = \Delta\varphi_U - \Delta\varphi_D, A' = (A_U - A_D) \times \left(\rho^i_{max} - \rho^i_{min}\right) \qquad (14)$$

where, $\Delta\varphi_U$ and A_U represents the phase difference and amplitude ratio obtained when the transmitting coil on the swept frequency dielectric logging tool works normally, $\Delta\varphi_D$ and A_D respectively represent the phase difference and amplitude ratio obtained when the lower transmitting ring of the swept frequency dielectric logging tool works, $\Delta\varphi'$ represents directional phase difference signal, A' represents the directional amplitude ratio signal. Generally, the resistivity of the downhole oil and gas reservoir is relatively high, and the reverse of the vector angle should be in the direction of the oil and gas reservoir. The direction of the swept frequency dielectric logging tool axis is defined as rectangular coordinates z axis direction, x axis is defined as the upward direction perpendicular to the ground, and the swept frequency dielectric logging tool xy plane is defined as the instrument face. When the transmitting antenna of the sweep frequency dielectric logging tool is axial and the receiving coil is transverse, the real and imaginary components of the geosteering vector on the surface of the sweep frequency dielectric logging tool can be expressed as follows:

$$s_b = (V^r_{xz}/\sqrt{(V^r_{xz})^2 + (V^r_{yz})^2}, V^r_{yz}/\sqrt{(V^r_{xz})^2 + (V^r_{yz})^2}, 0) \times \Delta\varphi' \qquad (15)$$

$$x_b = (V^i_{xz}/\sqrt{(V^i_{xz})^2 + (V^i_{yz})^2}, V^i_{yz}/\sqrt{(V^i_{xz})^2 + (V^i_{yz})^2}, 0) \times \Delta\varphi' \qquad (16)$$

Where, V^r_{xz} and V^i_{xz} represents the signal received on the swept frequency dielectric logging tool x real and imaginary parts of the component electromotive force, V^r_{yz} and V^i_{yz} respectively represent corresponding y real and imaginary parts of the component electromotive force, when the transmitting coil of the sweep frequency dielectric logging tool is inclined laterally and the receiving coil is axial, the downhole geological steering vector can be calculated using the following formula:

$$s'_b = (-V^r_{zx}/\sqrt{(V^r_{zx})^2 + (V^r_{zy})^2}, -V^r_{zy}/\sqrt{(V^r_{zx})^2 + (V^r_{zy})^2}, 0) \times s_b \qquad (17)$$

$$x'_b = (-V^i_{zx}/\sqrt{(V^i_{zx})^2 + (V^i_{zy})^2}, -V^i_{zy}/\sqrt{(V^i_{zx})^2 + (V^i_{zy})^2}, 0) \times x_b \qquad (18)$$

Where, V^r_{zx} and V^i_{zx} represents the transmitting coil of swept frequency dielectric logging tool x The real and imaginary parts of the electromotive force generated by the component magnetic field on the receiving coil, V^r_{zy} and V^i_{zy} respectively represent the corresponding transmitting coil y real and imaginary parts of the electromotive force generated by the component magnetic field of. According to the definition, the vector direction of the real part and imaginary part of the electromotive force both points to the more conductive area of the downhole formation, and the direction of its corresponding resultant vector indicates that the measured area of the swept frequency dielectric logging tool is more conductive in this direction, while the opposite direction of its resultant vector indicates the direction of higher formation resistivity.

3.2   Numerical Simulation and Anisotropy Measurement of Swept Frequency Dielectric Logging Geosteering Vector Angle

The space overlap algorithm is used to decompose the actual model, remove all the upper and lower surrounding rocks of the actual three-layer formation model, place the swept frequency dielectric logging tool in an infinite space, and after the above decomposition of the actual model, the relationship between the emf response tensor of the actual downhole three-layer formation model and the emf response tensor in the decomposed downhole three-layer formation model is $V = V_1 + V_2 - V_3$, V_1, V_2, V_3 represents the electromotive force tensor received in different models, V represents the total received electromotive force tensor of swept frequency dielectric logging tool, both of which are 3×3. Assuming that it is necessary to tilt the transmitting coil and receiving coil to detect the anisotropy of the downhole formation, define T_6 and T_2 represent swept frequency dielectric logging transmitting coils, R_4 and R_3 respectively represent the receiving coil and measure separately T_6 and T_2 at launch R_4 and R_3 phase difference and amplitude ratio between, where R_4 and R_3 electromotive force on is only taken as a_0 phase difference and amplitude ratio signals of swept frequency dielectric logging are subtracted and used as the phase difference and amplitude ratio signals for detecting the anisotropy of the downhole formation, which are expressed by the following formula:

$$Att_a = \log \frac{[\text{Re}(a_0(T_6R_4))]^2 + [\text{Im}(a_0(T_6R_4))]^2}{[\text{Re}(a_0(T_6R_3))]^2 + [\text{Im}(a_0(T_6R_3))]^2} \tag{19}$$

$$PS_a = \tan^{-1} \frac{\text{Im}(a_0(T_6R_4))}{\text{Re}(a_0(T_6R_4))} \tag{20}$$

In the downhole homogeneous anisotropic medium, the above definition can be used to detect the anisotropic characteristics of the downhole formation. Taking the 500 kHz transmission frequency as an example, the phase difference and amplitude ratio under different horizontal and vertical resistivity contrast are calculated at $0°$ tilt angle, and shown in Fig. 1.

It can be seen from the figure that in a vertical well, when the downhole formation is thick, the horizontal and vertical resistivity of the downhole formation can be obtained by using the intersection diagram. When the horizontal resistivity of the transmitting coil of the swept frequency dielectric logging tool exceeds 15 Ω·m, the sensitivity of the swept frequency dielectric logging tool to the formation anisotropy gradually decreases. When the well inclination angle gradually increases to $50°$, the intersection diagram is shown in Fig. 2. It can be seen that. After the well inclination angle is gradually increased, the sensitivity of the swept frequency dielectric logging tool to the formation anisotropy is gradually increased. When the horizontal resistivity of the transmission coil of the swept frequency dielectric logging tool is 15 Ω·m, the anisotropy of the downhole formation can be detected. Similarly, the intersection diagram under different inclination angles and arbitrary inclination frequencies can be drawn. In practice, according to the working mode of the sweep frequency dielectric logging tool and the inclination angle of the well, the intersection diagram is selected to obtain the anisotropy information of the downhole formation.

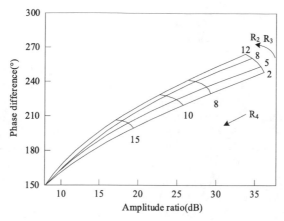

Fig. 1. Intersection Diagram of Phase Difference and Amplitude Ratio of Anisotropy at 500 kHz in Vertical Well

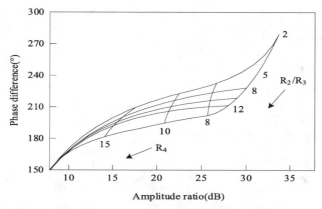

Fig. 2. Intersection Diagram of Phase Difference and Amplitude Ratio of Anisotropy at 400 kHz in 50° Relative Tilt Angle

4 Experimental Results and Analysis

Firstly, by comparing the numerical results of the swept frequency dielectric logging response, the validity of the numerical simulation model of the swept frequency dielectric logging response based on wireless communication proposed in this paper is verified. The response characteristics of the swept frequency dielectric logging are studied through the numerical simulation results on heterogeneous and anisotropic formations. During the numerical simulation of the swept frequency dielectric logging response, the long and short source distances of the coil system are 50 m, respectively. The short axis radius of the tilt coil is 30 cm, the transmission frequency is 500 kHz, and the borehole radius is 20 cm. Under the configuration of 2.4 GHz computer frequency and 4G memory, the sweep frequency dielectric logging response numerical simulation experiment is carried out. The experimental environment includes the electromagnetic field finite element analysis software COMSOL Multiphysics and the swept frequency dielectric logging system. The experimental parameters include frequency range, transmission power, receiver distance, burial depth and dielectric constant. Specifically, according to the experimental requirements, the frequency range is 1 Hz to 1 GHz, the transmission power range is 10 dBm to 30 dBm, the receiver distance range is 1 m to 100 m, and the burial depth range is 1 m to 50 m. The rock dielectric constant is set to 2, 5, 10, 15, 20, 30, 40, 50, and 60 for testing. During the experiment, several simulations were carried out for each group of parameters, and the wireless communication response data under different parameters were collected and analyzed. The data is selected from iResearch data, and the website of the database is http://www.iresearch.cn. Through the collection and analysis of experimental data, the application of wireless communication technology in swept frequency dielectric logging is deeply studied and verified.

In order to more intuitively observe the inversion results, the inversion simulation of the following formations was carried out. The calculation conditions in Fig. 3 are: the borehole diameter is 40cm, the mud resistivity is 20 Ω·m, the target layer thickness is 20m, the horizontal resistivity of the downhole formation is 0°, and the resistivity of the formation surrounding rock is 90°. Figure 3 shows the influence of the downhole formation anisotropy coefficient on the sweep frequency dielectric logging response.

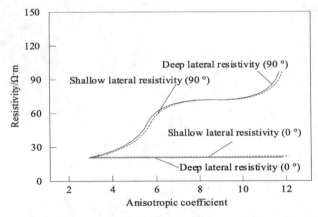

Fig. 3. Sweep frequency dielectric logging response changes with formation anisotropy coefficient at different well deflections

It can be seen from the analysis of Fig. 3 that the relationship between the swept frequency dielectric logging response and the variation of the layer anisotropy coefficient is quite different in the vertical well (0°) or horizontal well (90°) environment. In a vertical well, the current plane generated by the sweep frequency dielectric logging tool is parallel to the horizontal direction of the formation. The resistivity obtained by the sweep frequency dielectric logging tool basically comprehensively describes that the resistivity in the horizontal direction of the anisotropic formation changes little with the anisotropic formation coefficient, while in a horizontal well, the current plane generated by the sweep frequency dielectric logging tool is perpendicular to the horizontal downhole formation, Sweep frequency dielectric logging is obviously affected by vertical resistivity.

Figure 4 (a) The calculation model is the high resistivity of the target layer of the horizontal well. The resistivity in the horizontal direction is set as 50 Ω·m. The surrounding rock of the underground formation is isotropic. The resistivity of the surrounding rock of the formation is 20 Ω·m, the borehole diameter is 20cm, and the mud resistivity is 2 Ω·m. Figure 4 (b) The calculation model is the low resistivity of the target layer of the horizontal well. The corresponding horizontal resistivity is 5 Ω·m, and the surrounding rock resistivity of the underground formation is 50 Ω·m. As shown in Fig. 4, the sweep frequency dielectric logging response of the target layer with high resistance (a) and low resistance (b) of horizontal wells varies with the thickness of the layer under different downhole formation anisotropy conditions.

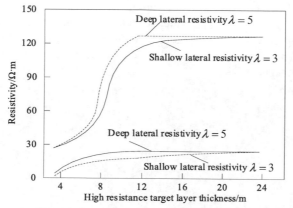

(a) Sweep frequency dielectric logging response of high resistivity target layer changes with layer thickness

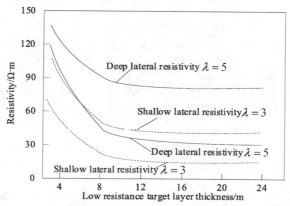

(b) Change of sweep frequency dielectric logging response with formation thickness in low resistivity target formation

Fig. 4. Sweep Frequency Dielectric Logging Response of Target Layer of Horizontal Well Changing with Layer Thickness under Different Formation Anisotropy Conditions

As shown in Fig. 4, when the underground strata have different high resistivity and low resistivity, the influence of underground surrounding rock layer thickness on the response of swept-frequency dielectric logging is obviously different, but when the target is thin, the deep and shallow lateral resistivity of underground strata change obviously with the target layer thickness. When the underground strata thickness reaches a certain value, the corresponding resistivity reaches a certain value, and the changing trend of underground isotropic strata and anisotropic strata is similar. The above characteristics are due to the anisotropy of underground strata, which increases the vertical resistivity of swept-frequency dielectric logging, thus increasing the difference between the resistivity of underground target layer and the resistivity of surrounding rock.

In the actual sweep frequency dielectric logging environment, mud invasion causes the radial resistivity of permeable formation along the receiving coil to change. Due to the different detection depths of sweep frequency dielectric logging resistivity, the obtained resistivity has amplitude difference. Figure 5 calculates the target layer thickness of 8m, and the measured horizontal resistivity of the target layer is 50 Ω·m, the invasion radius is 0.9m, and the mud resistivity is 2 Ω·m.

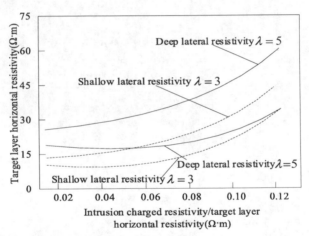

Fig. 5. Sweep frequency dielectric logging response changes with invasion resistivity under different formation anisotropy conditions

Analysis of Fig. 5 shows that mud invasion has obvious influence on the resistivity of swept-frequency dielectric logging. With the gradual increase of the resistivity of swept-frequency dielectric logging in invasion zone, the deep and shallow lateral resistivity of underground formation increases, while the amplitude difference gradually decreases, but the resistivity of anisotropic formation changes more sharply than that of isotropic formation, and the amplitude difference between deep and shallow lateral resistivity of swept-frequency dielectric logging increases gradually, which is more affected by mud invasion. Under the condition that the resistivity of swept-frequency dielectric logging in invasion zone is known,

5   Conclusion

In view of the interference of swept frequency dielectric logging signal caused by mud invasion in the current method, it can not meet the real-time requirements. Therefore, a numerical simulation model of swept frequency dielectric logging response based on wireless communication is proposed. According to the analysis of prior knowledge, the swept frequency dielectric logging response of anisotropic formation in horizontal wells is quite different from that in vertical wells. The results show that the swept frequency dielectric logging response of downhole anisotropic formation is mainly affected by the vertical resistivity of the formation. When the target formation has high resistivity, the anisotropic formation is greatly affected by the thickness of the surrounding rock under the well, and when the target formation has low resistivity, the downhole formation is greatly affected by the lateral formation. Compared with isotropic formation, mud invasion has more obvious and complex effects on the response of sweep frequency dielectric logging in downhole formation.

Acknowledgement. 2020 Teaching Research Project of Wuhan Institute of Design and Sciences: Case based "linear algebra" Hybrid Teaching Research and Practice (Project No.: 2020JY101).

References

1. Takahashi, A., Takyu, O., Fujiwara, H., et al.: Overloaded wireless MIMO switching for information exchanging through untrusted relay in secure wireless communication. IEICE Trans. Commun. **104**(10), 1249–1259 (2021)
2. Li, K., Li, Z., Xiong, Z., et al.: Thermal camouflaging MXene robotic skin with bio-inspired stimulus sensation and wireless communication. Adv. Funct. Mater. **32**(23), 1–10 (2022)
3. Yan, X., Pan, W., Zou, X., et al.: Optical frequency comb assisted denoising for multiple access and capacity enhancement of covert wireless communication. Opt. Lett. **47**(6), 1442–1445 (2022)
4. Zhang, F.F., Gao, R., Liu, J.: Acoustic wireless communication based on parameter modulation and complex Lorenz chaotic systems with complex parameters and parametric attractors. Chin. Phys. B **30**(8), 805–815 (2021)
5. Wang, A., Yang, L., Yi, X.J., et al.: Wireless communication applications of the variable inclination continuous transverse stub array for Ku-band applications. IET Microw. Antennas Propag. **15**(6), 644–652 (2021)
6. Okan, T.: High efficiency unslotted ultra-wideband microstrip antenna for sub-terahertz short range wireless communication systems. Optik **242**(1), 1–13 (2021)
7. Sharma, R.: RF analysis of double-gate junctionless tunnel FET for wireless communication systems: a non-quasi static approach. J. Electron. Mater. **50**(1), 1–17 (2021)
8. Huang, S., Safari, M.: SPAD-based optical wireless communication with signal pre-distortion and noise normalization. IEEE Trans. Commun. **70**(4), 2593–2605 (2022)
9. Liu, X., Chen, S., Song, L., et al.: Self-attention negative feedback network for real-time image super resolution. J. King Saud Univ. Comput. Inf. Sci. **34**(8B), 6179–6186 (2022)
10. Han, C.W., Zhou, H., Liu, L., et al.: Power and rate control for wireless communication networks based on data. Comput. Simul. **39**(2), 375–379+511 (2022)
11. Liu, S., Li, Y., Fu, W.: Human-centered attention-aware networks for action recognition. Int. J. Intell. Syst. **37**(12), 10968–10987 (2022)

12. Liu, S., Gao, P., Li, Y., Weina, F., Ding, W.: Multi-modal fusion network with complementarity and importance for emotion recognition. Inf. Sci. **619**, 679–694 (2023)

13. Hu, S., Liu, W.N., Liu, Y.M., Liu, K.: Acoustic logging response law in shales based on petrophysical model. Pet. Sci. **19**(5), 2120–2130 (2022)

14. Liu, S., et al.: Human inertial thinking strategy: a novel fuzzy reasoning mechanism for IoT-assisted visual monitoring. IEEE Internet Things J. **10**(5), 3735–3748 (2023)

15. Li, Y., Chen, S., Liu, X., et al.: Study on the logging response characteristics and the quantitative identification method of solid bitumen at different thermal evolution stages. Fuel **316**(5), 1–16 (2022)

16. Liu, S., Xiyu, X., Zhang, Y., Muhammad, K., Weina, F.: A reliable sample selection strategy for weakly-supervised visual tracking. IEEE Trans. Reliab. **72**(1), 15–26 (2023)

17. Bisbing, S.M., Buma, B.J., Vander Naald, B., et al.: Single-tree salvage logging as a response to Alaska yellow-cedar climate-induced mortality maintains ecological integrity with limited economic returns. For. Ecol. Manag. **503**, 1–9 (2022)

18. Shen, J., Shu, D., Shen, Y., et al.: Response of casing hoop and geometry factor to transient electromagnetic logging in cased wells. Geophys. Prospect. Pet. **58**(4), 613–624 (2022)

19. Sun, J., Cai, J., Feng, P., et al.: Study on nuclear magnetic resonance logging T2 spectrum shape correction of sandstone reservoirs in oil-based mud wells. Molecules **26**(19), 1–10 (2021)

20. Sun, Z., Du, M., Xu, R., et al.: Wireless communication utilizing berry-phase carriers. Laser Photonics Rev. **16**(4), 1–9 (2022)

21. Alset, U., Mehta, H., Kulkarni, A.: Evaluation of antenna dependent wireless communication based on LoRa for clear line of sight (CLOS) and non-clear line of sight (NC-CLOS) applications. J. Phys. Conf. Ser. **1964**(3), 32–44 (2021)

22. Gummineni, M., Polipalli, T.R.: Implementation of reconfigurable emergency wireless communication system through SDR relay. Mater. Today Proc. **8**(3), 1–10 (2021)

23. Jassim, A.K.: Microstrip patch antenna with metamaterial using superstrate technique for wireless communication. Bull. Electr. Eng. Inform. **10**(4), 2055–2061 (2021)

24. Li, W., Lee, Y.H., Chen, Y.L., et al.: Proposed model for performance analysis of fourth-generation mobile wireless communication system. Sens. Mater. Int. J. Sens. Technol. **33**(4), 1375–1385 (2021)

25. Eid, M.M.A., Seliem, A.S., Rashed, A.N.Z., et al.: High modulated soliton power propagation interaction with optical fiber and optical wireless communication channels. Indones. J. Electr. Eng. Comput. Sci. **21**(3), 1575–1583 (2021)

Sports Athlete Error Action Recognition System Based on Wireless Communication Network

Yanlan Huang[✉] and Lichun Wang

Guangxi College for Preschool Education, Nanning 530022, China
hyl5825@163.com

Abstract. Incorrect actions not only affect the training effect, but also cause certain harm to the athlete's body. A sports athlete incorrect action recognition system based on wireless communication network is proposed. Build a wireless communication network architecture, obtain sports athletes' sports videos, extract key frames, determine athletes' positions, extract sports athletes' action characteristics (STIP characteristics, Cuboids characteristics, enhanced dense trajectory characteristics and covariance characteristics), based on the extraction of sports athletes' action characteristics, build a Hyperplane based on support vector mechanism, and finally realize the recognition of sports athletes' wrong actions. The experimental data shows that after the application of the system in this article, the maximum recognition rate of incorrect movements of sports athletes obtained is 96.35%.

Keywords: Sports Athletes · Standard Action · Movement Characteristics · Action Recognition · Wireless Communication Network · Wrong Action

1 Introduction

Sports are various activities that are gradually carried out in the process of human development to consciously cultivate one's physical fitness. It adopts various forms of physical activities such as walking, running, jumping, throwing, and dancing. These activities are commonly referred to as physical exercise processes, with rich content, including athletics, ball games, swimming, martial arts, aerobics, mountain climbing, ice skating, weightlifting, wrestling, judo, cycling, and other events [1]. The state encourages and supports the exploration, organization, protection, promotion, and innovation of excellent ethnic, folk, and folk traditional sports projects, and regularly holds ethnic minority traditional sports games. The newly revised Sports Law makes clear provisions on the inheritance and development of traditional sports projects. Modern sports were first introduced to China's Christian school by Western missionaries in the late Qing Dynasty. The so-called Christian school is a new school established by missionaries in accordance with the western educational model and philosophy to spread Christian culture and scientific knowledge. Different from the traditional Chinese private schools or academies, Christian school, while focusing on intellectual education, also vigorously

L. Yun et al. (Eds.): ADHIP 2023, LNICST 548, pp. 375–388, 2024.
https://doi.org/10.1007/978-3-031-50546-1_25

promote sports; Not only are there specialized physical education courses in the curriculum, but also amateur sports teams and sports competitions have been established. All this originates from the Olympic tradition and belief in ancient Culture of Greece, that is, sound thoughts need healthy bodies to carry them. Sports have functions such as physical fitness, entertainment, and education, politics, and economy. It can also be said that sports have different functions in different historical stages, but since the emergence of sports, physical fitness and entertainment have always been the main functions of sports. Sports is a complex social and cultural phenomenon that uses physical activity as the basic means to enhance physical fitness, enhance health, and cultivate various psychological qualities of individuals. In recent years, sports competitions have become one of the world's key exchange activities, especially the Olympics. Athletes have also received widespread attention from various countries and the general public.

Athletes often face various technical requirements in training and competition. However, due to the limitations of human vision and training environments, coaches often find it difficult to accurately identify and correct athletes' incorrect movements. The traditional manual guidance method requires coaches to have rich experience and professional knowledge, and consumes a lot of time and energy. Developing a sports athlete error recognition system has important research significance and practical application value. Firstly, this system can help coaches more accurately identify and analyze athlete's incorrect actions, providing timely and effective guidance and correction [2]. Secondly, by automatically identifying and analyzing erroneous movements, the system can provide personalized training suggestions for athletes, helping them improve their technical skills faster. In addition, the system can also be used to evaluate and monitor the training progress of athletes, providing scientific basis for the formulation of training plans. In short, the research and development of a sports athlete error recognition system will greatly improve the guidance effect of coaches and the training efficiency of athletes, which is of great significance for promoting the development of sports.

In recent years, more and more coaches and researchers have begun to study how to conduct efficient replay analysis based on competition videos. The analysis method gradually transformed from direct observation of records to data analysis combined with computer technology. For example, the Chinese national team of table tennis has successively introduced table tennis technical and tactical analysis software such as Simi Scout developed by Germany and "Table Tennis Military" developed by the Software Architecture Laboratory of Northern Polytechnical University. These software achieve diversified needs such as technical and tactical data statistics and analysis, as well as the editing and synthesis of exciting fragments. They also recognize technical and tactical actions through manual processing based on the experience of coaches, This method can relatively accurately identify athlete movements, but with the sudden increase in the number of videos that need to be analyzed, manual processing alone will consume a lot of time and manpower, and will reduce accuracy. Therefore, how to use computer technology to efficiently and accurately identify sports athlete movements is a current research hotspot. The above methods can be applied to the training process of sports athletes to identify their erroneous movements, facilitate timely adjustment and change of training plans, and maximize the quality of athletes' training.

In order to achieve the above objectives, a design and research on a sports athlete error action recognition system based on wireless communication networks is proposed. The designed system is based on a wireless communication network architecture, extracting key frames from sports athlete motion videos and determining the athlete's position. After extracting multiple features of sports, support vector machines are used to identify incorrect actions of sports athletes.

2   Sports Athletes' Error Action Recognition System

2.1   Wireless Communication Network Architecture Building Module

Wireless communication network is the basis and premise of sports athletes' action data collection, so the first step of research is to build a wireless communication network architecture.

Due to the large number of sports athletes, the wireless communication network architecture has been set up as a hierarchical distributed topology structure. The topology structure of the wireless communication network is shown in Fig. 1.

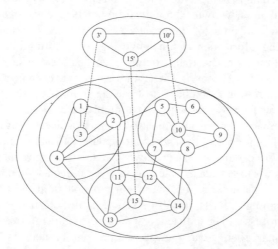

Fig. 1. Schematic diagram of wireless communication network topology

As shown in Fig. 1, the nodes in a hierarchical distributed topology are generally divided into cluster head nodes, gateway nodes, and ordinary nodes. From a structural perspective, the hierarchical distributed topology can be seen as a combination of the star-shaped structure and the fully distributed topology structure. It is commonly used in systems with a large number of nodes [3].

The topology structure of wireless communication networks has been studied earlier. Another important component in wireless communication networks is routing protocols. Routing, as the name suggests, is to choose the best path for data transmission in the network. A suitable routing protocol can improve the stability and access efficiency of the entire network system. For traditional routing protocols such as OSPF and RIP, they cannot respond in a timely manner to rapidly changing topologies, and are likely to encounter various unknown errors when updating routing information, such as the appearance of routing loops. According to the routing method, the routing of wireless communication networks can be divided into active routing and on-demand routing.

Among them, active routing requires nodes to actively exchange routing information regardless of whether the topology structure in the network has changed, in order to update the routing changes in the network in a timely manner. It can continuously monitor the topology structure changes in the network and track updates. The advantage of active routing is that for networks with high throughput, its routing is always in the latest state. This way, for a data transmission process, available routing information can be directly found without initiating the route discovery process. However, for networks with low throughput and topology that are not easily changed, frequent updates of routing are particularly wasteful of system resources when there is no data transmission. Common active routing protocols include DSDV, WRP, etc. The DSDV routing protocol distinguishes between new and old routes by managing the sequence number of the route. Each route is assigned a sequence number, and the quality of the route is judged by determining the size and parity of the sequence. The size of the sequence is directly proportional to the quality of the route. The main feature of DSDV protocol is that it can distinguish the old and new routes through serial numbers, and can also avoid routing loops. The WRP routing protocol requires nodes in the network to send routing packets to exchange routing information with neighboring nodes during idle time, in order to update routing information. Due to the routing exchange between adjacent nodes, routing loops can be avoided. However, WRP requires nodes to exchange information with neighboring nodes, resulting in a huge cost for the entire network in routing updates, which is not suitable for networks with a large number of nodes.

The main difference between on-demand routing and active routing is that on-demand routing does not require scheduled and active updates of routing information. It only updates the route in a certain way when network nodes exchange data. In general, when the source node needs to send data to the target node, on-demand routing will first check the routing table or local cache for routing information from the source node to the target node. Assuming it exists, it will be sent according to this route. Otherwise, a route discovery process will be initiated to update the routing information in the routing table [4]. On demand routing can alleviate the pressure on network system resources in situations or topologies where real-time requirements are not strong due to its on-demand update characteristics. Common on-demand routing protocols include DSR, AODV, TORA, ABR, SSR, etc. DSR is a typical on-demand routing protocol designed based on the concept of source routing. The source route contains all the information of the data transmission process route, and is also forwarded according to the routing information in the group when forwarded by the intermediate node. The DSR routing protocol can route as needed and update routes in a timely manner, but due to the complete routing information contained in packets, it can cause a lot of overhead. AODV is also an on-demand routing protocol that combines the advantages of DSR. The difference is that AODV abandons the complete information of routes in DSR packets and instead uses a hop by hop forwarding method for routing. AODV can quickly establish effective routes, but the establishment of routes has a long delay. The TORA routing protocol adopts an analogy to the "high mountains and flowing water" approach, comparing the destination nodes to other nodes in the network with specific routing heights. The data transmission path can only be transmitted in the direction of height from large to small, which can effectively avoid the generation of loops.

Based on the above description and combined with the needs of sports athlete action data collection, TORA is selected as the routing protocol for wireless communication network.

In order to ensure the stable operation of wireless communication networks, it is necessary to scientifically manage their resources, which are mainly divided into three stages: channel adjustment, power adjustment, and load balancing, as follows:

First, channel adjustment

Reasonable use of non overlapping channels for each AP within a certain wireless coverage range is crucial for wireless network management. For example, on a 2.4G network, there are only three non overlapping channels, and achieving intelligent channel allocation for APs is crucial for wireless applications. At the same time, a large number of possible interference sources, such as radar and microwave ovens, exist in the frequency band of wireless network operation, and the normal operation of the AP will be interfered by them [5]. To ensure that the optimal channel can be allocated to each AP and minimize or avoid adjacent channel interference, channel adjustment functions can be used to achieve this; At the same time, in order to enable the AP to avoid interference sources such as radar and microwave in real-time, real-time channel detection can be used to achieve this. Continuous communication is achieved through dynamic channel adjustment, providing reliable transmission for the network. The schematic diagram of channel adjustment is shown in Fig. 2.

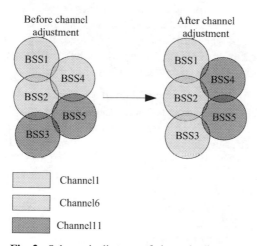

Fig. 2. Schematic diagram of channel adjustment

Second, power adjustment

The traditional RF power control method only statically sets the transmission power to the maximum value in order to simply pursue the signal coverage, but excessive power may cause unnecessary interference and affect other wireless devices. Therefore, it is necessary to select the best power to balance the coverage and system capacity. Power adjustment is to allocate power dynamically and reasonably in the whole wireless network operation process according to the real-time wireless environment. When the AP starts to run for the first time, it uses the maximum transmission power; When it detects other neighbor APs and gets reports from other neighbor APs, the power will increase or decrease according to the actual situation of the environment. The power increase or decrease ultimately depends on the detection conclusion:

1. When the neighbor AP is added, the power will decrease. The wireless signal covers the same area, and the default maximum number of neighbors is 3; After adding AP4, the maximum number of neighbors is 3, and each AP will run the power adjustment function. The power after adjustment is less than the power required by the first three APs.
2. When the neighbor AP is reduced, the power will increase. When AP3 is reduced and the number of neighbors is 2, each AP will run the power adjustment function to repair the black hole covered by the signal caused by the offline neighbor APs, and the adjusted power is greater than the power required by the first three APs.

Third, load balancing
The purpose of wireless communication network load balancing is to ensure the performance and bandwidth of each wireless user, and accurately balance the load of users in WLAN network according to the actual wireless network situation. WLAN load balancing can effectively ensure the reasonable access of wireless users, and is suitable for the wireless network environment of high-density applications [6]. The AC is responsible for performing load balancing when the client connects to the AP. The AC will periodically receive the information sent by the AP to the connected wireless client. The AC performs load balancing according to the neighbor information. When the client connects to the AP, AC will check whether it exceeds the set load; If the set responsibility is not exceeded, accept the current requested connection; Otherwise, the current connection will be accepted or rejected based on the configured load balancing:

1. When the configuration is in session mode, the operation of load balancing mainly depends on the number of users connected to the AP and RF. Client 1 is associated with AP1, and client 2 to client 6 are associated with AP2. To start user based load balancing on the AC, the threshold number of users is 5 and the maximum load difference is 4. When client 7 sends a connection request to AP2, the number of users on AP2 has simultaneously exceeded the threshold of 5 and the session difference threshold of 4. So client 7 is associated with AP1.
2. When configuring for traffic mode, load balancing mainly considers using traffic snapshots. Client 1 and Client 2 are associated with AP1. Start the traffic based load balancing on the AC. If the traffic of AP1 is greater than the set traffic threshold and the traffic difference threshold at the same time, the final client 3 will be associated to AP2.

   Through the above process, the wireless communication network was built, and its resources were scientifically managed, which laid a solid foundation for the follow-up research.

2.2 Movement Data Collection Module of Sports Athletes

Based on the wireless communication network built above, obtain sports video of sports athletes, extract key frames, determine the position of athletes, and provide support for the subsequent extraction of sports athletes' action characteristics.

The visual word bag model, being an image recognition algorithm for image processing, requires the conversion of video data into image data. To accomplish this, the key frame extraction method is employed to extract the significant frames from the video [7]. Nonlinear editing technology in Direct Show plays a major role in this extraction process. Given that sports involve high-speed movements and variations in poses for the same action, it is crucial to ensure the completeness of the action. Therefore, when extracting feature points, the key frame extraction rate is constrained based on the video's frame rate attribute. The key frame extraction algorithm is as follows:

Input: video data A
Output: keyframe image data
Begin:

    (1) Get the video frame rate as p
    (2) Get the total number of video frames, recorded as N
    (3) Calculate key frame capture rate $S = \frac{1000}{p}$
    (4) Extract keyframe images based on S and save them

End

The time complexity of this algorithm is $O(1)$, and only the frame rate of video data is used as the standard for keyframe extraction rate to complete keyframe extraction. After the keyframe extraction is completed, it is necessary to preprocess the keyframe data. The preprocessing operation first randomly selects 80% of the extracted keyframe images as the training set, and the remaining 20% keyframe images as the test set.

Based on the extracted keyframes mentioned above, a real-time and high recognition rate human detector, C4, was adopted. This method utilizes boundary signals from video frames and uses the CENTRIST visual descriptor, followed by a cascaded classifier for classification and recognition. This method can quickly recognize the human body in video frames, with a resolution of 640 × 480 images can achieve a detection effect of over 90% when the thread reaches 20 fps. Using C4 human detector for athlete position detection on the BSMDataset dataset. The schematic diagram of the position detection results for sports athletes is shown in Fig. 3.

To evaluate the detection accuracy of C4 on the BSMDataset dataset, this study utilizes the first frame image of each video segment as the experimental sample. Only the first frame is used for detector performance testing since subsequent frames can rely on the human body's position in the previous frame to determine its location. The C4 human body detector is employed to detect the position of the human body in these samples. The results indicate that the C4 human detector achieves a detection accuracy of 86.25% on these experimental samples, with a processing speed of 20 fps on the visual system's modest hardware. These findings meet the requirements of this study regarding the accuracy and speed of sports athlete position detection. Hence, this detector is selected for further detection purposes.

Fig. 3. Schematic diagram of sports athletes' position detection results

After detecting the position of the athlete in the video frame, a rectangular area containing the athlete is obtained. This article uses the rectangular area obtained by the human body detector as a mask, and by only detecting feature points within the area where the human body is located, it reduces the computational complexity of the algorithm while enhancing its robustness [8]. When human detection is not used, feature points need to be detected throughout the entire video frame, which not only consumes a lot of computational resources and time, but also includes many feature points in the background, causing interference to subsequent tracking and classification recognition. When using a human body detector to detect the human body, a large amount of useless calculations and background interference are eliminated. After the application of C4 detector, the detection range of feature points has increased from the original 800 × 450, dropped to 90 × 270, the detection range has been reduced by 93.25%, which not only reduces the computational complexity, but also avoids the interference of useless feature points on the recognition effect and increases the robustness of the algorithm.

The above process completed the collection of sports athlete videos, extracted key frames, and detected the position information of sports athletes, providing convenience for subsequent sports athlete motion feature extraction.

2.3 Sports Athletes' Action Feature Extraction Module

Based on the aforementioned collected data of sports athlete movements, the extracted features of sports athlete movements (including STIP features, Cuboids features, enhanced dense trajectory features, and covariance features) serve as a foundation for recognizing the final erroneous movements.

The STIP feature was proposed by Laptev, who extended the feature extraction method from two-dimensional spatial images to three-dimensional spatiotemporal videos. Firstly, a Gaussian mean function was used to calculate the first-order spatiotemporal derivative mean, which was used to construct the second-order spatiotemporal matrix used in the feature extraction process [9]. The Gaussian mean function expression is

$$\mu(\sigma, \tau) = g\left(\sigma^2, \tau^2\right) * \left[\nabla L(\sigma, \tau)(\nabla L(\sigma, \tau))^T\right] \tag{1}$$

In Eq. (1), $\mu(\sigma, \tau)$ represents the Gaussian mean function; σ and τ represent independent spatial and temporal parameters, and there is no mutual influence between the two; $g(\sigma^2, \tau^2)$ represents the time function; $\nabla L(\sigma, \tau)$ represents the spatial gradient value.

By utilizing the local maximum value, the position of the point of interest can be accurately determined, enabling the retrieval of the corresponding point of interest. Consequently, the time-space value of the point of interest can be calculated as a descriptor for further analysis.

The calculation formula of local maximum value is:

$$H = \det(\mu) - ktrace^3(\mu) \tag{2}$$

In Eq. (2), H represents the local maximum value; $\det(\mu)$ and $ktrace(\mu)$ represent the maximum and minimum values of the Gaussian mean function.

Cuboids was proposed by Dollar et al., who independently calculated the points of interest in space and time. For spatial dimension, 2D Gaussian kernel smoothing filter is used for detection, while for temporal dimension, 1D Gabor filter is used for detection. The response equation is:

$$R = (I * g * h_1)^2 + (I * g * h_2)^2 \tag{3}$$

In Eq. (3), R represents the characteristic value of Cuboids; I represents the size of the image; g represents the average grayscale value of the keyframe image; h_1 and h_2 represent an orthogonal pair of 1D Gabor filters.

The local maximum value of formula (3) represents the final point of interest detected. After detecting Cuboids' interest points, each pixel generated within the cube is considered as a center point for gradient calculation. These gradient values are then concatenated. To enhance computational efficiency, simplify feature data, and reduce dimensionality, linear discriminant analysis is employed as a dimensionality reduction method. Research analysis suggests that applying PCA can reduce the feature dimension to 100 dimensions, while extracting 200 cube feature blocks from each video.

The calculation method for dense trajectory features mainly involves calculating the optical flow field of the entire video that needs to extract features. The optical flow field is a two-dimensional instantaneous velocity field composed of all pixels in each frame of the video image. This two-dimensional velocity field is the projection of the three-dimensional velocity field of visible points in the image onto the two-dimensional imaging surface. After calculating the optical flow field of the entire image, pixel dense sampling is performed on the first frame of the video that requires feature extraction. Generally, sampling is set every W pixels, and this study sets $W = 5$. To enhance the dense trajectory features, the image is first densely sampled at intervals of W pixels in the spatial dimension. In order to ensure that the extracted feature points can express all the action information as much as possible and reduce the influence of the surrounding environment on the action information, independent sampling under different spatial scales is needed. In the study, sampling was carried out at most on 8 different spatial scales, and each sampling space was reduced $\sqrt{2}$ times in turn according to the resolution of the video. Then the sampled points are tracked to facilitate the tracking and calculation of the optical flow space ω_t of the image. In this study, a 3×3 kernel median filter is used to calculate the optical flow field, and the expression is

$$\omega_t = (\mu_t, \nu_t) \tag{4}$$

In Eq. (4), μ_t represents the horizontal component of the optical flow field; ν_t represents the vertical component of the optical flow field.

For a given sampling point P_t in one frame of image I_t, its tracking point in the next frame of image I_{t+1} is P_{t+1}. By connecting the points in the following image using this method, the trajectory feature of the point can be obtained: $\{P_t, P_{t+1}, P_{t+2}, \cdots\}$.

The covariance feature is obtained on the basis of enhancing dense trajectory features. Select the HOF feature portion from each frame of image features and calculate its corresponding covariance features, represented as:

$$C_i = \frac{1}{n-1} \sum_{i=1}^{n} (\sigma_i - \mu)(\sigma_i - \mu)^T \tag{5}$$

In Eq. (5), C_i represents the covariance feature; σ_i represents the standard deviation; μ represents variance.

In order to facilitate the application of movement characteristics of sports athletes in the future, they are integrated into set $X = \{x_1, x_2, \cdots, x_M\}$ to provide accurate data support for subsequent research.

2.4   Identification Module of Sports Athletes' Wrong Actions

Based on the extracted movement feature $X = \{x_1, x_2, \cdots, x_M\}$ of sports athletes, support vector machine is used to identify incorrect movements of sports athletes, providing certain assistance for sports athlete training.

Support Vector Machine (SVM) is a generalized linear classifier that employs the supervised learning method for binary classification of data. Its decision boundary is determined by maximizing the margin of the training samples. SVM utilizes the hinge loss function to calculate empirical risk and incorporates a regularization term into the solution system to optimize structural risk. As a classifier, SVM demonstrates sparsity and robustness. By utilizing kernel methods, SVM can also perform nonlinear classification, making it one of the commonly used kernel learning methods.

The properties of support vector machines are as follows:

Robustness and sparsity: The optimization problem of SVM considers both empirical risk and structural risk minimization, so it is stable. From the geometric point of view, the stability of SVM is reflected in that it requires the maximum margin when constructing the hyperplane Decision boundary, so there is sufficient space between the interval boundaries to contain the test samples [10]. SVM uses hinge loss function as proxy loss, and the value of hinge loss function makes SVM sparse, that is, its Decision boundary is only determined by the support vector, and the rest of the sample points do not participate in empirical risk minimization. In nonlinear learning using kernel method, the robustness and sparsity of SVM ensure reliable solution results while reducing the computation and memory overhead of the kernel matrix.

Relationship with other Linear classifier: SVM is a generalized Linear classifier, and other types of Linear classifier can be obtained by modifying the loss function and optimization problems under the algorithm framework of SVM. For example, replacing the loss function of SVM with a logistic loss function can get an optimization problem close to logistic regression. SVM and logistic regression are classifiers with similar functions. The difference between the two is that the output of logistic regression has probability significance and is easy to expand to multi classification problems. However, the sparsity and stability of SVM make it have good generalization ability and less computational complexity when using kernel methods.

As a property of kernel methods, SVM is not the only machine learning algorithm that can use kernel techniques. Logistic regression, ridge regression, and linear discriminant analysis (LDA) can also obtain kernel logistic regression, kernel ridge regression, and kernel linear discriminant analysis (KLDA) methods through kernel methods. Therefore, SVM is one of the implementations of generalized kernel learning.

The hyperplane and interval boundary of support vector machine are expressed as:

$$\begin{cases} \zeta^T X + \alpha = 0 \\ \begin{cases} \zeta^T X + \alpha \geq +1 \\ \zeta^T X + \alpha \leq +1 \end{cases} \end{cases} \tag{6}$$

In Eq. (6), ζ and α represent the normal vector and intercept of the hyperplane.

Load a clear set of sports athletes' correct action characteristics $Y = \{y_1, y_2, \cdots, y_N\}$ and wrong action characteristics $Z = \{z_1, z_2, \cdots, z_q\}$, which are distributed on both sides of the hyperplane. Calculate the similarity between sports athletes' action characteristics $X = \{x_1, x_2, \cdots, x_M\}$, correct action characteristics and wrong action characteristics, and the expression is:

$$\begin{cases} \psi(X, Y) = \frac{\text{cov}(X,Y)}{\sqrt{D(X)}\sqrt{D(Y)}} \\ \psi(X, Z) = \frac{\text{cov}(X,Z)}{\sqrt{D(X)}\sqrt{D(Z)}} \end{cases} \tag{7}$$

In Eq. (7), $\psi(X, Y)$ and $\psi(X, Z)$ represent the similarity between the movement characteristics of sports athletes and the correct and incorrect movement characteristics; $\text{cov}(\cdot)$ represents covariance; $D(\cdot)$ represents variance.

According to the calculation result of formula (7), the athletes' action judgment rules are formulated as follows:

$$\begin{cases} \psi(X, Y) \geq \xi \ \ right \\ \psi(X, Y) < \xi \ \ wrong \end{cases} \cup \begin{cases} \psi(X, Z) \geq \Psi \ \ right \\ \psi(X, Z) < \Psi \ \ wrong \end{cases} \tag{8}$$

In Eq. (8), ξ and Ψ represent the athlete's action judgment threshold, which needs to be set according to the actual situation.

Through the above process, the recognition of sports athletes' wrong actions is realized, which provides assistance for athletes' training and skill improvement.

3 System Performance Test

3.1 Experiment Preparation Stage

In order to ensure the smooth progress of the follow-up experiment, the badminton athletes were taken as the experimental objects, and their action data were accurately intercepted in the preparation stage of the experiment, providing certain convenience for the follow-up experiment.

Because the segmentation of experimental data is offline, the motion data can be observed well without considering the real-time. The most accurate window segmentation method is naturally the event based window segmentation method, followed by the action window based data segmentation method. However, the event based window segmentation method is relatively strict in determining the start point and end point, and this paper does not require too much experimental data, so the action window based segmentation method is selected.

In order to illustrate the process of capturing experimental data, take a section of action data in which the swing action type is flat gear as an example. For the experimental data used, the existence of the swing action is known, so the size and position of the action window can be determined only by determining the start and end positions of the action. First of all, we need to find the general position of the swing action. In this paper, we choose to determine the general position of the swing action according to the position of the hitting point, and then determine the starting point and ending point of the swing action. Through the comparison and observation of several experimental data, it is found that the position of the hitting point always appears near the peak of the action data.

Through a large number of tests, it can be seen that the average time spent on a badminton swing action is about 0.8 s, and the sampling frequency of the data acquisition system is 50 Hz, that is, the size of an action window should be 40 groups of data. Here, set the hitting point as the center point of an action window, take the 19 groups of data in front as the first half data, and the remaining 20 groups of data are the 20 groups of data after the hitting point.

In the actual situation, the swing action cannot be continuous all the time, and there must be non swing action data in the middle, and it needs to ensure real-time, so it is necessary to process the data immediately after collecting certain data. This paper uses the method of combining sliding window and action window to intercept real-time action data. First, judge whether there is swing action in the sliding window data. If there is no swing action, discard this part of data and directly read the next sliding window data; If there is swing action, find the hitting point first, and then intercept the swing action data according to the size of the action window determined by the hitting point.

3.2 Analysis of Experimental Results

Based on the contents of the above experimental preparation stage, the video action recognition system based on HOF-CNN and HOG features and the fitness action recognition system based on deep learning are selected as the comparison systems 1 and 2 to design the contrast experiment of sports athletes' wrong action recognition.

The design system is applied to identify the wrong actions of sports athletes, and the recognition results are shown in Fig. 4.

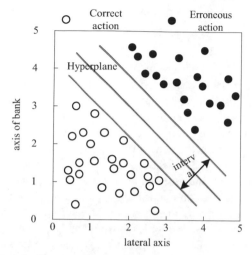

Fig. 4. Schematic diagram of sports athletes' error action recognition results

As shown in Fig. 4, the design system can effectively identify the wrong actions of sports athletes, but in order to highlight the advantages and disadvantages of the application performance of the design system, the recognition rate of wrong actions is calculated under different experimental conditions, as shown in Table 1.

Table 1. Results of error action recognition rate of athletes/%

Test conditions	design system	Comparison system 1	Comparison system 2
1	89.45	45.23	35.02
2	90.45	50.12	45.10
3	91.20	36.26	42.11
4	96.35	50.48	43.59
5	94.25	45.69	50.12
6	92.10	51.78	54.78
7	85.45	46.19	53.46
8	86.23	45.23	52.49
9	91.25	48.79	50.14
10	95.12	46.23	40.16

As shown in the data in Table 1, after the application of the design system, the error action recognition rate of sports athletes obtained is higher than that of the comparison systems 1 and 2, with the maximum value of 96.35%, which fully confirms that the application performance of the design system is better.

4 Conclusion

High-performance sport have various effective rules to prevent injustice, which is an artistic creation and gives people a feeling of intense, wonderful, harmonious and beautiful. With the continuous development of High-performance sport, sports athletes have also been widely concerned by the public. In order to continuously improve the skills of sports athletes, it is necessary to conduct scientific training and correct their incorrect actions. This requires precise identification of the correctness and error of sports athletes' actions. Therefore, a design and research on a sports athlete's incorrect action recognition system based on wireless communication networks is proposed. The design system has significantly improved the recognition rate of incorrect actions by sports athletes, providing more effective system support for sports athlete training. In future research work, multiple sensor data (such as images, videos, inertial sensors, etc.) can be fused to obtain more comprehensive and accurate motion information. By comprehensively analyzing the data of different sensors, we can better capture and identify wrong actions.

References

1. Hameed, I., Tuan, P.V., Camana, M.R., et al.: Optimal energy beamforming to minimize transmit power in a multi-antenna wireless powered communication network. Electronics **10**(4), 509 (2021)
2. Liu, W.: Simulation of human body local feature points recognition based on machine learning. Comput. Simul. **38**(06), 387–390+395 (2021)
3. Wu, Y., Meng, D., Wu, Z.G.: Disagreement and antagonism in signed networks: a survey. IEEE/CAA J. Automatica Sinica **9**(7), 1166–1187 (2022)
4. Liu, S., Li, Y., Fu, W.: Human-centered attention-aware networks for action recognition. Int. J. Intell. Syst. **37**(12), 10968–10987 (2022)
5. Shimada, S., Golyanik, V., Xu, W., et al.: Neural monocular 3D human motion capture with physical awareness. ACM Trans. Graph. **40**(4), 1–15 (2021)
6. Liu, X., Chen, S., Song, L., et al.: Self-attention negative feedback network for real-time image super-resolution. J. King Saud Univ. Comput. Inf. Sci. **34**(8B), 6179–6186 (2022)
7. Wang, L., Li, Y., Xiong, F., et al.: Gait recognition using optical motion capture: a decision fusion based method. Sensors **21**(10), 3496 (2021)
8. Miao, A., Liu, F.: Application of human motion recognition technology in extreme learning machine. Int. J. Adv. Rob. Syst. **18**(1), 4–18 (2021)
9. Zhang, H., Rao, P., Xia, H., et al.: Modeling and analysis of infrared radiation dynamic characteristics for space micromotion target recognition. Infrared Phys. Technol. **116**(3), 1289–1297 (2021)
10. Xiao, G.: Research on athlete's action recognition based on acceleration sensor and deep learning. J. Intell. Fuzzy Syst. **40**(2), 2229–2240 (2021)

Design of Adaptive Detection Algorithm for Mobile Social Network Security Vulnerability Based on Static Analysis

Fang Qian[✉], Qiang Chen, and Lincheng Li

Engineer of China Southern Power Grid Ultra High Voltage Transmission Company,
Guangzhou 510000, China
qianfang202304@163.com

Abstract. In order to improve the accuracy of adaptive detection of security vulnerabilities in mobile social networks and achieve the ideal effect of high-precision adaptive detection of vulnerabilities, static analysis is introduced and an adaptive detection algorithm design of security vulnerabilities in mobile social networks based on static analysis is developed. Use plug-in technology to scan mobile social network ports, databases, operating systems, Web, security baselines, weak passwords, and industrial control systems to obtain network data. The abnormal data propagation rules are used to preprocess the scanned data and extract the network abnormal data. The static analysis of the extracted abnormal data defines the corresponding rules of network security vulnerabilities by building an abstract simulation of network applications, extracts the corresponding relationship between abnormal data and network security vulnerabilities, calculates the final score of network security vulnerabilities according to the basic evaluation utilization factor, and identifies and detects the security vulnerabilities of mobile social networks. The experimental analysis results show that the designed algorithm has a vulnerability detection rate of more than 90% with and without security protection mechanism, and the adaptive vulnerability detection rate is high.

Keywords: Static Analysis · Mobile Social Network · Loophole · Adaptive · Testing

1 Introduction

Nowadays, people can't live and study without social networks for a long time. Social networks are the hubs connecting people. The purpose of information transmission through software or websites via the Internet is social networks (SN). Social networks generally allow users to have their own network space to save and share relevant information, such as email, online cloud disks, etc. With the progress and application of intelligent mobile terminal devices, network sensors and other devices, the use of mobile terminal devices to access social networks has become the mainstream. People have also moved from the original social model to mobile social networks, so mobile social networks are

L. Yun et al. (Eds.): ADHIP 2023, LNICST 548, pp. 389–402, 2024.
https://doi.org/10.1007/978-3-031-50546-1_26

developed from social networks [1]. With the wide spread and use of social software such as WeChat, Weibo and QQ, the popularity of mobile social networks among people has been further promoted. With the combination of mobile devices and communication technology, people can access massive information anytime and anywhere using the network, which has a huge impact on people's social activities and work learning. Mobile social networks collect and summarize the information of mobile devices to achieve communication through social networks. Mobile networks provide various network services for mobile users, including LBS. In the global population of more than 7 billion, more than 77% of users can enjoy the services provided by the network. On this basis, mobile social networks have contributed a lot. It is precisely because of the large number of mobile social network users and the huge amount of information transmitted on the network, many security problems will arise when people use mobile social networks to experience the services brought by LBS.

In network security attacks, mobile social network nodes are the main targets of attacks. Since mobile social network nodes can receive all information on mobile network devices, and can realize real-time detection and analysis of information, it is considered that node detection plays a vital role in network security. However, there are many types of network nodes in mobile social networks, and most of them exist in a highly decentralized state in the network environment. With the increasing attention paid to the security of mobile social networks, the detection of security vulnerabilities of mobile social networks has become an indispensable and crucial part of Internet security. With the continuous increase of mobile social network data scale, mobile social network attacks have become more complex and diversified. Traditional vulnerability detection algorithms cannot meet the security detection requirements of mobile social networks. How to use new computer algorithms to quickly and accurately detect known vulnerabilities under massive data, and achieve effective prevention and security precautions against mobile social network attacks, has become a hot topic of research institutes and researchers. In the work of in-depth research in this area, we found that static analysis is a method of using mathematical methods to process and analyze data. Compared with other data processing methods, this method can improve work efficiency. By collecting a large number of mobile social network data samples, we can judge and identify whether the samples are abnormal, and through appropriate processing of data, we can improve the node's computing ability. The purpose of simplifying the calculation. Therefore, this paper introduces static analysis in this research, and designs an adaptive detection algorithm for mobile social network security vulnerabilities, with a view to improving the security of mobile social network node transmission information, and improving the ability to detect mobile social network security vulnerabilities.

2   Design of Adaptive Detection Algorithm for Mobile Social Network Security Vulnerabilities Based on Static Analysis

2.1   Mobile Social Network Security Vulnerability Scanning

Obtaining data is the first step of adaptive detection of security vulnerabilities in mobile social networks, which is a basic work. Here, we use network vulnerability scanning to obtain abnormal data. The scanning objects mainly include mobile social network ports, databases, operating systems, Web, security baselines, weak passwords and industrial control systems. Considering the large scanning range and complex scanning object types, this time we use plug-in technology to obtain data, write plug-ins in C language, connect the plug-in interface with the network interface, and insert a plug-in for each scanning object, as shown in Table 1.

Table 1. Scan plug-in class table

Scan plug-in ID number	category	name	explain
1100	Port scanning	Port	Service analysis
1200	Database dictionary scanning	Dict	Oracle and MySQL vulnerability scanning
1300	system scan	NteBIOS	Windows vulnerability attack
1400	Web Scan	HTTP	Web Service Scan
1500	Security Baseline Scan	SNMP	Third party network access security analysis
1600	Weak password scanning	General	Scanning for weak passwords existing in the system
1700	Industrial control scanning	Overflow	Safety analysis of industrial control system

Table 1 shows that a plug-in file is constructed for each mobile social network vulnerability class. The plug-in file contains the vulnerability scanning function corresponding to the vulnerability of the mobile social network. The mobile social network can be scanned by calling the corresponding scanning function to avoid confusion of the plug-in during the mobile social network scanning [2]. Among them, plug-ins exist in the form of compiled Windows DLLs. Each plug-in has a corresponding plug-in identity document (ID), and a unique function name [3] is designed for each scanning function. The function name is the index value of the mobile social network vulnerability in the plug-in plus the plug-in ID. The prefix symbol of the function name is designed as "_" in the form of "_1 * 00 * *". The specific scanning process is shown in the following figure (Fig. 1).

Make the mobile social network running, and scan the network one by one according to the plug-in sequence in Table 1. First, scan the mobile social network ports, which are the entry end of mobile social network attacks. Use 1100 plug-ins to scan all network ports and analyze whether the network port service is normal. Second, scan and check the firewall, router and server in the network operating system environment [4]. Thirdly, use 1200 plug-in to scan database dictionary and detect system tables and fields in database dictionary. Fourth, in the development process, due to programming logic errors, the website has SQL injection, cross site, dark chain and other problems, which may cause the website and website data to be tampered with, core data such as account

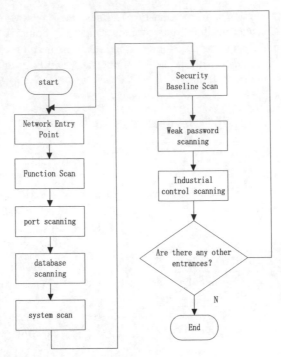

Fig. 1. Network Security Vulnerability Scanning Process

information to be stolen, the website server to become a dummy host and other problems. The 1400 plug-in can scan the website for Web vulnerabilities defined by OWASP Top 10. Fifth, security configuration verification is the basic work of security management and an important technical means of security operation and maintenance. Security configuration verification should first establish a baseline that meets the security configuration requirements of the organization's information security management system. Sixth, use the 1500 plug-in to scan the specific products of these support systems, such as Windows, Oracle, WebLogic, switches, firewalls, etc. Seventh, use the 1600 plug-in to detect the weak password existing in the system. The plug-in supports well-known protocols such as Telecom Network Work Protocol (TELNET), File Transfer Protocol (FTP), Secure Shell (SSH) protocol, Post Office Protocol Version 3 (POP3), Server Message Block (SMB)Simple Network Management Protocol (SNMP), Remote Display Protocol (RDP), Simple Mail Transfer Protocol (SMTP), Remote DictionaryServer (REDIS), Oracle, etc. The default dictionary library is built in and supports uploading customized dictionary library [5]. Eighth, use 1600 plug-ins to comprehensively scan industrial control systems, including Siemens, Schneider, AB, Rockwell, domestic Hollysys, central control and other mainstream programmable logic controllers (PLC), distributed control systems, DCS), Supervisory Control And Data Acquisition (SCADA) equipment, etc., to collect vulnerability attack data on current mainstream production business equipment and software.

Store all scanned data in the network vulnerability attack information database, edit the scanning file name according to different scanning objects, and store the scanned abnormal data by category for subsequent static analysis.

2.2 Exception Data Extraction

Abnormal data refers to data related to vulnerability attacks on mobile social networks. The mobile social network is in an abnormal state [6] when there is a vulnerability in the mobile social network, but it is not attacked by the mobile social network attacker. Therefore, the data obtained through scanning is not completely related to mobile social network vulnerability attacks, but may also include mobile social network vulnerability data, so it is necessary to extract valuable abnormal data from the scanned data. There are essential differences between the two types of data. When the mobile social network is attacked by vulnerabilities, the abnormal data in the mobile social network is dynamic. In order to achieve the attack task, abnormal data needs to be transferred from one location to another. The mobile social network has loopholes, but has not been attacked. At this time, the abnormal data in the mobile social network is static, and the abnormal data extraction is based on the data propagation rules. The abnormal data is extracted through data inspection. The following figure is the abnormal data extraction flow chart.

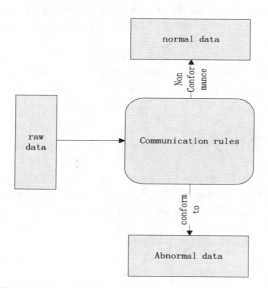

Fig. 2. Schematic Diagram of Abnormal Data Extraction

As shown in Fig. 2, the exception data propagation rule is designed according to the difference between the two, and the calculation formula of the rule is:

$$P_{uhd}(ins) = P_{uhd}(x, z) = isTaint(z) \qquad (1)$$

Where, $P_{uhd}(ins)$ Represents the mobile social network exception data propagation rules; P_{uhd} Abnormal data indicating that there are instructions for mobile social network access; x Indicates the memory address of read network access; z Indicates the address returned after the network instruction is executed; Use formula (1) to check the scanned data and whether it conforms to the rule [7]. If this rule is met, it will be determined as abnormal data generated by network vulnerability attacks, extracted and set up abnormal data sets.

2.3 Static Analysis of Abnormal Data

Using static analysis technology to analyze abnormal data, the mobile social network security vulnerability adaptive detection algorithm based on static analysis assumes that the abnormal data is contaminated data, and uses data flow to judge the value that can be controlled by the attacker. Therefore, it requires to know where the data enters the mobile social network program, how it is transmitted in the program, and finally the vulnerable operation. Static analysis is the key to identify many input validation flaws and represent flaws. In this way, if a mobile social network application contains an exploitable vulnerability, it almost always contains a path from receiving user input functions to vulnerable operations [8]. Therefore, vulnerabilities in mobile social networks can be detected through static analysis of abnormal data. The static analysis of abnormal data includes the following parts: (1) building an abstract simulation of the network; (2) Define the source rule, transmission rule, purification rule and receiving rule corresponding to the vulnerability; (3) Analyze mobile social network application vulnerabilities.

Since the mobile social network security vulnerability adaptive detection algorithm based on static analysis is a network data flow based detection algorithm, the network data flow model [9] must be established before vulnerability analysis. A mobile social network application can be represented as the mapping of HTTP requests and the current state of Web applications to replies, application dependency graphs, and new Web application states:

$$P_{uhd}(ins) : R \times S \to P \times E \times D \tag{2}$$

Where, R Represents HTTP requests submitted to mobile social network applications; S Indicates the current status of mobile social network applications; P Represents the HTTP reply of the mobile social network application; E Represents the control flow and data flow used by network applications to process a given HTTP request; D Indicates the status after the network application request is completed.

Exception data propagation includes a series of source rules, transmission rules, acceptance rules and purification rules, as shown in the following figure.

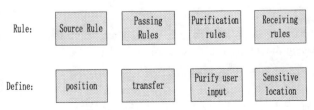

Fig. 3. Abnormal Data Rules

As shown in Fig. 3, these rules explain how the source function in the network generates pollution data, and how the receiving function is used. If the pollution data is not purified by the purification unit before being transmitted to them; It also explains how the pollution data is transferred through the transfer function. Source rules define where contaminated data enters the network application; The transfer rule defines the way in which the function manipulates the contaminated data, that is, how the contaminated data is transferred in the network; Purification rules define functions or modules used to purify user input in network applications to prevent potential cross site scripts. They are a form of transfer rules. The receiving rules define the sensitive locations of network applications that should not accept contaminated data. The source rules obtain data from HTTP requests. The receiving rules are used to return the obtained data to the client. In order to include other server information in the returned information, it is necessary to connect the string transmission rules [10]. Therefore, we can use the static analysis model to specifically describe the security vulnerability rules of mobile social networks.

According to the above description, the correspondence between the rules in static analysis and the mobile social network application in formula (2) can be formally defined:

$$P \to SS \cup DS \cup JS \cup RE = M \tag{3}$$

Where, SS Indicates the source rule of abnormal data; DS Indicates the transfer rules of abnormal data; JS Indicates the receiving rules of abnormal data; RE Represents the purification rule of abnormal data, M Represents the correspondence between mobile social network applications and exception data transmission rules [11]. The extracted exception data is statically analyzed according to the above process, and the corresponding relationship of the exception data rules in the mobile social network can be obtained.

2.4 Vulnerability Adaptive Detection

Use the above extracted correspondence to identify mobile social network security vulnerabilities, and the detection process is shown in the following figure.

As shown in Fig. 4, the vulnerability detection process reads the correspondence from the source file and parses it into automatic control rules. Match the scanned source code with the source code pattern in the correspondence. After successful matching, the pattern constraint will be checked [12]. When both pattern matching and pattern constraint are successful, trigger the pattern matching variable to bind the variable automatically to realize automatic state transfer. This process needs to analyze the environmental assessment indicators. Even in the same vulnerability, at the same time, in different environments, the vulnerability degree is different. Therefore, environmental assessment indicators must be analyzed during the transfer of bound variables. According to the environmental assessment index set, the basic assessment impact factors are calculated as follows:

$$Z = \min\{(M \times XR)(M \times YR)(M \times ZR)\} \tag{4}$$

Where, Z Indicates the security degree of mobile social network environment; XR Indicates confidentiality requirements; YR Represent integrity requirements; ZR Indicates availability requirements [13]. The basic detection results of mobile social network security vulnerabilities can be obtained by substituting formula (4) into formula (5):

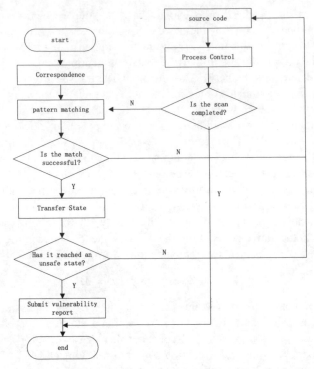

Fig. 4. Vulnerability Detection Flow Chart

$$H = (0.6Z + 0.4L) \times f \tag{5}$$

Where, H Represents the basic detection results of security vulnerabilities in mobile social networks; L Represents the basic detection utilization factor; f Represents the constant associated with the basic assessment impact factor. Basic detection utilization factor in the above formula L Is an unknown number, and its calculation formula is:

$$L = EV \times EC \times eu \tag{6}$$

Where, EV Indicates the propagation path of abnormal data; EC Indicates the complexity of abnormal data; eu Indicates the authentication strength of abnormal data. Substitute formula (5) into formula (7) to get the final detection results of mobile social network security vulnerabilities:

$$A = (H + H \times CDP) \times TD \tag{7}$$

Where, A Indicates the score of mobile social network security vulnerabilities; CDP Indicates measuring the potential impact of vulnerabilities; TD It indicates that vulnerability indicators exist in the potential harm of mobile social networks to the network [14]. Set the detection threshold according to the actual situation. If the calculated value of formula (7) is greater than the threshold, the output detection result is that the mobile social network has vulnerabilities; If the calculated value of formula (7) is less than the threshold value, the output detection result is that there is no vulnerability in the mobile social network. According to the comparison between formula (7) and the detection threshold value, a vulnerability adaptive detection report is generated. When the self motivation reaches the unsafe state, the vulnerability report is submitted to the user [15]. If it does not reach the unsafe state, it will continue to scan the subsequent source code until the scanning is completed, so as to complete the design of adaptive detection algorithm for mobile social network security vulnerabilities based on static analysis.

3   Experimental Analysis

3.1   Experiment Preparation

In order to verify the reliability and feasibility of the mobile social mobile social network security vulnerability adaptive detection algorithm designed in this paper based on static analysis, the experimental analysis is carried out as shown below. The operating platform used in the experiment is the 32-bit Win7 flagship version, the experiment uses C # language, and the programming experimental platform is the VS2010 flagship version. The experimental data uses the test page provided in the Wavsep program. With a mobile social mobile social network as the experimental environment, the advantages of the research method in vulnerability detection rate are verified by automatically expanding the simulation attack mode. The construction of simulation attack set mainly depends on the identification and judgment of attack types by security experts. Each attack is a separate individual for research. The artificially expanded simulation attack process is shown in the following figure (Fig. 5).

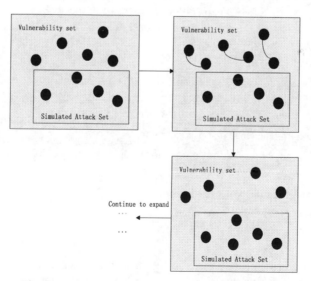

Fig. 5. Simulation attack process of manual expansion

As shown in the above figure, if the user enters a string when attacking the mobile social mobile social network, the final execution result will be obtained in the browser.

The process is as follows: enter the URL address in the browser address bar, add the refined string suffix, and the new page jump request in the string will be sent to the site business processor for processing. Send analysis request, and formulate jump strategy according to the analysis results. However, because the server does not deploy the corresponding security mechanism, the submitted URL string does not belong to the traditional jump request type. If there is a security vulnerability in the mobile social mobile social network application, the server will usually send a response, thus exposing sensitive information. By analyzing the response results returned to the client browser vulnerability, determine whether there is a corresponding vulnerability problem on the server side of the mobile social networking site. If the user constructs a string as the result of the jump request, when the customer browses the mobile social networking target website, the string in the browser address bar will be input and sent to the server, that is, an error query will be launched to the database, and the related information will be returned to the vulnerability injected by the external user through the browser.

In order to achieve the purpose of injection attack, the attacker must first find the vulnerability of the server side security mechanism. In terms of technical methods and formal rules of security scanning, their security verification mechanisms are similar, and they are all aimed at limited code, that is, the vulnerabilities covered by the simulation attack set used in security scanning are limited. The generation of different use cases is realized by deformation, so as to find the smallest set of simulated attacks. By studying the interrelationship between attack modes, a series of transformations and settings are carried out on the test case code to realize the vulnerability extension simulation process. In the experiment, the vulnerability data file of mobile social mobile social network includes 6 types of security vulnerabilities, namely NVD1.zip, Secunia2.zip, SecurityFocus3.zip, Cnvd4.zip, CNND5.zip, NSFocus6.zip. After the above design is completed, the initial parameters of the experiment are designed, and the specific parameters are shown in the table below.

Table 2. Table of Experimental Parameters

S/N	name	parameter
1	Vulnerability data load unit parameters	256
2	Initial value of vulnerability information sample mixing coefficient	1.25
3	Minimum carry value of vulnerability information	2
4	Number of security vulnerability types	6
5	Number of security vulnerabilities	90

During the experiment, the total transmission amount of vulnerability data annotation index in unit time is greater than or equal to 1, and the transmission order of vulnerability data does not change when the SQL annotation statement is executed, and the position of the encoding vector can only move from the first annotation node to the last annotation node.

3.2   Result Analysis

After the above experimental preparations are completed, the mobile social mobile social network security vulnerability adaptive detection experiment is carried out. The experimental execution process is as follows:

Step 1: Select the client PC: host as the experimental object, and connect it to the wireless LAN communication environment;
Step 2: Select the adaptive detection algorithm for security vulnerabilities of mobile social mobile social network based on static analysis as the application method of the experimental group;
Step 3: In order to make the experimental data and experimental results have a certain degree of explanation and reliability, two traditional algorithms are selected for comparison. The two traditional algorithms are the detection algorithm based on blockchain technology and the detection algorithm based on ant colony algorithm. The following are represented by traditional algorithm 1 and traditional algorithm 2 respectively, and the two traditional algorithms are used as the application methods of the control group;
Step 4: Design the experiment scenario, and the experiments are verified in this scenario.
Step 5: Explain the experimental equipment and parameters.
Step 6: Record the experimental values of the experimental group and the control group;
Step 7: Compare the experimental group and the control group to record the values and summarize the experimental rules. Next, set the evaluation indicators for this experimental test. In this experiment, the vulnerability detection rate is used as the evaluation index of three algorithms. In order to restore the vulnerability detection rate of mobile social mobile social network without security protection mechanism, no input URL filtering is added and no firewall is set, so as to test the vulnerability detection rate without security protection mechanism. Without adding any input URL filtering, the above three algorithms are used to analyze the vulnerability detection rate. The comparison results are shown in Table 2 (Table 3).

Table 3. Vulnerability detection rate of three algorithms without security mechanism (%)

Number of vulnerability samples	Design algorithm	Traditional algorithm 1	Traditional algorithm 2
10	98.8	65.4	68.4
20	97.6	62.5	64.2
30	96.8	58.4	60.2
40	96.3	56.3	55.6
50	96.4	55.1	52.4
60	95.2	51.4	50.3
70	95.1	48.5	47.5
80	94.3	45.3	44.4
90	94.1	42.7	41.2

From the data in the table above, we can see that the vulnerability detection rate of the algorithm designed in this paper is higher than that of the two traditional algorithms, and the number of detected vulnerabilities is basically consistent with the actual situation. The number of undetected vulnerability samples accounts for 6.9% of the total vulnerability samples, which is far lower than that of the two traditional algorithms. This is because the design algorithm utilizes plugin technology to scan mobile social networks and obtain network data. This technology can enhance the ability of vulnerability detection, enabling algorithms to better capture and identify vulnerabilities, and improve vulnerability detection rates. In contrast, traditional algorithms do not fully utilize plugin technology, resulting in a decrease in vulnerability detection rate.

In addition, in order to restore the vulnerability detection rate with security protection mechanism, it is necessary to use a dedicated line to connect the simulation software to the dedicated interface of the firewall to test the vulnerability detection rate of the three algorithms with security protection mechanism. Under the security protection mechanism, the above three algorithms are used to analyze the vulnerability detection rate. The comparison results are shown in the following figure (Fig. 6).

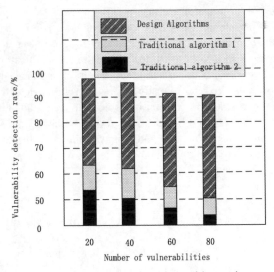

Fig. 6. Vulnerability detection rate of three algorithms with security protection mechanism

It can be seen from the above figure that under the security protection mechanism, for different vulnerability sample numbers, the comparison results of the vulnerability detection rate of the three algorithms are different. The vulnerability detection rate of the three algorithms decreases with the increase of the number of vulnerability samples, but the design algorithm is less affected by the number of vulnerability samples, and the vulnerability detection rate is always more than 90%, The vulnerability detection rate of both traditional algorithms decreases significantly with the increase of vulnerability samples. The vulnerability detection rate of traditional algorithm 1 can only be maintained at 50.3%–62.4%, and that of traditional algorithm 2 can only be maintained at 43.5%–54.6%.This is because the design algorithm in this paper adopts a static analysis method, which can more comprehensively analyze and detect security vulnerabilities in mobile social networks by building abstract simulation of Web application and defining vulnerability corresponding rules. In contrast, traditional algorithms cannot fully utilize the advantages of static analysis.

From this, it can be seen from this that the vulnerability detection rate of the algorithm designed in this paper is high, which can comprehensively detect the security vulnerabilities of mobile social mobile social networks. The algorithm designed in this paper has good feasibility and reliability, and is more suitable for adaptive detection of security holes in mobile social mobile social networks than traditional algorithms.

4 Conclusion

With the continuous updating of modern Internet technology, the data volume of mobile social network is increasing, and the operational efficiency and security of mobile social network nodes are increasingly concerned. In depth research on this aspect, it is found that mobile social mobile social network will involve a large amount of data related to cloud computing when carrying or transmitting information. If the node data vulnerability is not detected in time, the security risks of mobile social mobile social network will not be found in time, thus causing the mobile social mobile social network to fail to operate normally. Therefore, how to effectively solve the security problems and reduce the risks in mobile social networks has become one of the important topics in the current scientific research field. For this reason, this article uses plugin technology to scan mobile social networks, obtain network data, and preprocess using anomaly data propagation rules to extract network anomaly data. Then, the extracted abnormal data is statically analyzed, and the corresponding rules of network security vulnerabilities are defined by constructing an abstract simulation of Web application, and the corresponding relationship between abnormal data and network security vulnerabilities is extracted. Finally, the final rating of network security vulnerabilities is calculated based on the basic evaluation utilization factor to identify and detect security vulnerabilities in mobile social networks. After the completion of the design, this method has proved through comparative experiments that it can achieve accurate detection of vulnerabilities in mobile social mobile social network nodes, and can be popularized in the subsequent mobile social mobile social network engineering field using this method instead of traditional methods. The research in this article is of great significance for addressing security issues, reducing risks, and promoting the development of mobile social networks. By proposing an adaptive detection algorithm for security vulnerabilities in mobile social networks based on static analysis, precise detection of node vulnerabilities in mobile social networks can be achieved, providing effective guarantees for the safe operation of mobile social networks.

References

1. Aslan, Ö., Aktuğ, S.S., OzkanOkay, M., Yilmaz, A.A., Akin, E.: A comprehensive review of cyber security vulnerabilities, threats, attacks, and solutions. Electronics **12**(6) (2023)

2. Chen, G., Wang, H., Zhang, C.: Mobile cellular network security vulnerability detection using machine learning. Int. J. Inf. Commun. Technol. **22**(3) (2023)
3. Algarni, A., Thayananthan, V.: Autonomous vehicles: the cybersecurity vulnerabilities and countermeasures for big data communication. Symmetry **14**(12) (2022)
4. Jagannathan, J., Mohamed Parvees, M.Y.: Security breach prediction using artificial neural networks. Meas. Sens. **24** (2022)
5. Zheng, X.: Computer deep learning network security vulnerability detection based on virtual reality technology. Adv. Multimedia **2022** (2022)
6. Xing, Y.: Design of a network security audit system based on log data mining. Wirel. Commun. Mob. Comput. **2022** (2022)
7. Guoyu, L., Luo, G.: Research on network security vulnerability detection method based on artificial intelligence. J. Phys. Conf. Ser. **1651**(1) (2020)
8. Shaaban, R., Faruque, S.: Cyber security vulnerabilities for outdoor vehicular visible light communication in secure platoon network: review, power distribution, and signal to noise ratio analysis. Phys. Commun. 40 (2020). (prepublish)
9. Telecommunications - Physical Layer Communications; Study Data from University of North Dakota Provide New Insights into Physical Layer Communications (Cyber security vulnerabilities for outdoor vehicular visible light communication in secure platoon network: Review, power distribution, and ...). Comput. Netw. Commun. (2020)
10. Saudi Arabian Oil Company: "Cybersecurity vulnerability classification and remediation based on network utilization" in patent application approval process (USPTO 20200162498). Technol. Bus. J. (2020)
11. Afreen, S.S.: Analytical study on network security Breach's. J. Trend Sci. Res. Dev. **4**(3) (2020)
12. Xing, W.: Research on computer network security vulnerabilities and preventive measures based on multi-platform. IOP Conf. Ser. Mater. Sci. Eng. **740** (2020)
13. Syed, R.: Cybersecurity vulnerability management: a conceptual ontology and cyber intelligence alert system. Inf. Manag. **57**(6) (2020)
14. FDA warns of urgent cybersecurity vulnerabilities with medical devices. Biomed. Saf. Stand. **50**(1) (2020)
15. Alvarez Valenzuela, D., Hevia Angulo, A.: Legal protection for the search and notification of cybersecurity vulnerabilities in Chile. Revista Chilena de Derecho y Tecnologia **9**(2) (2020)
16. Sun, H., Cui, L., Li, L., et al.: VDSimilar: vulnerability detection based on code similarity of vulnerabilities and patches. Comput. Secur. **110**(5–6), 102417 (2021)
17. Alaaraji, Z., Ahmad, S.S.S., Abdullah, R.S.: Propose vulnerability metrics to measure network secure using attack graph. Int. J. Adv. Comput. Sci. Appl. **12**(5), 2021 (2021)

Dynamic Mining of Wireless Network Information Transmission Security Vulnerabilities Based on Spatiotemporal Dimension

Qiang Chen$^{(\boxtimes)}$, Fang Qian, and Yukang Liu

China Southern Power Grid Ultra High Voltage Transmission Company, Guangzhou 510000,
China
cgycgy8680@126.com

Abstract. In order to improve the efficiency of dynamic mining for wireless network information transmission security vulnerabilities and improve the accuracy of mining results, this paper proposes a dynamic mining method for wireless network information transmission security vulnerabilities based on the spatiotemporal dimension. Firstly, collect data on security vulnerabilities in wireless network data transmission; Secondly, wavelet transform is introduced to filter and process wireless network information transmission security vulnerability data; Then, in the deep neural network architecture, the instruction level word embedding method based on Word2vec obtains the feature attributes of wireless network information transmission security vulnerabilities; Finally, dynamically mine wireless network information transmission security vulnerabilities based on the spatiotemporal dimension. The experimental results show that the vulnerability dynamic mining method proposed in this paper takes 25.8 s, with an accuracy of 99.0% and a recall rate of 98.1%, which can improve the effectiveness of vulnerability dynamic mining.

Keywords: Spatiotemporal Dimension · Wavelet Transform · Deep Neural Network · Instruction Level Word Embedding · Dynamic Vulnerability Mining

1 Introduction

With the development of wireless internet and the penetration of information technology, there are currently more and more industrial control systems connected to computer networks [1]. Compared with classic information systems, industrial systems mostly use specialized software, devices, and protocols, so the security vulnerabilities of industrial control systems are different from those of information systems. Industrial vulnerabilities can be classified into device vulnerabilities, wireless protocol vulnerabilities, and industrial software vulnerabilities based on their location [2, 3]. Hackers or attackers exploit vulnerabilities to attack the system. If vulnerabilities are discovered and exploited by the attacker in a timely manner, corresponding remedial measures can be

© ICST Institute for Computer Sciences, Social Informatics and Telecommunications Engineering 2024
Published by Springer Nature Switzerland AG 2024. All Rights Reserved
L. Yun et al. (Eds.): ADHIP 2023, LNICST 548, pp. 403–417, 2024.
https://doi.org/10.1007/978-3-031-50546-1_27

taken to effectively reduce the likelihood of the system being attacked. Vulnerability detection and mining technology is used to discover system vulnerabilities in a timely manner. In addition, the vulnerabilities of industrial systems have their own characteristics, so the vulnerability detection and mining methods of traditional information systems may not be fully applicable to vulnerability discovery under the Industrial Internet [4]. Therefore, the study of vulnerability detection and mining technology combined with traditional information system and industrial control system will add a layer of barrier to protect modern industrial control system, escort the vigorous development of Industrial Internet, and even have important significance to protect people's livelihood and national stability. Reference [5] puts forward the method of network Computer security hidden trouble and vulnerability mining, which requires a comprehensive security hidden trouble analysis of the network computer system. Identify potential security risks by reviewing known vulnerability databases, analyzing malware samples, and network attack behaviors. Based on the existing security hazard analysis results, it is necessary to research and develop corresponding vulnerability mining methods. After discovering potential vulnerabilities, verification and repair are required. Trigger vulnerabilities by constructing input with malicious intent and detect whether the system is under attack. Repairs can include modifying source or binary code, adding security checks and error handling mechanisms, or updating system and application patches. Establish a security monitoring system, regularly scan and monitor networks and systems, and promptly identify new security risks and vulnerabilities. At the same time, it is also necessary to pay attention to the latest security technologies and research results, and continuously improve vulnerability mining methods and security protection measures. Reference [6] proposes a network vulnerability mining method based on Apriori risk data analysis to collect and prepare data related to network vulnerabilities. This includes vulnerability databases, network attack logs, system logs, etc. Ensure the integrity and accuracy of data, and carry out appropriate preprocessing, such as removing duplicate data, handling missing values, etc. Perform risk analysis on network vulnerability data using the Apriori algorithm. Based on the results of risk data analysis, specific vulnerabilities can be selected for mining and evaluation. This can be achieved through in-depth analysis of vulnerability combinations and frequent itemsets in association rules. Based on the mining results, evaluate the severity, scope of impact, and potential risks of vulnerabilities. After discovering specific vulnerabilities, corresponding repair and preventive measures need to be taken. This may include application patch updates, system configuration adjustments, network security policy improvements, etc. At the same time, based on the mining results, potential vulnerabilities can be predicted in the future, and proactive preventive measures can be taken to reduce the probability and impact of vulnerabilities occurring. However, the above methods have poor efficiency in dynamically mining vulnerabilities and improving the accuracy of mining results.

Therefore, this article proposes a dynamic mining of wireless network information transmission security vulnerabilities based on spatiotemporal dimensions.

2   Wireless Network Information Transmission Security Vulnerability Data Processing

2.1   Wireless Network Information Transmission Security Vulnerability Collection

Due to the sensitivity and aggressiveness of vulnerabilities, the vulnerability site data does not have a publicly available dataset, making it difficult to obtain. In the early stage, by searching and reading a large amount of literature online, only a portion of SQL injection attack statement data was collected and no vulnerability data was found. Therefore, the experiment requires collecting SQL injection vulnerability data on one's own. There are two ways to collect vulnerabilities, one is to scan specific websites, and the other is to use non-public data authorized by the internship company for research purposes. During the experiment, a total of eight weeks were spent using a scanner to scan over 1500 websites and identify over 1300 vulnerabilities.

The sources of vulnerability websites include a large number of testing stations built through dockers and online vulnerability testing stations publicly available for testing.

Each vulnerability data should contain the following basic information: URL, which is the website URL containing SQL injection vulnerabilities; The HTTP request method is related to the location of the injection point; Injection type, which is a detailed classification of SQL injection types; Vulnerability attack statement, which detects the presence of SQL injection vulnerabilities. By analyzing and extracting the scanning results, the following vulnerability related information was obtained, as shown in Table 1.

Table 1.   Scan Results Table

Vulnerability Information	content
URL	Request Target Address
Injection Point	Parameters that can be injected into SQL
HTTP request type	GET, POST Wait for request type
SQL Injecting attack statements	Message for implementing injection attacks
Attack Details	The final SQL injection attack statement and the SQL injection attack statement used during the attack process
HTTP request	Including request type, request header, and request data
HTTP Response information	HTTP requests return information, usually HTML pages or JSON data
Vulnerability classification	Vulnerability type, including CVE and CVSS information
Vulnerability Details	Include vulnerability details and solutions

The scanning results of the scanner, whether displayed through a web page or exported through a report, cannot be input into the model. Therefore, it is necessary to extract the required vulnerability information from the scanner and process it into a format that is convenient for input into the model. By analyzing the principle of the scanner, all information about the vulnerability was obtained through the API interface. Finally, the vulnerability data was obtained through constructing HTTP requests and stored in MySQL [8].

The internship company obtained approximately 2000 pieces of vulnerability information through authorized penetration vulnerability scanning, covering various databases and injection types of data. The vulnerability information includes URLs, injection points, database types, HTTP request types, and SQL injection attack statements, meeting the requirements of model training. The vulnerability information obtained through the above two methods is processed and the final stored data format is shown in Table 2:

Table 2. Vulnerability Information Data

field	data
domain	http://www.*****
url	http://www.*****
inject_type	SQL injection
request_type	POST
inject_key	q_year
payload	Page = 0&PageSelect = &q_day = &q_month = &q_year = %BF'%BF"

Complete the collection of wireless network information transmission security vulnerability information.

2.2 Wireless Network Information Transmission Security Vulnerability Data Filtering Processing

At present, mainstream wireless network protocols include Modbus, EtherCAT, Powerlink, Porfinet, Ethernet/IP, TSN (Time Sensitive Networks), etc. There are various methods to capture data packets from different wireless network protocols, among which the most direct method is to apply appropriate message packet capture tools to capture data packets generated by industrial control systems from real industrial control network environments as training data. After capturing a sufficient number of data packets from the wireless network environment, it is necessary to perform data preprocessing operations on these raw data [9].

Leopard Mobile data set selected in this chapter is a binary code file, which is disassembled into assembly code by IDA pro tool; The CWE119 dataset is a C language program code file that needs to be compiled into assembly code format. Batch format the compiled assembly code through Python scripts, remove textual description information from the code, and retain function and instruction information. In the above dataset, each sample has a corresponding 0 or 1 label, with 0 indicating vulnerability and 1 indicating no vulnerability. The ratio of sample size for training, validation, and testing sets is 3:1:1. Before each training and testing, the order of the data is disrupted.

In order to better complete the data mining and processing of wireless network information transmission security vulnerabilities, priority is given to introducing wavelet transform to filter and process wireless network information transmission security vulnerability data [10]; Wavelet transform is a high-performance denoising algorithm that has applications in both data and image fields. It achieves multi-scale refinement of wireless network information transmission security vulnerability data through operations such as scaling and translation, thereby obtaining high and low frequency parts. Further refinement of high and low frequency parts can obtain detailed information of corresponding data. Wavelet analysis theory has been widely applied in fields such as signal and speech analysis.

cIf a represents $\tau(x)$ square integrable function, then the Fourier transform $\beta(x, y)$ needs to satisfy the constraint condition ψ:

$$\psi = \int\limits_{-\infty}^{\infty} \frac{\beta(x, y)}{\tau(x)} - \vartheta_{(x,y)}(t) \tag{1}$$

In the above equation, $\vartheta_{(x,y)}(t)$ represents the wavelet mother function, which is expanded, scaled, and translated to obtain the corresponding wavelet basis function. Then, the continuous wavelet transform $\left| \vartheta_{(x,y)}(t) \cdot \tau(x) \right|$ of wireless network information transmission security vulnerability data is represented in the form of formula (2):

$$\left| \vartheta_{(x,y)}(t) \cdot \tau(x) \right| = \begin{cases} \frac{1}{\sqrt{x}} \int \vartheta_{(x,y)}(t) \times \left(\frac{\alpha - x}{\tau(x)} \right) \\ \left\langle \vartheta_{(x,y)}(t), \tau(x), \alpha \right\rangle \end{cases} \tag{2}$$

In the above equation, x represents the scale factor; α represents the translation factor.

In order to simplify the calculation flow of the computer, it is necessary to discretization all continuous wavelets. Discretization mainly deals with the above two different factors. Applying binary dynamic networks to the wavelet transform process [10, 11] yields the binary wavelet transform c, which corresponds to the following expression:

$$R_{(i,j)} = \vartheta_{(x,y)}(t) \cdot \tau_{ij}(u, v) \cdot \tau(x) \tag{3}$$

In the above equation, $\tau_{ij}(u, v)$ represents the sliding factor cocfficient.

At present, some denoising methods for data have certain limitations, mainly targeting partial noise. However, using wavelet transform for denoising can not only achieve satisfactory denoising results, but also run faster than other methods. Priority should be given to conducting wavelet transform operations on wireless network information transmission security vulnerability data containing noise, obtaining wavelet coefficients. Further processing of the wavelet coefficients can obtain the latest wavelet coefficients, and reconstructing the wavelet coefficients can obtain the denoised data. Among them, the wireless network information transmission security vulnerability data noise detection model $\rho(i)$ can be expressed in the form of formula (4):

$$\rho(i) = t(i) + \vartheta(i) \cdot \tau_{ij}(u, v) \cdot \tau(x) \tag{4}$$

In the above equation, $t(i)$ represents wireless network data transmission security vulnerability data containing noise; $\vartheta(i)$ represents real wireless network data transmission security vulnerability data.

Select a suitable wavelet basis and decompose the abnormal transmission data of the cloud platform using the wavelet basis. The decomposition process is shown in Fig. 1.

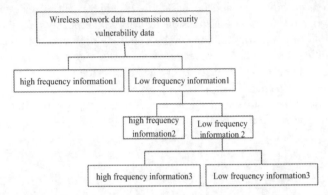

Fig. 1. Schematic diagram of data decomposition for wireless network information transmission security vulnerabilities

After completing the decomposition processing of wireless network information transmission security vulnerability data, select the appropriate wavelet basis to determine the decomposition level, and perform multi-level decomposition processing on the data. $H_{(n)}$ The corresponding calculation formula is:

$$H_{(n)} = \frac{\vartheta(i)}{t(i) - 1} \cdot \vartheta_{(x,y)}(t) \cdot \tau_{ij}(u, v) \tag{5}$$

After determining the threshold, perform soft threshold quantization processing $\varpi_{x,y}$ on all high-frequency coefficients using the selected threshold, and the corresponding calculation formula is as follows:

$$\varpi_{x,y} = \sqrt{\frac{\vartheta_{(x,y)}(t) \cdot \tau_{ij}(u, v)}{x - 1} - \frac{(x - R_{(i,j)})}{H_{(n)}}} \tag{6}$$

Calculate the sensitivity of all wavelet basis analysis wireless network information transmission security vulnerability data, sort and process them, select the least sensitive wavelet basis wireless network information transmission security vulnerability data for wavelet decomposition processing, and then reconstruct the high and low frequency parts, ultimately achieving wireless network information transmission security vulnerability data filtering processing.

2.3 Instruction Level Word Embedding Method Based on Word2vec

In the security leak detection of wireless network information transmission based on deep neural network, the original code data cannot be directly used as the input of neural network, so it is necessary to convert structured binary code into unstructured vector form. At present, the most common word embedding model is Word2vec, which is a language model that learns low dimensional word vectors rich in Semantic information from massive text corpora in an unsupervised manner. Word2vec word vector model maps words from the original space to the new low dimensional vector space. The similarity between each pair of words can be measured by vector similarity, so Word2vec word vector has good semantic mapping characteristics.

Word2vec includes two training models, namely, the Continuous bag-of-words model (CBOW) and the Skip gram model. CBOW is a model that calculates a word vector based on N words before or N words before and after a word, while Skip gram calculates a vector of several words that appear before and after a word.

Because the CBOW model and Skip gram model are mirror images of each other, a brief introduction to the CBOW model is sufficient here. The basic principle of the CBOW model is as follows:

The CBOW model utilizes contextual words to predict target words. The target word is, its contextual environment is $Context(w)$, and the word set of the entire corpus is w, so the model is transformed into calculating probability:

$$P(w|Context(w)) \tag{7}$$

When training the entire corpus to obtain word vectors, combined with the maximum logarithmic likelihood principle, the objective function form of the CBOW model can be obtained as follows:

$$L = \sum_{w \in W} \log P(w|Context(w)) \tag{8}$$

The specific steps are as follows:

Represent the input sentence as $v(Context(w)_1)$, $v(Context(w)_2)$, ..., $v(Context(w)_n) \in \Re^m$ context word vector based on the number of words, where m represents the dimension of the word vector and n represents the number of word vectors. Accumulate and sum the word vectors in the input layer, i.e. $X_w = \sum_{i=0}^{n} v(Context(w)_i) \in \Re^m$.

Build $N = |D|$ Huffman tree with the frequency of each word as the weight. The Huffman tree has a total of a leaf node and $N - 1$ non leaf nodes, where D is the number of all words in the corpus.

Using the Hierarchical Softmax method to solve the CBOW model, the objective function is transformed into:

$$L = \sum_{w \in W} \sum_{j=2}^{l^w} L(w, j) \tag{9}$$

$$L(w, j) = \{(1 - d_j^w) \log[X_w^T \theta_{j-1}^W] + d_j^w \cdot \log[1 - \sigma(X_w^T \theta_{j-1}^W)]\} \tag{10}$$

Among them, l^w represents the number of nodes covered in the route p^w of Huffman tree species from the root node to the corresponding leaf node of word w; The Huffman encoding of the f-th node in the path where the word '$d_j^w \in \{0, 1\}$ 'corresponds to' w', but the root node does not correspond. Code; The word vector of non leaf nodes in the path corresponding to the word $\theta_{j-1}^w \in w$; σ is the sigmoid function. At the same time, the CBOW model uses the random gradient ascent method to update parameters.

3 Fuzzy Testing Vulnerability Mining Algorithm Based on Spatiotemporal Dimension

In the algorithm architecture, there are two sub networks, namely generator network G and discriminator network D. One of our design concepts is to design a lightweight vulnerability mining model based on implementing a lightweight model. The model should have the characteristic of reducing computational resource consumption, making it easy to deploy to embedded devices and laying the foundation for the future online learning ability of the model. Therefore, on the premise of meeting the constraints of the spatiotemporal dimension architecture and the requirements mentioned earlier, we design a reasonable simplified architecture diagram based on the spatiotemporal dimension, as shown in Fig. 1. This framework corresponds to the minimax game between the generator and the decider, and the formula can be expressed as follows:

$$\min_G \max_D V(DG) = E_{x-Px}[\log(D(x))] + E_{z-Px}[-\log(D(G(z)))] \tag{11}$$

Among them, $D(x)$ represents the probability that the discriminator will determine the true data correctly, and $G(z)$ represents the weight matrix output by the generator after inputting noise data.

a. Generator Fig. 3.3 (a) depicts the network structure of the generator of a fuzzy testing vulnerability mining model based on the spatiotemporal dimension. The generator adopts a deconvolution neural network structure and is composed of multiple deconvolution layers. Specifically, unlike traditional convolutional networks, we replace the pooling layer in the generator with the deconvolution layer. Deconvolution, also known as transposed convolution or fractional step convolution, works by exchanging the forward and backward propagation of convolutions. Based on zero padding and non unit step size, the following formula formulates the output size of the generator's deconvolution in this study:

$$a = (i + 2p - k)\%s \tag{12}$$

$$o' = s(i' - 1) + a + k - 2p \tag{13}$$

Where $o'(o_1 = o_2 = o)$ represents the output size of the matrix, $i'(i'_1 = i'_2 = i')$ represents the input of the matrix, $k(k_1 = k_2 = k)$ represents the size of the convolutional kernel, s represents the step size along both axes, p represents the same filling along both axes, $i(i_1 = i_2 = i)$ represents the input size of the next convolutional layer, and d represents the amount of 0 added to the bottom and right of the input. We idealize parameter settings here, but please note that the formula here also extends to the case where the input matrix in the n dimension is not a square matrix.

Tanh is used in the output layer of the generator, and ReLU is used in the activation function of other layers, such as the input layer and the middle layer. The generator extracts noise data from a uniform noise distribution as input and outputs a two-dimensional matrix as input to the discriminator model. A two-dimensional matrix can be viewed as a sequence, transformed through a character embedding layer, with each row representing one character of the protocol message. It decodes the output of the generator into the generated test case. In order to decode the matrix generated by the generator, we also constructed a BLSTM network. It is considered here that the matrix generated by the generator is equivalence relation to the generated BLSTM network intermediate semantic vector. Characters are output through nonlinear transformation and softmax layer, and the matrix generated by the generator is input to the output part of the BLSTM network as input. The training and optimization strategies of the BLSTM network are introduced in the following text. The loss function of the generator is:

$$E_{z-P_z}[-\log(D(G(z)))] \tag{14}$$

b. Determinator In adversarial training, the discriminator network is designed to guide the training of the generator. Convert the byte representation of each preprocessed actual protocol message into one hot vector representation, and the matrix of the input message packet includes time step dimension and feature vector dimension. Most existing models only consider the time step dimension of the text to obtain a fixed length vector, while ignoring spatial structural features. However, the time step dimension and feature vector dimension are not mutually independent.

In order to fuse the features of two dimensions together, a combined model BLSTM-DCGAN based on BLSTM and CNN is proposed, which enables the discriminator to retain not only the time step dimension but also the feature vector dimension information. A one hot vector is a sequence of fixed length and dimension, which can be regarded as a matrix.

Convert to another vector through the character embedding layer as input to BLSTM. BLSTM not only has the ability to learn sequences forward from LSTM, but also benefits from having a reverse LSTM in its design structure, allowing it to learn sequence features from back to front. The second layer of the BLSTM network, BLSTM, serves as the encoder and generates an intermediate semantic vector as the input for the DCGAN decision maker the calculation of the a character in the BLSTM network, which combines forward and reverse transmission outputs, is as follows i_{th}:

$$h_i = [\overrightarrow{h_i} \oplus \overrightarrow{h_i}] \tag{15}$$

The intermediate semantic vectors $H = \{h_1 h_2 h_{l\_max}\}$ and $H \in R^{l\_max \times d}$ can also be seen as a matrix. l_max is the maximum frame length of the wireless network protocol, and d is the size of the embedded character in BLSTM. The l_max of different wireless network protocols is different, so the calculation steps of the discriminator are different. To simplify the operation here, make $d = l\_max$ obtain a square input size containing order information. For BLSTM network, we use the cross entropy function as the loss function, which is defined as:

$$H(pq) = -\sum_x p(x) \log q(x) \tag{16}$$

Among them, $p(x)$ is the true distribution of the sample, and $q(x)$ is the probability output by the model.

Dropout and L2 regularization terms are used in the above model to reduce the complexity of the model and eliminate the risk of overfitting as far as possible. Since the output of a BLSTM cell is binary, a BLSTM cell's loss function is obtained:

$$L_{BLSTM}^{<t>}(y^{<t>}\hat{y}^{<t>}\omega_1) = -\sum_{j=1}^{c} y_j^{<t>} \log \hat{y}_j^{<t>} + \lambda_1 \|\omega_1\|_2^2 \tag{17}$$

Where C is the size of the embedded character; y is a one hot vector of real characters; $\hat{y}$ is the probability of each class in the softmax layer; λ_1 is the weight of L2 regularization; ω_1 is the weight of the BLSTM layer and output layer. The loss function of BLSTM network to sequence S is:

$$L_{BLSTM}(y\hat{y}\omega_1) = \sum_{t=1}^{T_x} L_{BLSTM}^{<t>}(y^{<t>}\hat{y}^{<t>}\omega_1) \tag{18}$$

Among them, T_x represents the length of the input sequence.

Due to the BLSTM layer's ability to access forward and backward contexts, CNN is used to explore more meaningful information, such as the hierarchical structure of representations. The intermediate semantic vector adds positional features as input through the BLSTM network, and each filter can be regarded as a detector on CNN to detect whether the functional code positional features of the data frame are correct. This is beneficial for the model to quickly learn the format features of the wireless network protocol sequence data. Unlike generators, the discriminator uses batch normalization in all layers except for the input layer; Compared with ReLU, Leaky ReLU is used as the activation function in the model to avoid sample oscillation and model instability. In addition, unlike the generator using deconvolution layers instead of pooling layers, the discriminator uses step convolution layers to replace all pooling layers. In addition, we also use Z-Core to regularization H. Through a series of convolutional operations, we can obtain the output of the discriminator:

$$o' = \left[\frac{i + 2p - k}{s}\right] + 1 \tag{19}$$

The output value of the discriminator obtained at this point is the dynamic mining result of wireless network information transmission security vulnerabilities.

4 Experiment

4.1 Experimental Design

(1) Experimental environment

The specific environment for this experiment is as follows:
CPU: AMD® A8-7200P radeon r5 @2.40 GHz

Operating system: CentOS Linux release 7

Running memory: 8 GB

(2) Experimental testing process

Leopard Mobile data set selected in this chapter is a binary code file, which is disassembled into assembly code by IDA pro tool; The CWE119 dataset is a C language program code file that needs to be compiled into assembly code format. Batch format the compiled assembly code through Python scripts, remove textual description information from the code, and retain function and instruction information. In the above dataset, each sample has a corresponding 0 or 1 label, with 0 indicating vulnerability and 1 indicating no vulnerability. The ratio of sample size for training, validation, and testing sets is 3:1:1. Before each training and testing, the order of the data is disrupted.

In the instruction level word embedding model training based on Word2vec, this chapter uses an assembly code corpus, with the first column containing assembly operation instructions and the remaining columns containing operation data. Regular expressions are used to match the numbers in the operation data. The input of the instruction level word embedding model based on Word2vec is assembly code, and the output is a two-dimensional tensor. The input of the convolutional neural network model proposed in this chapter is the two-dimensional tensor output of the word embedding layer mentioned above, and the output is the probability of the code being classified as a bad sample. It is often observed in the training process of convolutional neural network model that, with the increase of the epoch of the training round, the training and verification error of the binary code vulnerability detection model will increase after reaching the local minimum, that is, the binary code vulnerability detection model training consumes more time and the latest parameters are not retained when the training is terminated. Therefore, the experiment in this chapter adopts the early termination strategy, that is, when the error of the binary code vulnerability detection model in the verification set does not further reduce within the epochs specified in advance, the training algorithm will terminate.

4.2 Experimental Result

4.2.1 Wireless Network Information Transmission Security Vulnerability Dynamic Mining Recall Rate

In order to verify the dynamic mining effect of wireless network information transmission security vulnerabilities using the method proposed in this article, the reference [5] method, the reference [6] method, and the method proposed in this article were used to verify the recall rate of wireless network information transmission security vulnerability dynamic mining. The results are shown in Table 3.

According to Table 3, when the resource level is 100 GB, the recall rate for dynamic mining of wireless network information transmission security vulnerabilities using reference [5] method is 76.0%, the recall rate for dynamic mining of wireless network information transmission security vulnerabilities using reference [6] method is 79.2%, and the recall rate for dynamic mining of wireless network information transmission security vulnerabilities using this method is 99.3%; When the resource amount is 500

Table 3. Dynamic Mining Recall Rate of Wireless Network Information Transmission Security Vulnerabilities

Resource quantity/GB	Dynamic Mining of Security Vulnerabilities in Wireless Network Information Transmission and Recall Rate/%		
	Reference [5] Method	Reference [6] Method	proposed method
100	76.0	79.2	99.3
200	79.1	68.3	99.2
300	60.2	79.5	95.6
400	68.8	80.8	96.8
500	79.2	63.6	96.3
600	66.8	68.5	98.1

GB, the recall rate of dynamic mining for wireless network information transmission security vulnerabilities in reference [5] method is 79.2%, the recall rate of dynamic mining for wireless network information transmission security vulnerabilities in reference [6] method is 63.6%, and the recall rate of dynamic mining for wireless network information transmission security vulnerabilities in this method is 96.3%; When the resource amount is 600 GB, the recall rate for dynamic mining of wireless network information transmission security vulnerabilities using reference [5] method is 66.8%, the recall rate for dynamic mining of wireless network information transmission security vulnerabilities using reference [6] method is 66.8%, and the recall rate for dynamic mining of wireless network information transmission security vulnerabilities using this method is 98.1%; The above results indicate that the method proposed in this paper can effectively improve the recall rate of dynamic mining for wireless network information transmission security vulnerabilities.

4.2.2 Dynamic Mining Accuracy of Wireless Network Information Transmission Security Vulnerabilities

In order to verify the efficiency of dynamic mining of wireless network information transmission security vulnerabilities using the method proposed in this paper, the methods of reference [5], reference [6], and the method proposed in this paper were used for English translation error recognition and time-consuming verification. The results are shown in Table 4.

According to Table 4, when the resource size is 100 GB, the accuracy of dynamic mining for wireless network information transmission security vulnerabilities using reference [5] method is 58.3%, the accuracy of dynamic mining for wireless network information transmission security vulnerabilities using reference [6] method is 55.8%, and the accuracy of dynamic mining for wireless network information transmission security vulnerabilities using this method is 98.0%; When the English resource is 500 GB, the accuracy of dynamic mining for wireless network information transmission security vulnerabilities using reference [5] method is 68.8%, the accuracy of dynamic mining for

Table 4. Dynamic Mining Accuracy of Security Vulnerabilities in Online Network Information Transmission

Resource quantity/GB	Dynamic Mining Accuracy of Wireless Network Information Transmission Security Vulnerability/%		
	Reference [5] Method	Reference [6] Method	proposed method
100	58.3	55.8	98.0
200	60.0	78.9	96.3
300	55.1	60.1	99.5
400	63.6	62.3	95.0
500	68.8	65.9	96.6
600	78.0	53.2	99.0

wireless network information transmission security vulnerabilities using reference [6] method is 65.9%, and the accuracy of dynamic mining for wireless network information transmission security vulnerabilities using this method is 96.6%; When the English resource is 600 GB, the accuracy of dynamic mining for wireless network information transmission security vulnerabilities using reference [5] method is 78.0%, the accuracy of dynamic mining for wireless network information transmission security vulnerabilities using reference [6] method is 53.2%, and the accuracy of dynamic mining for wireless network information transmission security vulnerabilities using this method is 99.0%; The above results indicate that the method proposed in this paper can effectively improve the accuracy of dynamic mining of wireless network information transmission security vulnerabilities.

4.2.3 Dynamic Mining of Security Vulnerabilities in Wireless Network Data Transmission Takes Time

In order to verify the efficiency of dynamic mining and identification of wireless network information transmission security vulnerabilities using the method proposed in this paper, reference [4], reference [5], reference [6], and the method proposed in this paper were used to verify the time consumption of dynamic mining of wireless network information transmission security vulnerabilities. The results are shown in Table 5.

According to Table 5, when the resource level is 100 GB, the dynamic mining time for wireless network information transmission security vulnerabilities using reference [4] method is 15.8 s, the dynamic mining time for wireless network information transmission security vulnerabilities using reference [5] method is 11.9 s, the dynamic mining time for wireless network information transmission security vulnerabilities using reference [6] method is 32.9 s, and the dynamic mining time for wireless network information transmission security vulnerabilities using this method is 2.9 s; When the English resource is 600 GB, the dynamic mining time for wireless network information transmission security vulnerabilities using reference [4] method is 282.8 s, the dynamic mining time for wireless network information transmission security vulnerabilities using reference

Table 5. Time consumption for dynamic mining of wireless network information transmission security vulnerabilities

Resource quantity/GB	Dynamic mining of security vulnerabilities in wireless network data transmission takes time/s			
	Reference [4] Method	Reference [5] Method	Reference [6] Method	proposed method
100	15.8	11.9	32.9	2.9
200	48.9	28.9	58.9	5.8
300	82.6	86.2	99.6	12.0
400	198.6	99.1	109.2	18.9
500	252.3	136.9	136.1	22.3
600	282.8	188.7	148.3	25.8

[5] method is 188.7 s, the dynamic mining time for wireless network information transmission security vulnerabilities using reference [6] method is 148.3 s, and the dynamic mining time for wireless network information transmission security vulnerabilities using this method is 25.8 s; The above results indicate that the proposed method can effectively improve the efficiency of dynamic mining of wireless network information transmission security vulnerabilities.

5 Conclusion

The paper proposes a dynamic mining method for wireless network information transmission security vulnerabilities based on spatiotemporal dimensions. Collect wireless network information transmission security vulnerability data, introduce wavelet transform to filter and process wireless network information transmission security vulnerability data, and in the deep neural network architecture, command level Word embedding method based on Word2vec is used to obtain wireless network information transmission security vulnerability feature attributes, and realize dynamic mining of wireless network information transmission security vulnerabilities based on space-time dimensions. The experimental results indicate that the dynamic mining method based on spatiotemporal dimension can effectively discover security vulnerabilities in wireless network information transmission. By real-time monitoring and analysis of wireless network data flow, potential security threats and attack behaviors can be captured. The accuracy of vulnerability dynamic mining can reach 99.0%, and dynamic mining methods can timely detect security vulnerabilities in wireless network information transmission and provide corresponding warnings. By identifying abnormal behaviors and patterns, we can quickly respond and take corresponding security measures to avoid or reduce potential security risks. The dynamic vulnerability mining takes 25.8 s, indicating that our method can improve the effectiveness of vulnerability dynamic mining.

References

1. Liu, Y.: Research on wireless communication network data security situation awareness based on deep learning. Inf. Rec. Mater. **23**(08), 182–185 (2022)
2. Lv, G., Ju, L.: A method for identifying network security vulnerabilities in power systems based on data mining. Electrotech. J. (02), 49–51 (2023)
3. Ding, J.: Research on software security vulnerability automatic mining method based on big data technology. J. Taiyuan Norm. Univ. (Nat. Sci. Ed.) **21**(01), 45–50 (2022)
4. Yin, Y.: Research on network protocol vulnerability mining technology based on fuzzy. Microcomput. Appl. **37**(09), 8–10+16 (2021)
5. Zhang, M.: Analysis of network computer security hazards and vulnerability mining technology. Wirel. Internet Technol. **19**(10), 13–15 (2022)
6. Guan, J., Shi, G., Chen, H.: Research on network vulnerability mining based on Apriori risk data analysis. Comput. Simul. **39**(01), 343–347 (2022)
7. Gu, M., et al.: Software security vulnerability mining based on deep learning. Comput. Res. Dev. **58**(10), 2140–2162 (2021)
8. Li, M., Zhu, M.: A security vulnerability detection method for wireless communication networks based on ant colony algorithm. Comput. Meas. Control **30**(10), 51–56 (2022)
9. Wang, X., Wang, C., Li, Q., Ren, T.: A method for mining network security vulnerabilities in power systems based on black box genetic algorithm. J. Shenyang Univ. Technol. **43**(05), 500–504 (2021)
10. Jiang, Z., Fan, L.: Intelligent monitoring system for wireless communication network security vulnerability based on machine learning. Electron. Des. Eng. **29**(15), 115–119 (2021)

A Method for Identity Feature Recognition in Wireless Visual Sensing Networks Based on Convolutional Neural Networks

Chenyang Li[✉] and Zhiyu Huang

Shenyang Institute of Technology, Shenyang 113122, China
liklu22@163.com

Abstract. Due to the problems of low recognition accuracy and long recognition time in traditional wireless visual sensing network identity feature recognition methods, a convolutional neural network-based wireless visual senscto the operation results, the global threshold method is used to obtain the binary image sequence and perform morphological processing. Based on the processing results, Extract target regions from video image sequences of wireless visual sensing networks, detect human targets, and construct a Softmax classifier using convolutional neural networks to classify human targets in video image sequences of wireless visual sensing networks, in order to identify identity features. The simulation results show that the proposed method has high accuracy and short recognition time for identity feature recognition in wireless visual sensing networks.

Keywords: Convolutional Neural Network · Wireless Visual Sensing Network · Identity Feature Recognition · Image Sequence · Mean Method

1 Introduction

The wireless vision sensor network is composed of multiple wireless nodes integrated with miniature vision sensors, which can transmit the acquired visual perception information to the Sink node (Sink node) through the cooperative way of multiple nodes, and then send it to the application server for subsequent processing and analysis. Wireless vision sensor networks not only have the advantages of traditional wireless sensor networks, such as self-organization, self-healing, flexible configuration, fast coverage and low-cost deployment, but also have the characteristics of the traditional vision application system with rich information, which can support a wider range of intelligent applications, such as traffic monitoring and traffic statistics, assisted living, public behavior analysis and modeling, and virtual reality. However, the resources of unlicensed frequency band suitable for wireless multi-hop transmission of visual information are limited, which limits the scale and performance of the network.Using wireless technology to access idle authorized frequency bands is one of the feasible ways to enhance the performance of wireless visual sensing networks and achieve large-scale applications [1]. The main challenge of wireless visual sensing networks is the randomness of available spectra

L. Yun et al. (Eds.): ADHIP 2023, LNICST 548, pp. 418–434, 2024.
https://doi.org/10.1007/978-3-031-50546-1_28

or channels. Due to the opportunistic access of wireless nodes to the idle authorized spectrum, their underlying link transmission capacity is dynamically changing. Under the traditional design principles of layered network protocols, the transmission of upper level visual information does not adaptively match changes in the transmission capacity of the lower level, thus unable to fully utilize the benefits brought by radio. This requires the use of cross layer design methods to enable upper layer applications to perceive potential transmission opportunities on lower layer links in real-time, in order to enhance the end-to-end service quality of visual information [2].

With the rapid development of information technology, wireless visual sensor network technology has been widely used in finance, e-commerce and other fields. The information of wireless visual sensor network is increasing rapidly, and the security of sensor network information has become a key issue in this field. How to effectively increase the security performance of wireless visual sensor network has become an urgent problem to be solved, and identity feature recognition is a necessary prerequisite to ensure the security of visual sensor network. How to effectively identify the user's identity and protect the security of visual sensor network information has been widely paid attention by experts and scholars in related fields. At the same time, there are also some good adaptive identity feature recognition algorithms [3]. Literature [4] proposed that the difference of footstep induced structural vibration signals in the walking process was used to identify personnel. Based on the energy threshold method, footstep events and non-footstep events were detected. A total of 16 footstep characteristic parameters of a single footstep event for different test personnel were compared and analyzed in the time domain and frequency domain. It is found that the parameter difference under different feature combinations can be used as the basis for identity recognition.In order to verify the effectiveness of the method, support vector machine (SVM) was used as a classification tool. With a test population of 10 people and 500 data samples, 16 foot feature parameters were selected with an average recognition rate of 79.21%. The Pearson correlation coefficient method was used to screen out 10 unrelated foot feature parameters with an average recognition rate of 91%, which was 11.79% higher than the average recognition rate using 16 foot feature parameters, We compared the impact of classification tools on the average recognition rate of 10 selected foot feature parameters under different SVM kernel functions, and found that the highest average recognition rate was 96% under linear kernel functions. The results indicate that an effective combination of foot feature parameters is suitable for identity recognition in small samples. However, the accuracy of the above methods for identity feature recognition is relatively low, resulting in poor recognition performance. Literature [5] uses BGN semi homomorphic encryption algorithm and Shamir secret sharing to design a threshold identity scheme based on biometric identification, which mainly uses BGN homomorphic encryption algorithm on bilinear pairs for data protection, uses a third-party authentication center for secret segmentation, and the server authenticates the user's identity in the ciphertext state to achieve threshold identity authentication.Literature [6] proposes an online detection and automatic identification technology for network video monitoring devices. Stateless scanning technology is used to carry out online detection of network terminal devices, extract BANNER and HTML page information from HTTP header information returned from specific ports of terminal devices, and construct Web identity features of devices through rough set attribute reduction. The cosine distance is used to calculate the similarity between the Web identity features of online devices and the sample of the known device signature database to realize the detection and identification of online devices. However, it takes a long time for the above two methods to recognize the identity features, resulting in low efficiency.

In response to the problems existing in the above methods, this paper proposes a wireless visual sensing network identity feature recognition method based on convolutional neural networks. Firstly, the architecture of wireless visual sensor networks is analyzed. Then, the background image of the application scene is processed by background subtraction to realize the target detection in the video image sequence. Using convolutional neural network, the Softmax classifier is constructed to classify the human body in the video image sequence of the wireless visual sensor network, and finally realize the recognition of identity features. Simulation experiments have verified that this method can quickly and accurately recognize the identity features of wireless visual sensing networks, laying a certain foundation for the safe operation of wireless visual sensing networks.

2 Wireless Visual Sensor Network Identity Feature Recognition Method

2.1 Analysis of Wireless Vision Sensor Network Architecture

The wireless vision sensor network system preloads part of AI computing power to the edge computing unit through the network architecture of the cloud side, and then completes a certain degree of accurate and lossless video data selection in the sensing front end through the completion of massive unstructured video data. This network architecture scheme can not only effectively reduce the transmission pressure of network bandwidth. It will also save system storage and computing resources, improve the real-time response speed and analysis accuracy of the system, reduce the system delay, and achieve efficient and timely response [7].

The security cloud edge end architecture scheme is shown in Fig. 1. The image and video data are collected through the camera, and then the target face frame in the photo is detected through local edge computing and features are extracted. The recognition results are fed back to the camera for output through feature matching with the cloud end.

2.2 Target Detection in Video Image Sequence of Wireless Visual Sensor Networks

The application scenario of the wireless visual sensing network identity feature recognition method proposed in this article is mainly video surveillance systems. Video surveillance systems have the characteristics of simple background and high real-time requirements. The original data of the dataset used in this article is also videos with relatively simple background, and the background image has not changed much, whether it is the experimental environment or the actual application environment of this article, The

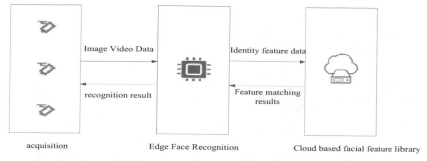

Fig. 1. Security Cloud Edge Architecture

background subtraction method is a good choice due to its performance advantages [8]. Therefore, this article chooses the background subtraction method as the basic algorithm for extracting human targets. For each image in the dataset, a rectangular box is used to mark the area where the human target is located and extract it.

The process of human object extraction is shown in Fig. 2, which can be divided into seven steps, namely image sequence acquisition, background modeling, background difference operation, binarization, open operation, connected domain analysis and target region extraction. Among them, open operation and connected domain analysis can be selected according to the actual situation.

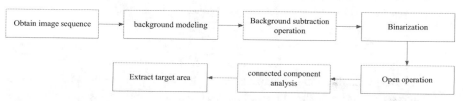

Fig. 2. Process of background difference method

(1) Obtain image sequence

The data obtained from surveillance videos are all in video format, and the data set adopted in this paper is also in video format. The image sequence corresponding to each video can be obtained by extracting video frames from the wireless visual sensor network. For the background video provided in the data set, frame by frame extraction is adopted in this paper to ensure that higher quality background images can be obtained in the process of background modeling. For the video of wireless vision sensor network containing human objects, extraction is carried out according to the number of 5 frames per second, so that the image sequence can contain the main gestures in the walking process. In terms of quantity, it can also meet the training needs of convolutional neural network, and at the same time avoid excessive redundant data leading to increased computation and reduced training efficiency [9].

(2) Background modeling

After obtaining a wireless visual sensing network video image sequence through video, the desired human target area is the foreground, while other areas of the image are processed as backgrounds. The purpose of background modeling is to obtain a background image. Only by obtaining an accurate background image can the desired human target be obtained through background difference calculation. Therefore, the quality of the background image directly determines whether the foreground target can be accurately extracted. At present, there are four commonly used background modeling methods: Mean method, Median method, Single Gaussian model and Mixed Gaussian model [10].

In this paper, according to the actual situation, we choose the Mean method for background modeling. The process of the mean method is: take n consecutive frames of images in an image sequence, calculate the average gray value of the corresponding pixel, and the average value is used as the final gray value of the pixel at the same position in the background image. The reason for choosing this method is that the quality of the background modeling averaging method can meet the requirements of the algorithm, the calculation speed is fast, and it is more in line with the real-time requirements of the video surveillance system.

Taking a background video from the CASIA (Chinese Academy of Sciences Institute of Automation) database Dataset B, Dataset SURF and Dataset CBD as an example, this article uses the mean method to calculate the background modeling process as follows:

After obtaining the image sequence of the background video frame by frame, grayscale each frame to obtain the image sequence P_i and i as the frame numbers. Calculate the average grayscale values of all corresponding points in the image sequence P_i, and establish the background image. The calculation formula is:

$$B(x, y) = \frac{1}{n} \sum_{i=1}^{n} P_i(x, y) \tag{1}$$

where, $B(x, y)$ is the background image, n is the total number of images in image sequence P_i, $P_i(x, y)$ is the grayscale image of frame i, and i is the frame number.

Through calculation, a background image can be obtained for the image sequence of each background video, that is, each person has a grayscale background image in each perspective, and the same background image is used for different dressing or walking posture under the same perspective.

(3) Background subtraction operation

For an image sequence containing human targets, first convert each image in the wireless visual sensing network video image sequence into a grayscale image to obtain image sequence G_i, and then perform a difference operation with the corresponding background image of the image sequence to obtain a new image F_i. The calculation formula is:

$$F_i(x, y) = |G_i(x, y) - B(x, y)| \tag{2}$$

where, F_i is the video image sequence of wireless visual sensor network obtained after background difference calculation, G_i is the video image sequence of grayscale wireless visual sensor network, and B is the background image.

As shown in Fig. 3, this figure is a new image obtained by background error calculation. It can be seen that the perfect foreground target cannot be obtained only by simple background error calculation.

Fig. 3. Image obtained by background subtraction method

(4) binarization

Binarization is a method often used in the process of image processing. Through binarization, an image can be divided into two areas that are either black or white, and the part of interest is usually set as white. The method of setting a threshold for the whole image is called the global threshold method. The image can also be divided into multiple areas, and each area sets a threshold, which is called the local threshold method.

In this paper, the global threshold method is used to set a global threshold T. If the gray value of a pixel is greater than T, the color of the pixel is set to white. If the gray value of a pixel is less than T, the color of the pixel is set to black, and the binary image sequence R_i is obtained. The image more accurately selects the area where the foreground target is located and displays it as white. The calculation formula is:

$$R_i(x, y) = \begin{cases} 1 \; F_i(x, y) \geq T \\ 0 \; F_i(x, y) < T \end{cases} \tag{3}$$

After many experiments, for Dataset B, Dataset SURF and Dataset CBD of CASIA database, threshold T was set to 40, which had the most ideal effect. Figure 4 shows the image after binarization. It can be seen that some noises in the figure need further processing, and some details of the human body are also missing.

(5) Open operation

After binarization processing, the image still cannot fully meet the requirements, because there will be foreground empty points and background noise points, especially the background noise points have a great impact on the subsequent work of this paper, so it is necessary to carry out morphological processing on the binary image obtained before.

Morphological processing of image is to Improve the quality of binary image by logical operation between structural elements and binary image. The two most basic operations of morphological processing are corrosion and expansion. Corrosion can eliminate noise points smaller than structural elements, and expansion can fill the

Fig. 4. Binary Diagram

holes in the target, but both corrosion and expansion will obviously change the area of the target region.

The combination of open and closed operations through corrosion and expansion solves this problem. In this paper, the binary image is opened, that is, the erosion operation is carried out first, and then the expansion operation is carried out. The main purpose is to remove some small noises in the binary image.

First, the corrosion operation is carried out, and the set obtained by etching the binary image R_i through the structural element S is the set of the origin position of S when the structural element S is completely included in the binary image R_i. The corrosion calculation is as follows:

$$D_i = R_i \odot S \tag{4}$$

where, D_i is the image obtained after corrosion operation, R_i is the binary graph, and S is the structural element.

After the expansion operation, the set obtained by the expansion of the binary image D_i through the structural element S is the set of the origin position of S when the displacement of S' intersects with at least one non-zero element in the binary image D_i. The expansion calculation formula is:

$$E_i = D_i \oplus S \tag{5}$$

In the equation, E_i represents the image after expansion operation.

Figure 5. (a) shows the image after corrosion operation on Fig. 4, and Fig. 5 (b) shows the image after expansion operation.

(6) Connected domain analysis

The open operation can remove the noise with relatively small area, but the noise with relatively large area can be removed by connected domain analysis. Connected domain refers to the area formed by the connection of points with adjacent positions and equal pixel values in the image, and connected domain analysis refers to marking the white area in the binary image so that each single connected area has a unique mark, so that geometric parameters such as contour, centroid and external rectangle

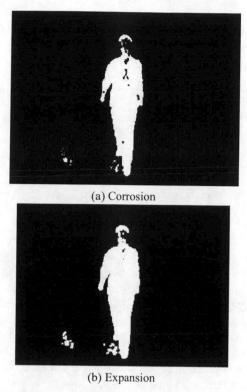

(a) Corrosion

(b) Expansion

Fig. 5. Corrosion and Expansion of Binary Graph

of these blocks can be obtained further. In this paper, pixel labeling method, which is widely used in connected domain analysis, is used.

Scan all pixels in the binary graph and assign a unique label to the set of pixels located in the same connected region. During the scanning process, there may be multiple small connected regions assigned different labels, but these small connected regions belong to a larger connected region. Write these small connected region labels into equivalent pairs and record their equality relationship.

Merges equal connected domains into one connected domain and assigns a new tag. The area of each connected domain is calculated, and the connected domain whose area is less than a certain threshold is regarded as noise for removal. Through many experiments, it is found that for the data set used in this paper, the effect of setting the threshold as 125 pixels is ideal.

The image obtained after connected domain processing is shown in Fig. 6. It can be seen that the noise with a relatively large area is also removed. Although some details are lost in the foreground area, it can meet the need of extracting the human body target in the video image sequence of wireless visual sensor network.

(7) Extract target area

Fig. 6. Connected Domain Analysis

After background subtraction, binarization, open operation, and connected domain analysis, the background noise of the color image containing the target character is eliminated, resulting in a binary image containing the complete target, where the white area is the target area. If the white part is used as the selection area to extract the target, as shown in Fig. 7 (a), there will be some loss in the details, especially in the facial details, which will have a certain negative impact on the accuracy of the recognition results. Therefore, the method adopted in this article is to calculate the outer rectangle of the target character area through the outline of the white area, and use the internal area contained in the rectangle as the selection area of the target character, Through this selection area, a rectangular area can be captured from the original color image, which includes both the human target and some background images, as shown in Fig. 7 (b). The rectangular area is used as the final human target to be extracted.

2.3   Recognition Model Based on Convolutional Neural Network

Convolutional neural networks (CNN) have evolved from traditional neural networks, with the main difference being that the feature extractor of convolutional neural networks is mainly composed of convolutional feature extractors, while traditional neural networks are mainly composed of fully connected layers. The structure of CNN network used in this paper includes three main parts: convolution layer, activation layer and pooling layer. Among them, the convolutional layer is one of the core components in CNN, which extracts the local features of the input image through the convolution operation. The convolutional layer consists of multiple kernels, each of which multiplies element-wise with a local region of the input image and sums the results to obtain an element of the output feature map. The activation layer introduces nonlinear transformation to increase the expression ability of the network. Commonly used activation functions include ReLU (Rectified Linear Unit), sigmoid, and tanh. The activation layer usually follows the convolutional layer and performs element-wise activation function computation on the output of the convolutional layer. The pooling layer reduces the size of the feature map by downsampling operation while retaining important feature information, and maps the

(a) Human targets

(b) Rectangular area

Fig. 7. Human target area

pooling results to the output layer to realize the identification of identity features. In this paper, convolutional neural networks are used to construct Softmax classifier to classify human body targets in video image sequences of wireless visual sensor networks, so as to identify identity features.

2.3.1 Convolutional Layer

In the traditional fully connected network, each node will connect all nodes in the upper layer, which will lead to excessive parameters, difficulty in model training, and overfitting. In convolutional neural networks, the convolutional layer implements local connections, which can effectively reduce the number of model parameters. Each neuron in the convolution layer is only connected to the local Receptive field of the previous layer. The size of the local Receptive field depends on the size of the convolutional nucleus. In CNN convolution, two-dimensional convolution is defined as follows:

$$s(i, j) = (X \times W)(i, j) \tag{6}$$

It can be seen from the above equation that the convolution result of the human object image in the video of wireless visual sensor network is the result of multiplying the local region of the image and the elements of each position of the convolution kernel matrix, and then adding. It is assumed that the size of the feature mapping image input at layer l is $W \times H \times D$, where W and H are the width and height of the human object identity feature image in the two-dimensional wireless visual sensor network video, and D determines the depth of the feature image. Parameter P, the number of 0 elements filled around the image, is used to adjust the size of the output identity feature map. If the size of the two-dimensional convolution kernel is $W_c \times H_c$, the number of output channels set is k, and the step size is S, then the size of the feature map obtained after convolution is $W' \times H' \times D'$. We can control the size of the feature image after convolution by using the convolution kernel size, step size S, and zero padding P. For example, if the convolution kernel size is set to 3, P is set to 1, and step size is set to 1, the feature image remains the same size after passing through the convolution layer. If the size of the convolution kernel is set to 2, P is set to 0, and the step size is set to 2, the feature image will be reduced to 1/4 of the original to achieve the effect of Downsampling.

2.3.2 Activation Layer

Convolutional layers are essentially linear, but for sample data that needs to be learned, their distribution may not necessarily be linearly separable. In order to learn the nonlinear part, an activation layer is usually connected behind the convolutional layer. The activation function in the activation layer transforms the data nonlinearly, which is the key for neural networks to solve nonlinear problems. Several common activation function are sigmoid, tanh, Re LU.

The sigmod activation function has two main disadvantages: (1) It is easy to be saturated. When the input is very large or very small, the gradient is approximately 0, which will lead to the gradient dispersion in the back propagation. (2) The non zero mean output of sigmoid will affect the output of the back layer, thereby affecting the update of the gradient: if the inputs of the back layer neurons are all positive, the local gradient obtained is positive. During the backpropagation process, the parameters will always update in the positive direction, and vice versa, the parameters will always update in the negative direction, resulting in slower convergence speed.

tanh activation function compresses the output to the range of -1 1, and its output basically follows the mean value distribution of 0, but the problem of gradient saturation still exists in tanh function.

The Re LU activation function does not require exponential calculation, and its computational complexity is relatively low. It only inhibits negative input values, and its output is sparse. Re LU solves the problem of gradient saturation in sigmod function. In this paper, Re LU is selected as the activation function, and its expression is as follows:

$$f(x) = \max(x, 0) \tag{7}$$

2.3.3 Pooling Layer

In the deep convolutional neural network, a huge amount of parameters will be generated with the deepening of the network, and the hardware conditions limit the infinite increase of parameters in the convolutional neural network, which requires the control of the number of parameters in the network. In addition, the image is static, and the same feature may apply to different areas in the image, which indicates that there must be redundancy in the original parameters. The essence of the pooling layer is to aggregate statistics of features of different locations and compress the input wireless visual sensor network identity feature map, that is, to conduct downsampling of the feature map. Pooling layer can effectively reduce the parameters required by subsequent layers, reduce the possibility of overfitting, and make the convolutional neural network translation invariant, that is, when the pixels in the feature map have a small displacement in the neighborhood, the pooling layer can keep the output unchanged, which enhances the robustness of the network. The commonly used pooling layer downsampling methods include mean pooling and max pooling. Pooling of regional average values can preserve the characteristics of the overall data and highlight the background information. Pooling the maximum value of a region can better preserve the features on the texture. The pooling methods used in this article are regional maximum pooling and mean pooling.

2.3.4 Softmax Classifier

Softmax is a multinomial logic regression model. Logistic regression model belongs to log-linear model and is also a probabilistic model, which is generally used for binary classification problems. Multinomial logistic regression model is the extension of logistic regression and can be used to solve multi-classification problems. The calculation process of Softmax is as follows:

$$P(Y|x) = \frac{\exp(w_k \times x)}{k = 1 \sum^{K-1} \exp(w_k \times x)} \tag{8}$$

In the equation, w is the desired parameter model, x is the input vector, and $P(Y|x)$ is the probability value of predicting the category of x as category Y.

Input human target identity features from wireless visual sensing network videos as samples into Softmax classifier to obtain classification results for identifying wireless visual sensing network identity features:

$$\theta_t = -\eta \times g_t \tag{9}$$

where, η is the global learning rate initially set, and g_t is the gradient.

3 Experimental Analysis

3.1 Preparation for Experiment

In order to verify the effectiveness of the wireless visual sensing network identity feature recognition method based on convolutional neural networks proposed in this article in practical applications, the accuracy and recognition time of the wireless visual sensing network identity feature recognition were selected as experimental indicators, and the methods of reference [4] and reference [5] were used as comparative methods for experimental testing.

This article also used GPU for acceleration during the experimental process. The GPU brand model is NVIDIA Tesla K40c, which has 2880 Cuda cores and 12G graphics memory. Its parallel computing and storage capabilities can well meet the needs of convolutional neural network training. Using M3001 robot module camera, capturing image size 1920 × 1080, which can set parameters such as automatic exposure, automatic white balance, color correction, brightness, contrast, saturation, sharpness, etc. It supports a variety of protocols such as common TCP/IP, ICMP, HTTP, FTP, DHCP, DNS, DDNS, RTP, RTSP, etc. It can access the network through RJ45 10M/100M adaptive Ethernet port, and transmit the photos taken to the local processing unit. The camera and processor are shown in Fig. 8.

Fig. 8. Camera and processor

The Hiss 3516DV300 neural network processing chip used in the terminal computing unit in this paper NNIE, short for Neural Network Inference Engine, is a hardware unit specializing in accelerated processing of neural networks, especially deep learning convolutional neural networks in So C of Hesis Media. It supports most existing public networks. For example, classification networks such as Alex Net, VGG16, Res Net18 and Res Net50, detection networks such as Faster R-CNN, YOLO and SSD, and scene segmentation networks such as Seg Net and FCN.

3.2 Data Set Construction and Training

(1) Building a dataset

Obtaining multi view video data of all objects to be identified through the reasonable arrangement of monitoring equipment. From a practical application perspective, it is reasonable to have no more than three views. Then, the video data is converted into an image sequence and target detection is performed using background subtraction to extract the rectangular area where the human target is located as the dataset. If image data from three different perspectives is obtained, the dataset is divided into three subsets according to three different perspectives, and each subset is further divided into a training set, a validation set, a single perspective testing set, and a multi perspective testing set.

(2) Training model

The multi-network identity model is trained and tested on the multi-view data set according to the training and testing process. Firstly, an appropriate convolutional neural network should be built as the subnet for training each perspective, and the training set and verification set of each subset should be input into the corresponding subnet for training, so as to obtain the model of each subnet. During the training process, each subnet should be adjusted independently, so that each subnet can obtain the best recognition result on its corresponding data set. The accuracy of each subnet is tested by the single view test set, and the weight of each subnet is calculated according to the weight calculation formula, and the weighted fusion identity model is obtained. The final identity model is tested through the multi-view test set to verify its comprehensive performance and further optimize the model.

(3) Identity recognition

Input the multi view image of the object to be recognized into the corresponding subnet for recognition, obtain the recognition results of each subnet, and then calculate the final identity recognition result through weighting.

CASIA database, created by the Institute of Automation of the Chinese Academy of Sciences, consists of three data sets, of which Dataset A is a small-scale database with 20 people, each with three shooting angles ($0°$, $45°$, $90°$), a total of 240 image sequences, and the acquisition environment is outdoor. Select Dataset B, Dataset SURF and Dataset CBD of the CASIA database as the dataset for this article. Each subject has eleven perspectives, and the images from different perspectives are shown in Fig. 9.

From the different perspectives mentioned above, the methods of this article, reference [4], and reference [5] were used to compare and analyze the accuracy of identity feature recognition in wireless visual sensing networks. The comparison results are shown in Table 1.

According to Table 1, the accuracy of the method used in this paper for identity feature recognition in wireless visual sensing networks can reach up to 99.5%. The accuracy of the method used in reference [4] for identity feature recognition in wireless visual sensing networks can reach up to 84.1%. The accuracy of the method used in reference [4] for identity feature recognition in wireless visual sensing networks can reach up to 70.5%. The accuracy of the method used in this paper for identity feature recognition in wireless visual sensing networks is the highest, and the recognition effect is the best.

Using the methods of this article, reference [4], and reference [5], a comparative analysis was conducted on the time required for identity feature recognition in wireless visual sensing networks. The comparison results are shown in Table 2.

(a) 0 ° viewing angle

(b) 90 ° viewing angle

(c) 180 ° viewing angle

Fig. 9. Images from Different Perspectives

As can be seen from Table 2, the time used for the identification of wireless visual sensor network identity features by the method in this paper is within 6.2 s, the time used for the identification of wireless visual sensor network identity features by the method in reference [4] is within 16.5 s, and the time used for the identification of wireless visual sensor network identity features by the method in reference [4] is within 26.4 s. The method presented in this paper has the shortest time and the highest recognition efficiency for wireless visual sensor network identity feature recognition.

Table 1. Comparison results of identification accuracy of wireless visual sensor networks /%

Number of experiments/times	Textual method	Method of reference [4]	Method of reference [5]
10	94.2	80.1	64.5
20	94.9	80.6	65.2
30	95.6	81.2	66.2
40	96.8	81.6	67.4
50	97.5	82.3	68.1
60	98.2	83.5	69.2
70	99.5	84.1	70.5

Table 2. Comparison results of identity feature recognition time in wireless visual sensor network /s

Number of experiments/times	Textual method	Method of reference [4]	Method of reference [5]
10	5.2	15.2	22.2
20	5.3	15.6	22.6
30	5.4	15.7	23.4
40	5.5	15.9	23.5
50	5.5	15.9	24.9
60	5.8	16.2	25.8
70	6.2	16.5	26.4

4 Conclusion

In recent years, with the gradual development and widespread application of wireless visual sensing networks, people's requirements for information security have become increasingly high. Identity verification, as one of the important means to ensure information security, can advantageously ensure that system users have corresponding application rights. Therefore, studying adaptive recognition algorithms for identity features is of great significance and has become a key research topic for relevant scholars, receiving increasingly widespread attention. As a key issue in the development of network security, identity recognition technology has received increasing attention from scholars. Commonly used identity recognition technologies mainly include face recognition, iris recognition, fingerprint recognition, and related algorithm research has achieved certain results. However, in the research of visual optimization identification, due to the influence of posture, light, expression and other factors, it is impossible to accurately identify the identity in the wireless visual sensor network under the uncontrollable environment. In traditional visual identity recognition, uncontrollable factors need to be transformed into controllable and stable characteristic factors in an uncontrollable environment with relatively complex node distribution before identity recognition. The conversion process leads to long recognition time and low efficiency. In this paper, an identity feature recognition method based on convolutional neural networks is proposed for wireless visual sensor networks, and the experimental verification shows that the proposed method has good identification effect and high recognition efficiency. Future research should also focus on the use of identity feature recognition techniques for privacy protection and security. This paper explores how to fully consider the needs of user privacy and information security while ensuring high accuracy.

References

1. Zheng, Y.L., Burns, J.H., Wang, R.F., et al.: Identity recognition and the invasion of exotic plant. Flora - Morphology Distribution Functional Ecology of Plants **280**, 151828 (2021)
2. Yang, W.-H., Dai, D.-Q.: Two-dimensional maximum margin feature extraction for face recognition. IEEE Transactions on Systems, Man, and Cybernetics. Part B, Cybernetics : a Publication of the IEEE Systems, Man, and Cybernetics Society (4), 1002–1012 (2019)
3. Li, D., Huang, L.: Reweighted sparse principal component analysis algorithm and its application in face recognition. J. Comp. Appl. **40**(3), 717–722 (2020)
4. Hou, X., Li, R., Zhang, Y.: Personnel characteristics identification based on foot induced structural vibration. J. Vibration and Shock **41**(23), 241–248, 292 (2022)
5. Liu, Y., Guo, S., Yang, X.: Threshold identity authentication scheme based on biometrics. Appl. Res. Comput. **39**(4), 1224–1227 (2022)
6. Ding, W.: Network video surveillance equipment identification based on web identity characteristics. J. Shenyang University of Technol. **42**(4), 427–431 (2020)
7. Zhao, D., Lu, Y., Liu, X., et al.: Design of emergency UAV network identity authentication protocol based on Beidou. MATEC Web of Conferences **336**, 04004 (2021)
8. Zhang, Y., Sun, Z.: Identity authentication for smart phones based on an optimized convolutional deep belief network. Laser & Optoelectronics Progress **57**(8), 081009 (2020)
9. Tian, Z., Yan, B., Guo, Q., et al.: Feasibility of identity authentication for IoT based on Blockchain. Procedia Computer Sci. **174**, 328–332 (2020)
10. Liu, Y.N., Lv, S.Z., Xie, M., et al.: Dynamic anonymous identity authentication (DAIA) scheme for VANET. Int. J. Communication Syst. **32**(5), e3892.1-e3892.13 (2019)

Research on Image Super Resolution Reconstruction Based on Deep Learning

Zhiwen Chen, Qiong Hao[✉], and Liwen Liu

Wuhan Railway Vocational College of Technology, Wuhan 430205, China
wruqhao@163.com

Abstract. To enhance the precision and clarity of graphic and image depictions, we propose a super-resolution image reconstruction method driven by the power of deep learning. This method initiates by obtaining the reconstruction object from graphics and images, subsequently simulating their degradation process. The preprocessing of initial images is accomplished via registration and expansion, setting a solid foundation for the subsequent stages. Deep learning algorithms are employed to interrogate and dissect the inherent features of the graphics and images. Subsequently, a lineup of techniques including feature fusion and bilinear interpolation are deployed to gain super-resolution reconstruction results of the graphics and images. Upon examining and juxtaposing our deep learning-based method with conventional techniques, we discerned a noticeable advantage of the former. Intriguingly, the resolution deviation within the image reconstruction results derived via our idealized strategy has been remarkably minimized. Concurrently, peak signal-to-noise ratio and structural similarity attributes have been substantially augmented. This unique confluence of improvements as embodied in our approach places it squarely as a potential game-changer in the domain of super-resolution image reconstruction.

Keywords: Deep Learning · Image Reconstruction · Super-Resolution Image

1 Introduction

A key index to evaluate the resolution of an image acquisition system is the spatial resolution of the image. Super resolution reconstruction is the process of obtaining the highest quality image from one or more low resolution images through signal processing and image processing methods. Super resolution reconstruction is widely used in image coding, image processing, high-definition television, image synthesis, face recognition and monitoring, medical diagnosis and other fields. Due to camera cost, limited bandwidth, limited storage space, limited computing power and other reasons, image resolution is often compressed [1]. An intuitive example is that due to the limitation of network bandwidth, compressed low resolution images are often transmitted in the network. Super resolution reconstruction method has become an important part of the network receiver.

© ICST Institute for Computer Sciences, Social Informatics and Telecommunications Engineering 2024
Published by Springer Nature Switzerland AG 2024. All Rights Reserved
L. Yun et al. (Eds.): ADHIP 2023, LNICST 548, pp. 435–450, 2024.
https://doi.org/10.1007/978-3-031-50546-1_29

In the context of super-resolution reconstruction, the pursuit of increased spatial resolution relies on acquiring multiple low-resolution images of the same scene. These images must not only be undersampled and aliased but also possess sub-pixel displacement relative to one another. When the displacement is by integer pixels, the information provided on the same low-resolution grid sampling points is identical across all images. Consequently, integrating these images for reconstruction would be futile, as no new information about the scene would be gained. However, when there is sub-pixel displacement between low-resolution images, along with aliasing, each image no longer represents a simple transformation or duplication of another. Each low-resolution image captures a specific recording outcome for the same scene. In this scenario, the new scene information contained within each low-resolution image can be harnessed for super-resolution image reconstruction. Current reconstruction methods predominantly focus on identifying easily extractable new features and extracting image details for training purposes. This approach, unfortunately, exhibits several drawbacks, including heavy reliance on predefined features, high computational complexity, limited robustness, and fixed input image requirements. These limitations ultimately hinder the quality of super-resolution reconstruction for graphical images. Acknowledging these limitations, there is a need for more advanced approaches that address these challenges. This research aims to overcome the shortcomings by proposing a novel reconstruction method that leverages deep learning techniques. By utilizing deep learning algorithms, the proposed method aims to enhance feature extraction capabilities, reduce computational complexity, improve robustness, and flexibly adapt to varying input image conditions. The objective is to ultimately enhance the quality and fidelity of super-resolution reconstruction for graphic images.

The proposed approach contributes to the broader field of super-resolution reconstruction by addressing key limitations of existing methods. By combining the power of deep learning with the unique characteristics of sub-pixel displacement and aliasing, this research demonstrates the potential to significantly improve the quality and robustness of reconstructed images. The outcomes of this study will provide valuable insights and advancements towards more effective super-resolution reconstruction techniques, benefiting fields such as medical imaging, remote sensing, and computer vision.

Deep learning inSearch technology, data mining, machine learningMT, natural language processing, Multimedia learning, voice, recommendation and personalization technology, as well as other related fields have made many achievements. In order to improve the final image reconstruction quality, the deep learning algorithm is applied to optimize the image super-resolution reconstruction method.

2   Design of Super-Resolution Reconstruction Method for Graphics and Images

Graphic image reconstruction is a technology to obtain the shape information of three-dimensional objects through digital processing from the external measured data of objects. With the support of deep learning algorithm, the operation process of optimizing the design of super-resolution reconstruction method for graphics and images is shown in Fig. 1.

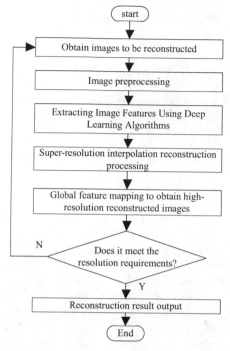

Fig. 1. Flow chart of super-resolution reconstruction of graphic image

The optimized super-resolution reconstruction method takes graphics as the basic content of the image, and obtains the super-resolution reconstruction results of graphics and images through feature fusion, feature reconstruction and other steps.

2.1 Acquire Graphic Image Reconstruction Object

Let the graphic image be sampled separately on the two coordinate systems x and y k_1 and k_2. Second, the Dirac distribution function is used to represent the sampling function of the grid image:

$$f(x, y) = \sum_{k_1=1}^{M} \sum_{k_2=1}^{N} \chi(x - k_1 \Delta x, y - k_2 \Delta y) \tag{1}$$

where M and N Respectively represents the length and width of the image to be reconstructed, $\chi(x, y)$ Is Dirac distribution function, Δx and Δy The sampling interval [2] on the corresponding real coordinate axes x and y. Where Dirac distribution function $\chi(x, y)$. The expression of is as follows:

$$\chi(\text{x}, \text{y}) = \begin{cases} 1, x = y = 1 \\ 0, \text{ else} \end{cases} \tag{2}$$

The Dirac distribution function expressed by formula 2 meets the following conditions:

$$\int_{-\infty-\infty}^{\infty}\int_{-\infty}^{\infty} \chi(x, y) = 1 \tag{3}$$

Substitute the solution result of formula 2 into formula 1 to get the sampling result of graphic image.

2.2 Simulate the Degradation Process of Graphics and Images

In the process of natural image imaging, the original natural scene and target image are a continuous signal. The imaging system samples the input continuous signal to obtain the observed digital image. Because the original image will be affected by motion and deformation, blur, downsampling, noise and other factors, the actually acquired low resolution image will be degraded to a large extent compared with the original high resolution image [3]. Usually, the linear shift invariant or linear shift variant model is used to model the fuzzy interference. Downsampling is the discretization and quantization of the input analog continuous signal. The commonly used downsampling strategies are average sampling or ideal sampling, and generally assume average sampling;The quantization process is to convert analog signals into digital signals. In addition, the image imaging process will also be polluted by noise, resulting in errors, mainly including sensor, sampling and quantization, and model errors. The degradation model of graphics and images is shown in Fig. 2.

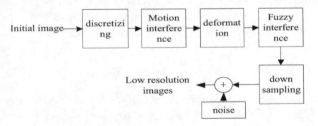

Fig. 2. Graph Image Degradation Model

According to the representation principle in Fig. 2, it can be concluded that the degradation result of graphics and images is:

$$g(x, y) = D \cdot B \cdot H_{atm}M \cdot f(x, y) + n_{\text{additive}} \tag{4}$$

Among $f(x, y)$ is the sample object of the initially collected graphics and images, D, B, H_{atm} and M the corresponding real down sampling matrix expression, optical ambiguity matrix expression, atmospheric disturbance matrix and deformation matrix expression, n_{additive} Is the additive noise in the spatial domain, and the final result is obtained $g(x, y)$ is the low resolution image to be reconstructed.

2.3 Initial Graphic Image Pre-processing

Before super-resolution reconstruction of graphic images, the initial image needs to be preprocessed first. The preprocessing steps mainly include registration and expansion. Image registration is the estimation of image motion information. Image registration is based on the scene, and it needs to use each point to estimate [4]. That is, it is necessary to find a highly accurate point-to-point matching model between input images: suppose two different images are taken for the same scene, and the corresponding points in the other image can be found for one point in one image, which are exactly the same points of the real scene. If the image registration is not accurate, then the information used in the image reconstruction is not correct, can not express the information of the real image, and can not make the image add more details than the original. Therefore, motion estimation is a very critical step in the super-resolution reconstruction algorithm, and if the estimation accuracy cannot reach the sub-pixel level, it will greatly affect the quality of the reconstructed image. The specific registration process in the graphic image preprocessing can be quantified as:

$$g'(x, y) = h_{\text{Registration}}\big[g(x, y)\big] \tag{5}$$

Among $h_{\text{Registration}}()$ represents the registration transformation function, which includes three parts: rigid body transformation registration, affine transformation registration and projection transformation registration. Rigid body transformation registration refers to the coordinate transformation of the three relations of translation, rotation and mirror image, that is, the image frames must be from the same perspective and taken by the same sensor. The rigid body transformation registration function can be expressed as:

$$h_{\text{Registration}}^{\text{rigid body}} = \begin{bmatrix} \varepsilon_x \\ \varepsilon_y \end{bmatrix} + \begin{bmatrix} \cos\theta & -\sin\theta \\ \sin\theta & \cos\theta \end{bmatrix} \begin{bmatrix} x \\ y \end{bmatrix} \tag{6}$$

Variables in Formula 6 ε_x and ε_y Is the offset in the horizontal and vertical directions respectively, θ is the rotation angle. Affine transformation registration is a coordinate transformation of translation, rotation, scaling and mirroring, which is slightly more general than rigid body transformation model and increases the ability to handle scaling transformation [5]. The projection transformation registration can complete the linear transformation of two-dimensional and three-dimensional planes. The affine transformation registration and projection transformation registration functions are as follows:

$$\begin{cases} h_{\text{Registration}}^{\text{affine}} = \begin{bmatrix} \varepsilon_x \\ \varepsilon_y \end{bmatrix} + \begin{bmatrix} \lambda_{11} & \lambda_{12} \\ \lambda_{21} & \lambda_{22} \end{bmatrix} \begin{bmatrix} x \\ y \end{bmatrix} \\ h_{\text{Registration}}^{\text{projection}} = \begin{bmatrix} \lambda_{11} & \lambda_{12} & \lambda_{13} \\ \lambda_{21} & \lambda_{22} & \lambda_{23} \\ \lambda_{31} & \lambda_{32} & \lambda_{33} \end{bmatrix} \begin{bmatrix} x \\ y \\ z \end{bmatrix} \end{cases} \tag{7}$$

among λ_{ij} Is a nonsingular matrix element. The registration result of the initial graphic image can be obtained by weighted fusion of Formula 6 and Formula 7 and substituting

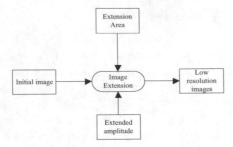

Fig. 3. Schematic diagram of graphic image expansion processing

them into Formula 5. In addition, the expansion principle of graphics and images is shown in Fig. 3.

According to the principle shown in Fig. 3, the result of image expansion processing is:

$$g_{extend}(x, y) = g(x, y) \times \kappa_{extend} \tag{8}$$

Parameters in the above formula κ_{extend} Is the image expansion coefficient. All pixels in the initial graphic image are processed according to the above process to obtain the preprocessing result of the initial graphic image.

2.4 Using Deep Learning Algorithm to Extract Image Features

In order to ensure the quality of image reconstruction, convolutional neural network algorithm is used for image reconstruction. The basic structure of the network is shown in Fig. 4.

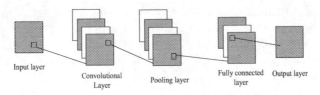

Fig. 4. Structure of deep convolution neural network

As can be seen from Fig. 4, the deep convolution neural network is composed of convolution layer, pooling layer, full connection layer, etc. The convolution layer is the most important part of the convolution neural network, and its main function is to extract various features in the image. Input a feature map. The convolution core in the convolution layer and the feature map are convolutioned and added with the offset weight value. Repeat this process until the whole map is traversed. A new feature map is obtained and used as the input of the next network layer [6]. The convolution process of the convolution layer is shown in Fig. 5.

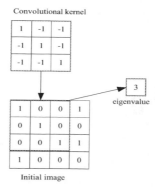

Fig. 5. Schematic diagram of convolution process of convolution layer

For a training process, more than one feature needs to be extracted. The convolution process of a convolution kernel can be regarded as a feature extraction. In order to get more features, we need to use multiple convolution kernels. Each convolution kernel is equivalent to a feature. The feature map of the previous layer in the convolution layer is convolved with the learnable kernel function, and the feature map of this layer is output after the activation function. A trainable parameter will be added after convolution b At this time, the output result of the convolution layer is:

$$x^l_{\text{convolution}} = f_{\text{activation}}\left(\sum_{i\in M_j} x^{l-1}_i * u^l_{ij} + b^l_{\text{convolution}}\right) \tag{9}$$

among $x^{l-1}_{\text{convolution}}$ by $l-1$ Image input value of layer, u^l_{ij} Is the convolutional kernel, $f_{\text{activation}}()$ Is the activation function, and its expression is:

$$f_{\text{activation}}(x) = \frac{1}{1+e^{-x}} \tag{10}$$

Each output will give a deviation b However, for some special output characteristic graphs, the input characteristic graphs will be convolved with some special kernel functions. The pooling layer can reduce the dimension of the feature map processed by the convolution layer. After being processed by pooling layer, the size of feature map tends to become smaller, which is also the result of reducing redundant information of feature map in pooling layer. The pooling layer reduces the size of the image and correspondingly reduces the number of feature parameters, which also speeds up the convergence of the model and improves the efficiency of network training [7]. Most importantly, the pooling layer can keep the most important features of the image from being lost after processing the feature map, that is, feature invariance. The pooling layer only removes some irrelevant information, while the important feature information is still not broken and can be expressed. The specific pooling process is shown in Fig. 6.

According to the above process, the output result of pooling layer is:

$$x^l_{\text{pooling}} = f_{\text{activation}}\left(\beta^l_{\text{pooling}}\text{down}\left(x^{l-1}_j\right) + b^l_{\text{pooling}}\right) \tag{11}$$

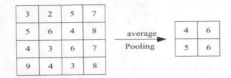

Fig. 6. Diagram of average pooling process of pooling layer

Each output characteristic graph has a multiple deviation value β^l_{pooling} and an additional deviation value b^l_{pooling}. Among, $\text{down}\left(x^{l-1}_j\right)$ is a down sampling function, which determines the pooling method. The features of the input image are obtained through the convolution layer and pooling layer, and then can be input into any classifier with differentiable weights. Generally speaking, the full connection layer is equivalent to the hidden layer of a multi-layer perceptron, which is usually located at the last part of the network model. As a classifier, the classification results are obtained through the feature vectors obtained in the convolution layer and pooling layer. Its principle is similar to that of the artificial neural network [8]. In the actual image feature extraction process, in addition to forward propagation, backpropagation is also required to improve the quality of feature extraction. In the backpropagation process of deep learning, first assign a random value in the interval (-1,1) to the network's convolution kernel weight parameter, randomly select the training set, including k input samples and the expected output corresponding to the input samples, and mark it as x_{in} and E_{out}. Through this process, the input and output of each neuron in the hidden layer are calculated and the partial derivatives of the error function in each layer to each neuron in the output layer are computed based on the actual output and expected output of the last layer of the network. The connection weight value is then adjusted using the derivatives of each neuron in the output layer and the output of each neuron in the hidden layer. Similarly, the derivatives of each neuron in the hidden layer and the input of each neuron in the input layer are used to modify the connection weight value. The global error is evaluated using the error formula to determine if it meets the requirements. If not, the next learning sample and its expected output are selected, and the process returns to the third step to commence the next round of learning. This continues until the error reaches the preset precision or the maximum number of learning times is reached [9]. Through the forward and backward propagation of the convolutional neural network, the extraction results of graphic image features are obtained, in which the extraction results of graphic image texture features are:

$$\tau_{\text{texture,infra-red}} = \frac{P(i,j|d,\varphi)}{\sum_i \sum_j P(i,j|d,\varphi)} \tag{15}$$

Among d and φ respectively represents the distance and direction vector between the target pixel and the image center, $P(i,j|d,\varphi)$ is the gray level co-occurrence matrix. Similarly, we can get the extraction results of other features such as color, shape and so on.

2.5 Realize Super-Resolution Reconstruction of Graphics and Images

The image super-resolution reconstruction is carried out through bilinear interpolation. The basic idea of bilinear interpolation is as follows: the output pixel value at the interpolation point after interpolation is its size in the input image 2×2 is the average value of four vertex samples in the field, and the output pixel interpolates the gray value of four pixels around the corresponding position of the input image in both horizontal and vertical directions [10]. The pixel value of any point in the square is obtained by interpolation, and the bilinear interpolation result is:

$$f_{interpolation}(x, y) = \alpha x + \psi y + \gamma xy + \zeta \qquad (16)$$

Equation 16 represents a hyperbolic paraboloid that will be fitted through four known points. Coefficient α, ψ, γ and ζ. It is selected by the known pixel values of four vertices. The final image super-resolution reconstruction results are:

$$F = \max\left(0, \tau * f_{interpolation}(x, y)\right) \qquad (17)$$

Repeat the above process to obtain the super-resolution reconstruction results of graphics images.

3 Experimental Analysis of Image Super-Resolution Reconstruction Quality Testing

To evaluate the efficacy of the deep learning-based image super-resolution reconstruction method optimized for enhanced performance, we conducted quality test experiments employing a contrast testing approach. The experiment aimed to assess the reconstruction quality by selecting multiple sets of graphic images that met the resolution requirements as experimental samples. Low-resolution images were generated through the process of degradation and downsampling, and served as inputs for the super-resolution reconstruction algorithm under investigation. Employing the configured development environment, we implemented the optimized reconstruction method and supplied the prepared image samples to obtain the corresponding reconstruction results. To measure the quality of the reconstructed images, we employed multiple quality indicators. Furthermore, we compared the performance of our optimized design method with traditional reconstruction results to highlight the advantages and efficacy of our approach.

3.1 Configuration and Reconstruction Method Development Environment

To fulfill the operational requirements of deep learning algorithms and the super-resolution reconstruction method for graphics and images, a comprehensive experimental development environment is established encompassing both hardware and software components. The hardware environment comprises two distinct components: a high-performance server employed in processing the neural network training process, and a low-configuration personal computer dedicated to software UI design and neural network testing. The neural network undergoes training and testing phases, which are facilitated by Ubuntu 16.04, a 64-bit operating system providing essential support. Anaconda serves as the package management tool for efficient virtual environment management, while PyCharm acts as the integrated development environment (IDE) platform to stream-line the development and coding processes. Software design, including UI aspects, is conducted utilizing PyQt5, PyQt5 tools, and other relevant packages on the Windows 10 operating system. The UI design phase is executed within the QtDesigner applica-tion software on macOS Catalina. Developing the image super-resolution reconstruction method relies on the PyTorch framework, celebrated as one of the most prominent frame-works among deep learning researchers. The framework's dynamic graph computation capabilities facilitate seamless model building and training, while also affording users the option to leverage GPU acceleration for enhanced computational performance.

3.2 Prepare Graphic Image Samples

In this experiment, a graphic image data set is used as the training set and test set of the reconstruction method. The data set contains 2000 training data and 200 test data. The resolution of the initial graphic image samples is 72 dpi. The initial preparation of some graphic image samples is shown in Fig. 7.

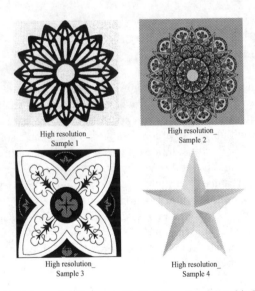

Fig. 7. Schematic Diagram of Initial Sample of Graphic Image

The OpenCV Bicubic interpolation method is used to preprocess the data set for 4-fold down sampling to generate a low resolution image, that is, a graphic image to be reconstructed. The processing result is shown in Fig. 8.

Similarly, the preparation results of all reconstructed object samples in the experiment can be obtained.

Fig. 8. Schematic diagram of image samples to be reconstructed

3.3 Input Operating Parameters of Deep Learning Algorithm

The convolutional neural network algorithm in the deep learning algorithm is selected as the technical support for the optimization design method, and the operation parameters of the algorithm need to be set from both the structure and implementation aspects (Table 1).

Table 1. Convolutional Neural Network Table

Convolutional Layer	Convolutional kernel size	Convolutional Kernel Number
1	9 * 9	64
2	1 * 1	32
3	5 * 5	1
4	9 * 9	64
5	1 * 1	32
6	5 * 5	1

The activation function is sigmoid, and the step size of the network is set to 1. The learning rate is set to 10-4, the learning rate is reduced to half of the original 250000 iterations, a total of 1000000 iterations, and the image batch size is set to 64.

3.4   Describe the Reconstruction Quality Test Experiment Process

Under the configured experimental environment, the development of the super-resolution reconstruction method of graphics and images based on deep learning for optimal design is realized, and the set operating parameters of the deep learning algorithm are input into the running program of the reconstruction method. In order to reflect the advantages of optimization design method in reconstruction quality, the traditional super-resolution reconstruction method based on maximum a posteriori estimation and super-resolution reconstruction method based on multiple sparse representation are set as the two comparison methods of the experiment. The comparison method is developed under the same experimental environment. In the actual operation process, the comparison method is not allowed to call the deep learning algorithm,The parallel switching mode is adopted to ensure the complementary influence between reconstruction methods. Select a reconstruction object from the prepared graphic image samples, and obtain the corresponding reconstruction results by running the reconstruction method. Figure 8 shows the reconstruction output results of some graphic images obtained by the optimization design method (Fig. 9).

According to the above method, the reconstruction result output by the traditional reconstruction method can be obtained, and the resolution information of the reconstructed image can be recorded.

3.5   Setting Super Resolution Reconstruction Quality Test Indicators

Resolution deviation, peak signal to noise ratio and structural similarity are set as test indicators to measure the quality of super-resolution reconstruction of graphic images. The numerical results of resolution deviation indicators are:

$$\Delta \eta = \eta_{\text{target}} - \eta_{\text{rebuild}} \tag{18}$$

among η_{target} and η_{rebuild} They are reconstruction target resolution and actual reconstruction resolution. In addition, the test results of peak signal to noise ratio index and structural similarity index are as follows:

$$\begin{cases} \sigma = \dfrac{N(\text{image})}{N(\text{noise})} \\ \mu = \displaystyle\sum_{i=1}^{n_{\text{pixel}}} \left(\left| x_{\text{original}}(i) - x_{\text{rebuild}}(i) \right| + \left| y_{\text{original}}(i) - y_{\text{rebuild}}(i) \right| \right) \end{cases} \tag{19}$$

Among, $N(\text{image})$ and $N(\text{noise})$ It respectively represents the number of effective pixels and noise points in the reconstruction result image, $(x_{\text{original}}(i), y_{\text{original}}(i))$ It is the No i Pixel value of pixels, $(x_{\text{rebuild}}(i), y_{\text{rebuild}}(i))$ Is the pixel value of the reconstructed image.

3.6   Experimental Results of Image Super-Resolution Reconstruction Quality Test

Through the statistics of relevant data, the test results of the reconstructed image resolution deviation index are obtained, as shown in Table 2.

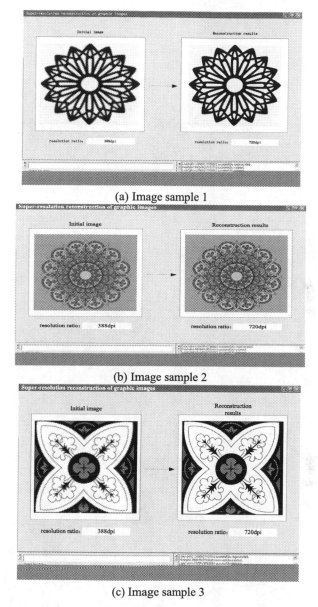

(a) Image sample 1

(b) Image sample 2

(c) Image sample 3

Fig. 9. Super resolution reconstruction results of graphics and images

By substituting the data in Table 2 into Formula 18, it is calculated that the average resolution deviation of the two comparison methods is 112.5 dpi and 50 dpi respectively, and the average value of the reconstructed image resolution deviation of the graphic image super-resolution reconstruction method based on deep learning optimized design is 2.5 dpi. In addition, the peak signal to noise ratio and structure similarity test results

Table 2. Reconstructed image resolution deviation test data table

Image sample number	Reconstruction target resolution/dpi	Super-resolution reconstruction method based on maximum a posteriori estimation	Super-resolution reconstruction method based on multiple sparse representations	Image super-resolution reconstruction method based on deep learning
1	720	650	680	720
2	720	660	650	715
3	720	540	680	715
4	720	550	680	720
5	720	600	660	720
6	720	650	650	720
7	720	660	680	720
8	720	550	680	710

of the reconstructed image are obtained through the calculation of Formula 19, as shown in Fig. 10.

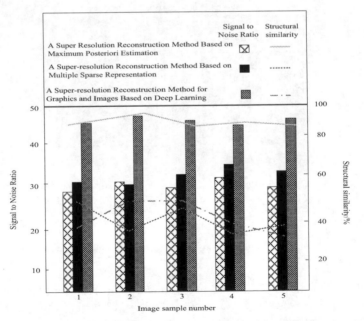

Fig. 10. Image SNR and Structure Similarity Test Results

From Fig. 10, it can be intuitively seen that for the image sample numbered 2, the maximum posterior estimation method has a signal-to-noise ratio of 31 and a structural similarity of 90.2%; The signal-to-noise ratio of the multiple sparse representation method is 29.2, and the structural similarity is 38.6%; The signal-to-noise ratio of the deep learning method is 46.8, and the structural similarity is 48.5%; Compared with two traditional super-resolution reconstruction methods, the optimized design method has higher peak signal-to-noise ratio and structural similarity.

4 Conclusion

In recent years, the utilization of image super-resolution reconstruction has seen a significant rise across various domains. For instance, in gaming, somatosensory peripheral devices enable the capture of players' body movements, enhancing human-computer interaction. By employing super-resolution reconstruction technology to improve the quality of depth data, the accuracy of posture recognition can be substantially elevated, greatly enhancing players' overall gaming experience. Super-resolution reconstruction techniques also find application in 3D reconstruction. By enhancing the density and accuracy of point cloud data obtained from depth cameras, a more realistic surface model of 3D objects can be achieved. This advancement promotes the utilization of 3D reconstruction in diverse domains like biomedicine, video surveillance, criminal case analysis, augmented reality, and more. In the field of unmanned driving, depth information plays a crucial role in determining the 3D position of the vehicle. Implementing super-resolution reconstruction on depth images enables unmanned vehicles to achieve higher positioning accuracy. Simultaneously, it enhances the vehicle's capability for accurate environmental description and obstacle avoidance operations. The development and maturation of depth image super-resolution reconstruction technologies have significantly propelled their widespread adoption in the realm of computer vision. Consequently, this progress has presented new and heightened requirements for future research endeavors. By leveraging deep learning algorithms, the reconstruction quality of graphics and images can be substantially improved. This advancement holds immense research significance and practical value. In summary, deep learning-based super-resolution reconstruction methods for graphic images demonstrate immense potential across various applications, including image enhancement, video compression, and medical image processing. Nevertheless, further research is necessary to enhance the method's robustness and generalization ability, while also addressing challenges such as computational speed and model size.

References

1. Kholil, M., Ismanto, I., Fu'Ad, M.N.: 3D reconstruction using structure from motion (SFM) algorithm and multi view stereo (MVS) based on computer vision. IOP Conference Series: Materials Science and Eng. **1073**(1), 12066–12072 (2021)
2. Lee, H., Chon, B.H., Ahn, H.K.: Rapid misalignment correction method in reflective fourier ptychographic microscopy for full field of view reconstruction. Opt. Lasers Eng. **16**(5), 138–145 (2021)

3. Inam, O., Qureshi, M., Laraib, Z., et al.: GPU accelerated Cartesian GRAPPA reconstruction using CUDA. J. Magn. Reson. **337**(21), 107175–107186 (2022)
4. Zhang, J., Xu, T., Zhang, Y., et al.: Multiplex Fourier ptychographic reconstruction with model-based neural network for Internet of Things. Ad Hoc Netw. **111**(22), 102350–102359 (2021)
5. Shi, Q., Hui, W., Huang, K., et al.: Under-sampling reconstruction with total variational optimization for Fourier ptychographic microscopy. Optics Communications **10**, 126986–126993 (2021)
6. Zhang, J., Tao, X., Yang, L., et al.: The integration of neural network and physical reconstruction model for Fourier ptychographic microscopy. Optics Communications **504**(22), 127470–127483 (2022)
7. Pan, B., Betcke, M.M., Arridge, S.R., et al.: Photoacoustic reconstruction using sparsity in curvelet frame: image versus data domain. IEEE Trans. Computational Imaging **26**(9), 8–15 (2021)
8. Zhang, Y., Zhang, Z., Li, T.: Research on image super-resolution reconstruction based on deep learning. J. Phys. Conf. Ser. **1802**(4), 42034–42043 (2021)
9. Jia, R., Wang, X.: Research on super-resolution reconstruction algorithm of image based on generative adversarial network. J. Phys. Conf. Ser. **1944**(1), 12014–12019 (2021)
10. Agarwal, V., Chitkariya, P., Miglani, A., et al.: Deep learning-based image processing for analyzing combustion behavior of gel fuel droplets. Smart Electrical and Mechanical Syst. **16**(22), 65–85 (2022)
11. Meng, Z., Zhang, J., Qiu, J.: Multi supervised loss function smoothing image super-resolution reconstruction. Chinese J. Image Graphics **27**(10), 2972–2983 (2022)
12. Ni, R., Zhou, L.: Face image super-resolution reconstruction method based on Convolutional neural network. Computer and Digital Eng. **50**(01), 195–200 (2022)
13. Ge, P., You, Y.: Super-resolution reconstruction of light field images based on sparse representation. Progress in Laser and Optoelectronics **59**(02), 94–100 (2022)
14. Liang, M., Wang, H., Zhang, Y., Li, J.: Image super-resolution reconstruction method based on accelerated residual network. Computer Appl. **41**(05), 1438–1444 (2021)
15. Wang, H.Y., Zhang, K.X., Guan, W.Z.: Single image super-resolution reconstruction method based on dense Inception. Computer Appl. **41**(12), 3666–3671 (2021)

Classification of Hyperspectral Remote Sensing Images Based on Three-Dimensional Convolutional Neural Network Model

Pan Zhao[1]([✉]), Xiaoling Yin[1], and Shida Chen[2]

[1] Chizhou University, Chizhou 247000, China
zhaopan0827@126.com
[2] Shanghai Urban Construction Vocational College, Shanghai 200438, China

Abstract. In response to the problems of low accuracy and long time consumption in traditional hyperspectral remote sensing image classification methods, this paper proposes a hyperspectral remote sensing image classification method based on a three-dimensional convolutional neural network model. Firstly, the image data is preprocessed and normalized. Based on this, a three-dimensional convolutional neural network is introduced into the learning of image data. On this basis, by optimizing the overall connectivity parameters of the convolutional kernel function, hyperspectral remote sensing image classification based on the convolutional kernel function was achieved. Experiments have shown that the algorithm proposed in this article can accurately classify hyperspectral images and achieve good results.

Keywords: 3D Convolutional Neural Network Model · Normalization Processing · Hyperspectral Remote Sensing Images · Support Vector Machine

1 Introduction

Hyperspectral remote sensing is a means of analyzing and identifying the reflectance characteristics of the measured substance at different wavelengths. It has been widely used in environmental monitoring, geological exploration, agricultural production, and other fields. The acquisition, processing, and application of hyperspectral remote sensing images have been a research hotspot internationally in recent years. Hyperspectral remote sensing images are the process of synthesizing the information of ground objects in different bands such as visible light and infrared to form multi band images. Compared to conventional remote sensing images, hyperspectral images have higher spatiotemporal resolution and can provide more ground information. Hyperspectral remote sensing image classification is an important research topic in hyperspectral remote sensing images [1]. It is an important aspect of hyperspectral remote sensing image data, and its accuracy will directly affect the practical application of hyperspectral remote sensing images. Hyperspectral remote sensing Image can provide abundant surface information due to its multi band, high spatial resolution, high Spectral resolution and other characteristics, and has important application value in land use, vegetation coverage, water

L. Yun et al. (Eds.): ADHIP 2023, LNICST 548, pp. 451–462, 2024.
https://doi.org/10.1007/978-3-031-50546-1_30

resources monitoring and other aspects. Hyperspectral remote sensing image classification refers to the classification of pixels in the image, including vegetation, water bodies, buildings, bare land, etc. The research results will provide scientific basis for land use, land cover, environmental monitoring, and resource management, as well as for agriculture, urban planning, and disaster risk assessment. However, currently hyperspectral remote sensing image classification still faces many challenges, such as high scene complexity and high image noise. Due to the similarity and confusion of a large number of pixels in the image, classification algorithms become more complex [2]. However, due to the frequent interference of clouds, shadows, and atmospheric disturbances in image data, the accuracy of image classification is low. In the context of massive hyperspectral data, achieving precise classification of remote sensing images remains a challenging issue. How to efficiently use hyperspectral data and classify it is crucial to improve.

In the classification of hyperspectral remote sensing images, researchers have used various methods to improve their classification accuracy. Reference [3] proposes a classification method for hyperspectral remote sensing images based on EMP and mixed kernel SVM. This method first effectively extracts spatial information through EMP, and then uses different kernel functions to process spatial and spectral information, ultimately completing hyperspectral image classification using mixed kernel SVM. This method has high classification accuracy for hyperspectral remote sensing images, but poor classification efficiency. Reference [4] proposes a hyperspectral remote sensing image classification algorithm based on multi-dimensional CNN. This algorithm integrates CNN from different dimensions and combines spatial and spectral information for hyperspectral remote sensing image classification. The comprehensive utilization of the abstract expression ability of three types of CNN for hyperspectral spatial spectral joint information has effectively promoted the application of CNN in the field of hyperspectral remote sensing image classification. This algorithm has good classification efficiency for hyperspectral remote sensing images, but poor classification accuracy.

In response to the problems of the above methods, this article intends to conduct research on hyperspectral remote sensing image classification based on convolutional neural networks. Compared with the traditional hyperspectral remote sensing image classification methods, this method introduces a three-dimensional Convolutional neural network model. Three dimensional Convolutional neural network can learn the spatial and spectral information of images at the same time, so as to capture image features more comprehensively. This method further improves classification accuracy by optimizing the overall connectivity parameters of the convolutional kernel function. The optimization of connectivity parameters can enhance the perceptual ability of convolutional kernels in hyperspectral remote sensing images, enabling them to better adapt to the feature distribution and classification requirements of images.

2   Hyperspectral Remote Sensing Image Preprocessing and Normalization Processing

The data preprocessing of hyperspectral remote sensing images is an important aspect that cannot be ignored in remote sensing data analysis. Firstly, hyperspectral images are composed of thousands of spectral bands, making data processing and storage difficult.

Moreover, hyperspectral image data often contains factors such as noise, atmospheric interference, and surface shadows, which can cause errors and affect the accuracy of the data. Therefore, it is necessary to preprocess hyperspectral remote sensing images to eliminate these effects and obtain more accurate and reliable information.

The preprocessing and normalization of hyperspectral remote sensing images are two important steps in hyperspectral data analysis. Preprocessing refers to the process of spatial correction, spectral correction, and denoising of hyperspectral images before analysis, in order to reduce the impact of errors and noise on the data and improve the quality and reliability of the data. Normalization processing is aimed at solving the problem of the difference in intensity values between different bands in hyperspectral data. A series of mathematical methods are used to convert hyperspectral image data into data at the same scale, making comparison and analysis between different bands more convenient and accurate.

2.1 Hyperspectral Image Data Preprocessing

For hyperspectral image classification algorithms, in order to achieve high classification accuracy while preserving the internal structure of the original data, it is necessary to find a low dimensional subspace composed of several end element spectral images, where each single pixel represents an actual feature. Using the dimensionality reduction method to construct linear graph M, ensuring that each adjacent pixel in the initial image maintains a relatively close projection distance and is always in the subspace. Thus, relevant information in local areas can be well preserved. For complex hyperspectral image classification structures, this can play a protective and constraining role, and effectively protect the diverse internal structure of the initial image [5]. Assuming the initial data of the training sample is:

Class $(x_{i,1}, x_{i,2}, ..., x_{i,j}..., x_{i,n})$, $x_{i,j} \in R^d$, is labeled as $y \in \{1, 2, ..., c\}$, n represents the total number of training samples, c represents the classification quantity value of training samples, and assuming n_l represents the l training sample number, there is $\sum_{l=1}^{c} n_l = n$. The intimate relationship between training sample x_i and training sample x_j is:

$$A_{i,j} = \exp\left(-\frac{\|x_i - x_j\|^2}{y_i - y_j}\right) \tag{1}$$

$$y_i = \|x_i - x_{i,m}\| \tag{2}$$

In the formula, y_i, y_j represents the number of samples, and $x_i x_j$ represents the m adjacent sample within local scale $x_{i,m}$.

On this basis, a new method that can effectively protect local information between pixels is proposed. It is necessary to use local dimensionality reduction methods to obtain a linear graph M through projection. Linear graph M not only effectively protects local information but also separates graph categories [6–8]. The formula for the scatter matrix between local information of adjacent pixels and the intra class scatter matrix is:

$$L_{lb} = \frac{1}{2} \sum_{i,j=1}^{n} W_{i,j}(x_i - x_j)(x_i - x_j)^T \tag{3}$$

$$L_{ho} = \frac{1}{2} \sum_{i,j=1}^{n} W_{i,j'} (x_i - x_j)(x_i - x_j)^T \tag{4}$$

In the formula, $W_{i,j}$ represents the close connection value between the training sample categories, and $W_{i,j'}$ represents the close connection value between the local categories of the training sample. Among them, the local inter category scatter matrix L_{ho} and the overall inter category scatter matrix L_{lb} are both $n \times n$-squared matrices.

According to formula (1), the close connection value between each adjacent pixel can be obtained [9], and the ratio of the maximum value obtained through local dispersion matrix is *Fisher*:

$$L_{lb}^T = \delta L_{ho}^T \tag{5}$$

In the equation, δ represents the diagonal characteristic value of the matrix, and the transformation matrix is:

$$T_{LFDA} = \arg \max_{T \in R^{d \times r}} tr\left[\left(T_{LFDA}^T L_{lb} T_{LFDA} \right)^{-1} T_{LFDA}^T L_{lb} T_{LFDA} \right] \tag{6}$$

In the equation, $tr(.)$ represents the motion trajectory function of the matrix.

This data preprocessing model has protective constraints and does not damage the diversified structure of the initial image. This model measures its constraint by the similarity between local adjacent pixels, utilizing the variable characteristics of matrix T_{LFDA} to make each individual adjacent pixel in the same category close to each other, while those in different categories are separated from each other. This not only preserves the local features of the initial data but also obtains a more compact distribution structure, while reducing Bayesian error, ultimately achieving the goal of improving classification accuracy.

2.2 Normalization Processing of Hyperspectral Remote Sensing Images

To achieve the desired effect, a basic preparation work needs to be carried out before classification, which is to perform standard normalization processing on the initial data. In classification models based on convolutional neural networks, using standardized methods to input data into different classification spaces can result in significant differences in classification results [10–12]. The spectral reflectance values of each individual pixel in the initial image may vary from several hundred to several thousand, and the numerical span is relatively large during the separation process, which increases the computational difficulty. However, using the standardized mean difference method can relatively improve. Firstly, each trajectory segment of the hyperspectral image is standardized. This operation can make the spectral curve fluctuation trajectory of each individual pixel more obvious and easier to determine, while increasing the variation of trajectory differences and relatively reducing the complexity [13–15]. Thus achieving an effect of improving classification training speed and accuracy. Assuming all pixels are taken as column vector x_i, the formula is as follows:

$$X_i = \frac{X_i - \mu}{\lambda} - 1 \tag{7}$$

In the equation, μ represents the average pixel value of the initial image, and λ represents the pixel standard deviation of the image at i curve band. After standardizing the initial data, it will lay a good foundation for subsequent image classification and greatly improve the classification effect [16, 17].

3 Image Classification Method Based on Three-Dimensional Convolutional Neural Network Model

3.1 Image Recognition

The CNN model is used to complete hyperspectral remote sensing image recognition, which includes an input layer, a hidden layer, and an output layer. The hidden layer serves as the neural structure layer in the model, including convolution, pooling, and a single-layer perceptron. This layer mainly realizes image recognition [5, 19].

Using processed hyperspectral remote sensing images as input samples for the CNN model, with a quantity of m, the sample set $\{(x^{(1)}, y^{(1)}), ..., (x^{(m)}, y^{(m)})\}$ consists of n categories. Taking sample $x^{(i)}$ as a Reference, the corresponding category labels are represented by j. The basic cost function calculation formula for the network model is:

$$\varphi(W, b) = R(W, b) = \left[\frac{1}{m} \sum_{i=1}^{m} \varphi\left(W, b; x^{(i)}, j\right) \right]$$

$$= \left[\frac{1}{m} \sum_{i=1}^{m} \left(\frac{1}{2} \left\| h_{W,b}\left(x^{(i)}\right) - j \right\|^2 \right) \right] \tag{8}$$

In the formula, the weight value is represented by W, which is used to complete the connection of each layer in the model; The offset term is represented by b; $h_{W,b}(x^{(i)})$ represents the output result and is completed at the last level of the model.

The training purpose of the model is to obtain the minimum value of $\varphi(W, b)$, using parameters W and b as Reference. The optimization objective function is completed through the gradient descent method, and the iterative equation is:

$$W_{ij}^{(l)} = W_{ij}^{(l)} - \alpha \frac{\partial}{\partial W_{ij}^{(l)}} \varphi(W, b) \tag{9}$$

$$b_{ij}^{(l)} = b_{ij}^{(l)} - \alpha \frac{\partial}{\partial b_{ij}^{(l)}} \varphi(W, b) \tag{10}$$

In the equation, the learning rate is represented by α. The BP algorithm is used to solve the partial derivative of formulas (9) and (10). The acquisition of $h_{W,b}(x^{(i)})$ is completed through the forward propagation algorithm. The difference between this value and the actual value is represented by $\delta_i^{(nl)}$, and the solution of $\delta_i^{(nl)}$ and nl represents the model output layer; The residual error of each layer of the model is solved based on the residual error of 6, and the partial derivative of formula (9) and (10) is solved.

The residual solution formula for the last layer of the network is:

$$\delta_i^{(nl)} = \frac{\partial \varphi_1}{\partial Z_i^{(nl)}} = \frac{\partial}{\partial Z_i^{(nl)}} \frac{1}{2} \left\| h_{W,b}\left(x^{(i)}\right) - j \right\|^2 \tag{11}$$

In the formula, both $Z_i^{(l)}$ and $Z_i^{(nl)}$ represent weighted sums, and both belong to the i unit, with the former located in the l layer; The latter is located on the last layer of the model. Based on the above steps, complete hyperspectral remote sensing image recognition.

3.2 Model Optimization Image Classification

During the recognition process of the CNN model, the full connection process of the single-layer perceptron determines the recognition output of the model. Therefore, in order to improve the recognition effect and convergence efficiency of the model, double optimization is performed on it.

If k represents the number of convolutional layers and $l_{img} \times l_{img}$ represents the size, it belongs to the convolutional kernel; The matrix of $l_{img} \times l_{img}$ represents the input image size, while both n_{in} and n_{out} represent the number of images and correspond to the input and output respectively; Using iterative methods, the feature matrix S with the smallest objective function value is processed to obtain a matrix of Mat_1. The convolution kernel optimization is completed through convolution coefficients, and the convolution results are analyzed using the binary method; Establish a function expression and complete it based on the interpolation principle; μ represents the coefficient, which belongs to dynamic convolution, and its calculation formula is:

$$\mu = \frac{n_{in} n_{out} l_{img}^2}{2^k} + \theta_1 \tag{12}$$

In the formula, the correction Error term is represented by θ_1. The formula for solving the number of parameters is represented by formula (13), which corresponds to the convolutional kernel for both input and output data. The formula is:

$$f_{in} = n_{in} l_{ker}^2, f_{out} = n_{out} l_{ker}^2 \tag{13}$$

The optimized convolution kernel calculation formula (14) represents and is initialized:

$$M_{at_2} = 2\sqrt{\frac{\mu}{f_{in} + f_{out}}} \cdot M_{at_1} \tag{14}$$

In the equation, the convolutional kernel parameter matrix is represented by M_{at_2} and is optimized.

Let ρ represent the optimization coefficient, which is used to optimize the fully connected parameters. The formula is:

$$\rho = \frac{n_{cag}}{2}\left(\gamma - \varepsilon k^{\varepsilon - 1}\right) \tag{15}$$

In the equation, γ represents a factor that is related to the optimization coefficient. If θ_2 indicates the correction of Error term, then:

$$\gamma = \left[\lambda\left(\sqrt{n_f l_{img}^2 - n_{cag}}\right) - k^\varepsilon\right] + \theta_2 \tag{16}$$

$$\lambda = k + \sum_{i=0}^{\varepsilon-1} i \tag{17}$$

The optimized fully connected layer parameter formula is:

$$Mat_4 = 2\sqrt{\frac{\rho}{n_{cag} + n_f l_{img}^2}} \cdot Mat_3 \tag{18}$$

Based on the above optimization steps, the calculation formula for the optimization coefficient η of the model is:

$$\eta = \frac{\mu_0}{\mu_0 + \mu}\rho + \frac{\mu}{\mu_0 + \mu}\left(\gamma - k^\varepsilon\right) \tag{19}$$

Update and solve formula (18):

$$Mat_4 = 2\sqrt{\frac{\eta}{n_{cag} + n_f l_{img}^2}} \cdot Mat_3 \tag{20}$$

On this basis, an optimized 3D convolutional neural network model is used to introduce a support vector machine to construct an image classifier:

$$g(x_j) = aign\left(\sum_{i=1}^{Num} a_i y_i \sum_{m=1}^{M} \beta_m K_m(x_i, x_j) + b\right) \tag{21}$$

In the formula, $g(x_j)$ represents the j expected image label parameter in $MKL-SVM$, and Num represents the number of collected parameters. Input Num has different expressions in different situations.

a_i、y_i represents the optimized parameters and the number of samples collected during training, while $K_m(x_i, x_j)$ represents the kernel function in this formula. In the test set and training set, it is $N \times (n - m)$、$N \times m$ respectively. To obtain the results of image classification, the features of the fused test set images are input into an $MKL-SVM$ high-resolution image classifier to complete image classification.

4   Experimental Methods

4.1   Experimental Parameter Settings

Using hyperspectral remote sensing images as experimental samples. 15% of the labeled samples will be randomly selected as test samples, and the remaining 85% will be used as training samples. Firstly, the preprocessed initial data is transferred into the input layer as raw data. Then, take 1 × 100 as the column vector of each pixel for preliminary data input, go through the second convolution layer, set the image feature to 30, and the convolution kernel to size 1 × 4, and use the actual effect activation function RELU for activation. At this time, take the column vector of 30 1 × 100 for each pixel as the initial data input into the convolution layer. Take the data output from the convolution layer as the input data of the next pooling layer. The window size of the pooling layer is 1 × 2. Each pixel passing through this layer will output 25 1 × 50 column vector. After this operation, good classification results can be achieved. Using Reference [3] and Reference [4] as experimental comparison methods to verify the effectiveness of the algorithm.

4.2   Experimental Results

Table 1. Comparison of Training Times and Coefficients of Three Methods

Index	Reference [3] Method	Reference [4] Method	Proposed method
training time/s	6.11	7.89	6.62
Test time/s	15.23	16.99	12.36
overall time/s	25.67	26.82	20.44
kappa coefficient	76.52	82.34	91.44

Table 1 shows the comparison of the time and *Kappa* coefficient required for the three classification algorithms in the classification experiment process. From the table, it can be seen that although the method in Reference [2] has slightly lower training time than the algorithm in this paper, the classification accuracy is the lowest among these methods. This is because in order to achieve ideal classification results during the training process, a large amount of training time is required. However, compared to other algorithms, this method has a relatively fast testing and training process, and has achieved the optimal classification experimental results. It can be said that there is a good balance between accuracy and time selection. The coefficient in this article *Kappa* is also the best among all methods, indicating that the optimal classification can be achieved in any aspect.

Classification of remote sensing images with different resolutions is an important research field in the application of remote sensing technology. Classification accuracy is the main indicator for measuring the quality of classification results. This project aims to study high-resolution and low-resolution data, and compare and analyze the image classification results under two different resolutions. The comparison results of classification accuracy obtained are shown in Figs. 1 and 2.

From Fig. 1, it can be seen that as the number of iterations increases, the three classification methods show an upward trend. However, compared to the other two methods, the overall classification accuracy of the classification method designed in this article is higher. When the number of experimental iterations reached 65000, the method

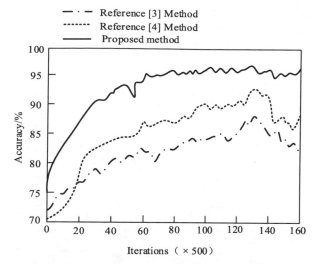

Fig. 1. Comparison results of classification accuracy for high-resolution remote sensing images

in this paper achieved the highest classification accuracy of 96.62% in the test dataset, and the classification effect was good. The reason for obtaining the experimental results is that the test set images selected in the experiment contain a large number of background regions, and the remote sensing image features are not obvious. The classification model construction module used in the literature method is relatively simple, so the accuracy of discriminative region positioning is lower, so the classification accuracy is lower. This article adopts a three-dimensional convolutional neural network model as the training model for hyperspectral remote sensing images. By optimizing the convolutional kernel and fully connected parameters, accurate classification of hyperspectral remote sensing images can be achieved.

Due to the blurry edge and target features of low resolution remote sensing images, as well as the partial loss of pixels in remote sensing images, it increases the difficulty of remote sensing image classification. From Fig. 2, it can be seen that as the number of samples increases, the overall classification accuracy of the classification method designed in this article is higher compared to the other two methods. When the sample size reached 90, the method in this paper achieved the highest classification accuracy of 95.5% in the test dataset, and the classification effect was good. The reason for obtaining the experimental results is that the method used in this paper preprocesses and normalizes the data of hyperspectral remote sensing images, eliminating the influence of noise and interference in hyperspectral image data, and thus obtaining more accurate and reliable information. This can prove that the proposed method can achieve better classification results for remote sensing images in both high-resolution and low-resolution scenarios.

Figure 3 shows the comparison of classification and recognition time between our method and traditional methods, with six datasets set in the experiment. From the figure, it can be seen that the classification and recognition time of our method is relatively short, while the Reference method has poor data convergence ability due to not removing redundant information, and cannot quickly achieve classification and recognition of

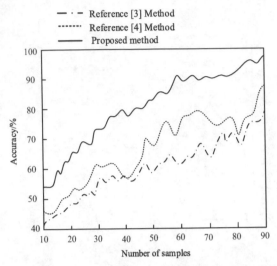

Fig. 2. Comparison Results of Classification Accuracy of Low Resolution Remote Sensing Images

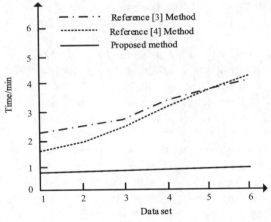

Fig. 3. Comparison of classification and recognition time for hyperspectral remote sensing images

hyperspectral remote sensing images. This also proves that the method proposed in this paper has good advantages and can adapt to precise image classification and recognition under various conditions. Three dimensional Convolutional neural network has good parallel computing ability, which can learn spatial and spectral information at the same time, effectively reducing processing time, thus accelerating the speed of classification.

On this basis, the recall rates of three methods for hyperspectral remote sensing image classification were tested, and the experimental comparison results are shown in Fig. 4.

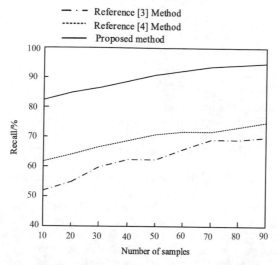

Fig. 4. Recall rate of hyperspectral remote sensing image classification

Analyzing Fig. 4, it can be seen that the recall rate of hyperspectral remote sensing image classification for the method in reference [3] is 68%, the recall rate of hyperspectral remote sensing image classification for the method in reference [4] is 73%, and the recommended English online teaching video resources for the method in this article take 94% of the time. Three dimensional Convolutional neural network model can automatically extract representative features from hyperspectral images by multiple convolution and pooling operations. These features typically have a higher level of abstraction and can better distinguish different categories. By feature extraction and dimensionality reduction, the dimensionality of input data can be reduced, thereby improving the classification recall rate of hyperspectral remote sensing images.

5 Conclusion

This article establishes a new classification method for hyperspectral remote sensing images. On this basis, further optimize the structure and parameters of the neural network, improve its learning and classification efficiency and stability, and enhance its ability to extract complex pixel distributions and semantics from remote sensing images. Experiments have shown that this algorithm can accurately classify hyperspectral images, better capture image feature information, improve classification accuracy, and have a relatively fast classification speed, effectively solving the classification problem of hyperspectral images.

In future research, it is planned to further improve and improve the classification algorithm of hyperspectral remote sensing images, improve their classification accuracy and robustness through in-depth research on convolutional neural network models based on previous work. At the same time, this project will also use convolutional neural networks to interpret hyperspectral remote sensing data and explore deep information within it.

Aknowledgement. 2020 Anhui Province University Excellent and Top notch Talent Cultivation (gxgnfx2020112)

References

1. Sadovnychiy, S.: Gabor features extraction and land-cover classification of urban hyperspectral images for remote sensing applications. Remote Sensing **13**(15), 2914 (2021)
2. Jijon-Palma, M.E., Kern, J., Amisse, C., et al.: Improving stacked-autoencoders with 1D convolutional-nets for hyperspectral image land-cover classification. J. Appl. Remote. Sens. **2**, 15 (2021)
3. Cao, H., Han, X., Li, J., et al.: Hyperspectral remote sensing image classification based on EMP and SVM with composite kernel. Geospatial Inf. **19**(11), 14–18 (2021)
4. Liu, J., Ban, W., Chen, Y., et al.: Multi-dimensional CNN fused algorithm for hyperspectral remote sensing image classification. Chinese Journal of Lasers **48**(16), 153–163 (2021)
5. Peng, Y., Wang, X., Zhang, J., et al.: Pre-training of gated convolution neural network for remote sensing image super-resolution. IET Image Processing **15**(5), 11791188 (2021)
6. Anand, R., Veni, S., Aravinth, J.: Robust classification technique for hyperspectral images based on 3D-discrete wavelet transform. Remote Sensing **13**(7), 1255 (2021)
7. Gong, H., Li, Q., Li, C., et al.: Multiscale information fusion for hyperspectral image classification based on hybrid 2D–3D CNN. Remote Sensing **13**(12), 2268 (2021)
8. Xi, J., Ersoy, O.K., Fang, J., et al.: Wide sliding window and subsampling network for hyperspectral image classification. Remote Sensing **13**(7), 1290 (2021)
9. Liu, X., Chen, S., Song, L., et al.: Self-attention negative feedback network for real-time image super-resolution. J. King Saud University - Computer and Information Sci. **34**(8B), 6179–6186 (2022)
10. Liu, M., Tan, J., Tian, Y.: Decoding auditory attentional states by a 3D convolutional neural network model. Int. J. Psychophysiol. **168**, 134–135 (2021)
11. Alameddine, J., Chehdi, K., Cariou, C.: Hierarchical unsupervised partitioning of large size data and its application to hyperspectral images. Remote Sensing **13**(23), 4874 (2021)
12. Hawkesford, M.J.: A neural network method for classification of sunlit and shaded components of wheat canopies in the field using high-resolution hyperspectral imagery. Remote Sensing **13**(5), 898 (2021)
13. Deshmukh, R.R., Ghule, A.: Wavelength selection and classification of hyperspectral non-imagery data to discriminate healthy and unhealthy vegetable leaves. Curr. Sci. **120**(5), 932–936 (2021)
14. Macfarlane, F., Murray, P., Marshall, S., et al.: Investigating the effects of a combined spatial and spectral dimensionality reduction approach for aerial hyperspectral target detection applications. Remote Sensing **13**(9), 1647 (2021)
15. Rossello, J.C., Graham, P., Prakash, A., et al.: Airborne hyperspectral data acquisition and processing in the arctic: a pilot study using the hyspex imaging spectrometer for wetland mapping. Remote Sensing **13**(6), 1178 (2021)
16. Zhang, Y., Du, J., Pi, W., et al.: Deep learning classification of grassland desertification in china via low-altitude UAV hyperspectral remote sensing. Spectroscopy **37**(1), 28+3035 (2022)
17. Ping, W., Fu, J., Qiao, W., et al.: Decision support system for hyperspectral remote-sensing data of Yellow River Estuary, China. Scientific Programming **2121**(9), 1–17 (2021)
18. Li, K., Xu, J., Zhao, T., et al.: A fuzzy spectral clustering algorithm for hyperspectral image classification. IET Image Processing **15**(12), 2810–2817 (2021)
19. Li, J., Shen, H., Li, H., et al.: Radiometric quality improvement of hyperspectral remote sensing images: a technical tutorial on variational framework. J. Applied Remote Sensing **15**(3), 1–33 (2021)

Texture Image Feature Enhancement Processing Method Based on Visual Saliency Model

Yuan Wang[✉]

Wuhan Institute of Design and Sciences, Wuhan 430205, China
yinwar822172@163.com

Abstract. To improve the feature visualization effect of texture images, a texture image feature enhancement processing method based on visual saliency model is proposed. After collecting texture images, use soft and hard threshold denoising algorithms to denoise the texture images. Extract and decompose the features of the denoised image based on the visual saliency model. Based on the results of feature decomposition, the resolution of the texture image is reconstructed using deep learning technology, and then the texture image is described using shear wave transformation method to enhance the expression of the image's feature information. According to the experiment, it can be seen that after applying this method, the distortion coefficient of the texture image is smaller and the clarity is higher, indicating the feasibility of this method.

Keywords: Texture Images · Feature Enhancement · Noise Reduction Processing · Visual Saliency Model · Feature Extraction

1 Introduction

As one of the visual foundations for human perception of the world, images are an important means for people to obtain, express, and transmit information. In the face of massive image data, how to enable computers to mine useful information to complete image analysis and understanding and provide effective decision making has become a current research focus [1].

Texture feature, as one of the underlying features of image, has been playing an important role in the fields of image analysis, machine vision and pattern recognition. No matter for natural images, medical images or remote sensing images, the extraction and analysis of texture features are the primary and basic problems to be solved [2]. Therefore, how to effectively obtain the representational texture features is the key to image analysis and understanding.

In response to this issue, relevant scholars have proposed enhancement processing methods for image features. A low illumination image enhancement method based on visual communication was proposed in reference [3], and an image enhancement method based on global and local multiple features was proposed in reference [4].

L. Yun et al. (Eds.): ADHIP 2023, LNICST 548, pp. 463–476, 2024.
https://doi.org/10.1007/978-3-031-50546-1_31

The visual saliency model is a visual attention model designed based on the visual nervous system of early primates [5]. This model first uses Gaussian sampling method to construct a Gaussian pyramid for the color, brightness, and direction of the image. Then, the Gaussian pyramid is used to calculate the brightness feature map, color feature map, and direction feature map of the image. Finally, by combining feature maps of different scales, brightness, color, and direction saliency maps can be obtained, and the final visual saliency map can be obtained by adding them [6]. This method does not require the process of training and learning, and can only complete the calculation of saliency maps through pure mathematical methods.

Therefore, based on the traditional research mentioned above, this paper proposes a new texture image feature enhancement processing method based on the visual saliency model.

2   Texture Image Acquisition

Firstly, the texture 3D imaging system is established, and its structure is shown in Fig. 1.

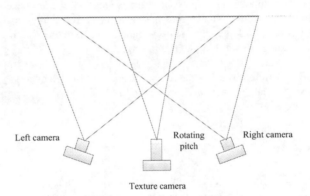

Fig. 1. Structure diagram of texture 3D imaging system

The texture 3D imaging system consists of three cameras, with the left and right cameras forming a binocular stereo vision system for reconstructing 3D point cloud models; The middle is a texture camera used to capture high-resolution texture images of objects. The texture camera is a telephoto camera that captures texture images from multiple perspectives by rotating and pitching.

Based on the texture 3D imaging system shown in Fig. 3, collect texture image information.

Before using multiple texture information to construct image information texture set, it is necessary to form local contour information feature vector of image first, including local edge contour information feature vector, image texture feature factor and enhanced information region coordinates, so as to construct image information texture set Q. Then, forward unification is used to decompose image texture feature information, and the feature equation of image edge state information feature matching can be obtained. The calculation formula is as follows:

$$a = \int_0^2 \rho^* \cos\theta^* \mathrm{d}\frac{\rho^*}{Q}, \rho^* \in [0, h] \tag{1}$$

In formula (1), h represents the set of edge contour pixels for low illumination images, θ represents the edge feature angle of the image.

Utilize the three-dimensional central coordinates of low illumination image information for feature aggregation processing of image information. Using $\rho^* - R$ as the displacement center point, the image information is optimized and sparsely identified, and then local contour image information is aggregated to obtain the texture image information collection model as follows:

$$b = \beta(g_l + f) \tag{2}$$

In Formula (2), b represents the acquisition model of low illumination image information; l represents the displacement distance of image information features, and the value range is $[1, R]$; f represents the aggregation model of image information; g_l represents the regional offset of image information; β stands for local contour scale.

3 Texture Image Denoising

After collecting the texture image mentioned above, the texture image is denoised using a soft and hard threshold denoising algorithm, and then the obtained wavelet coefficients are subjected to threshold processing. The soft threshold method needs to retract wavelet coefficients larger than the threshold to 0, and coefficients smaller than the threshold to 0; The hard threshold rule preserves coefficients that are greater than the threshold, and treats coefficients that are less than or equal to the threshold as 0. The constructed model is as follows:

$$\begin{cases} \hat{\delta}_{T2} = \begin{cases} sign(\delta)(|\delta| - T), |\delta| > T \\ 0, |\delta| \leq T \end{cases} \\ \hat{\delta}_{T1} = \begin{cases} \delta, |\delta| > T \\ 0, |\delta| \leq T \end{cases} \end{cases} \tag{3}$$

In Formula (3), $\hat{\delta}_{T1}$ and $\hat{\delta}_{T2}$ represent soft threshold and hard threshold respectively, and the sum of the deviations between them is E; $\hat{\delta}_T$ represents the wavelet coefficient after threshold processing; the wavelet coefficient of the original texture image is δ, and the selected threshold is T.

In the process of image denoising, if there are discontinuities or deviations in wavelet coefficients in the soft and hard threshold method during processing, it will cause vibration or blurring of the texture image [7]. Therefore, in order to avoid the occurrence of the above problems, the selected threshold function should have strong protection and scalability for edge coefficients, and also meet the requirement that the selected threshold is within the range of $\min E = \sum_k^n \left| \delta_{j,k} - \hat{\delta}_{j,k} \right|$. Among them, for wavelet coefficients with absolute values not less than the threshold, the above soft and hard threshold models are integrated and transformed as follows:

$$
\begin{cases}
\hat{\delta}_T = \begin{cases} 0, |\delta| \leq T_1 \\ sign(\delta)\dfrac{T_2(|\delta| - T_1)}{T_2 - T_1}, T_1 \leq |\delta| \leq T_2 \\ \delta, |\delta| < T_2 \end{cases} \\
\hat{\delta}_T = \begin{cases} sign\delta\left(|\delta| - \dfrac{2T}{1+\exp(m\delta^2)}\right), |\delta| \geq T \\ 0, |\delta| < T \end{cases}
\end{cases}
\tag{4}
$$

In Formula (4), the adjustment coefficient in the model is m, and T_1 and T_2 represent the upper and lower limits of the threshold respectively. When $m = 0$, the threshold values in the threshold function can maintain continuity, and there are high order differentiability in the interval $\delta \geq T$, so as to solve the problem of threshold deviation. The specific denoising process is shown as follows:

Step 1: Set the wavelet bases and their decomposition levels in the image based on the type of noise in the image to obtain the wavelet coefficient $\delta_{j,k}$ of the texture image.

Step 2: Based on the characteristics of the obtained wavelet coefficients, select an appropriate threshold to calculate the wavelet coefficients in the image, and obtain the wavelet estimation coefficient $\hat{\delta}_{j,k}$ of the texture image, while keeping the difference between $\hat{\delta}_{j,k}$ and $\delta_{j,k}$ in a minimum state.

Step 3: Use the obtained wavelet estimation coefficients to perform wavelet reconstruction on the texture image, thereby completing the denoising process of the texture image.

4 Feature Extraction and Decomposition of Texture Image Based on Visual Saliency Model

In this paper, a significance analysis algorithm based on graph theory is used to extract the visual significance of texture image features. Saliency analysis algorithm extracts visual saliency through Markov chain, whose properties are:

$$
P(X_{n+1} = x|X_0, X_1, \cdots, X_n) = P(X_{n+1} = x|X_n)
\tag{5}
$$

In Formula (5), x represents the process of a state, and X_n represents the state of time n [8].

Markov chain is used in graph theory significance extraction, texture image features can be written in the form of $M : [n]^2 \rightarrow R$. Define $M(i, j)$ and $M(p, q)$ as feature vectors, and the difference between the two vector values is represented by $d((i, j)\|(p, q))$, thus:

$$
d((i, j)\|(p, q)) \triangleq \log\left|\frac{M(i, j)}{M(p, q)}\right|
\tag{6}
$$

Therefore, texture images can serve as directed graphs with interconnected pixels, as the nodes of the image are represented by pixel points, and the two adjacent nodes in the graph are $M(i, j)$ and $M(p, q)$. Set the two nodes $M(i, j)$ to $M(p, q)$ as weights ω_1, which can be expressed as:

$$\omega_1((i, j), (p, q)) \underline{\underline{\Delta}} d((i, j) \| (p, q)) \cdot F(i - p, j - q) \tag{7}$$

$$F(a, b) \underline{\underline{\Delta}} \exp\left(-\frac{a^2 + b^2}{2\sigma^2}\right) \tag{8}$$

where σ is a free parameter in the algorithm. Therefore, the distance from node $M(i, j)$ to $M(p, q)$ is proportional to the weight ω_1 of node $M(i, j)$ to $M(p, q)$ and the difference between them. The initial texture image is represented by the pixel value of each node, and the node mode is used to represent the pixel of each node. All the weighted values and original significance values are added together respectively to calculate the probability of transfer. The original significance graph of the node is represented by the newly obtained significance value, and all nodes in the Markov chain are normalized. The excitation information obtained converges to many main places, thus creating a graph GN (including n^2 nodes), setting the two adjacent nodes in the graph as the new weight ω_2, which can be expressed as:

$$\omega_2((i, j), (p, q)) \underline{\underline{\Delta}} A(p, q) \cdot F(i - p, j - q) \tag{9}$$

In formula (9), $A(p, q)$ represents the original saliency map.

The nodes of Markov chains correspond to the states and have limited characteristics. Markov chains are also constrained by higher excitation nodes, resulting in a saliency map that further normalizes $A(p, q)$.

After extracting texture image feature $M : [n]^2 \rightarrow R$, the feature information is decomposed. The texture image information fusion model is generated by decomposing the information features of the image, and the multi-scale high frequency of the image information is decomposed by the Gaussian filter pixel detection method.

The sequence of nonlinear feature distribution chaotic function of image information is:

$$P = \sum_{k=1}^{n} I_{(k)}(x, y) \times 2^l / b \tag{10}$$

In formula (10), I represents the scale decomposition function of texture image information; k represents the edge grayscale pixel feature set of texture image information; b represents the collection model of texture image information; (x, y) represents a multi-scale input pixel chaotic sequence.

By fusing the distribution sequence of texture information, the feature fusion function obtained is:

$$W = P[2(x - 1) + u, 2(y - 1) + v] \tag{11}$$

In Formula (11), u represents the information intensity of image information edge extraction; v represents the information integration degree of image information distribution.

The amount of information distributed in adjacent domains of image information is:

$$r_\lambda^* = \begin{cases} \dfrac{W}{cs_k}, 1 \leq \lambda \leq x - y \\ \varepsilon_\lambda, \text{others} \end{cases} \tag{12}$$

In formula (12), λ represents the blur factor for image information enhancement; c represents the standardized component set of image information; s_k represents the grayscale information of the image; ε_λ represents the Gaussian filter coefficient of digital images of different levels.

Based on the above content, establish a set of regional feature decomposition vectors for image information as follows:

$$E = \frac{r_\lambda^* \sum_{i=1}^{p} \rho^* \times r_\lambda^* \sum_{j=1}^{p} \rho^*}{\sqrt{\int_1^p \left(\rho_i^* - R \right) \mathrm{d}\rho^* + \int_1^p \left(\rho_j^* - R \right) \mathrm{d}\rho^*}} \tag{13}$$

5 Using Deep Learning Technology to Reconstruct the Resolution of Texture Images

After the feature extraction and decomposition of the texture image is completed based on the visual saliency model, the resolution of the texture image is reconstructed by using deep learning technology, which lays a foundation for the subsequent feature enhancement processing.

Deep learning evolved from traditional neural networks, but it has deeper network results and more efficient training methods [9]. In the process of deep learning, loss functions can be optimized as far as possible to mine learning rules, and learning results can be obtained by directly inputting data, without manual design of learning rules. Deep learning can apply the learned rules in deeper network combinations to build complex models, distinguish and simplify different categories of problems, and further extract more complex features to replace artificial feature extraction [10]. Therefore, this study utilizes deep learning processes to construct a numerical imaging model for texture images and reconstruct image resolution.

The process of deep learning mainly includes preprocessing, feature learning, model training, and model construction. The specific content is as follows:

(a) Pre processing. Due to factors such as improper human operation and system errors, if noise data from some data samples is input into the deep learning network without processing, it will lead to phenomena such as unclear data features in the output results. Therefore, the process of processing data is very important. In the previous study, based on the processing of texture image features, a structure parallel to the linear shape was selected to process the corrosion reflection component, thereby effectively eliminating interference information in the image.

(b) Feature learning. The main goal of feature learning is to make the model more accurate, improve operational efficiency, and master more complex texture image features.

(c) Model training. After applying deep learning to construct a numerical imaging model for texture images, the optimal imaging model that meets the expectations and goals is ultimately generated through multiple training and calculation of data.

(d) Model construction. Using the feature regions obtained from the above processing, a numerical imaging model is constructed using deep learning as follows:

$$F(x, y) = -\frac{(x^2 + y^2)}{2\pi\lambda^2} \tag{14}$$

In Formula (14), $F(x, y)$ represents the constructed numerical imaging model, λ represents the range compression parameters in the texture image region, and x and y respectively represent the pixel points in the imaging region.

There is a compromise processing method of spatial proximity of pixels in the texture image region, which leads to the blurring of image edge details in the reconstruction of the resolution in the image. Therefore, guided filtering is used to establish the linear constraint relationship between pixels, forming the linear transfer of edge pixels. The local window is used to guide pixels to generate edge gradient, control pixels to form forward rotation, and control artifacts generated by image edges.

After the edge artifacts of the texture image are processed, the local window function is used to calculate the image cost generated by forward rotation [11, 12]. The reverse imaging process of the visual image is used to construct a downsampling numerical model to minimize the image cost formed by the processing. When reconstructing image resolution, the LR image reconstruction method is used to set interpolation functions between texture image pixels, randomly select a pixel, use a cubic polynomial to construct a correlation function between neighboring pixels of the pixel, and use the back projection method to project the pixel interpolation into the numerical imaging model constructed above. The changes in projection parameters during the projection process are shown in Fig. 2.

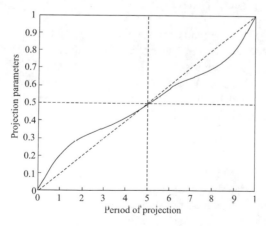

Fig. 2. Changes in projection parameters

As can be seen from the parameter changes shown in Fig. 2, the projection parameters show a changing process of increasing continuously within the range of calibrated projection cycle. In order to ensure the reconstruction accuracy of texture image resolution, SRCNN model is used to extract image block features and process them into corresponding high-dimensional vectors. The numerical relationship can be expressed as:

$$F(Y) = \frac{\max(Y)}{B_1} \tag{15}$$

In formula (15), $F(Y)$ represents the constructed high-dimensional vector, $\max(Y)$ represents the preprocessed image block, and B_1 represents the Bicubic preprocessing parameter. By integrating the high-dimensional vectors constructed above into the constructed numerical imaging model, the resolution in the visual image can be reconstructed.

6   Texture Image Feature Enhancement Processing

Since wavelet, curve wave and other transform methods can not best describe the texture image information [13], this study adopts the Shearlet transform method to describe the texture image. This method is roughly the same as curve wave and contour wave, and has the characteristics of multi-resolution and multi-direction, and has the best nonlinear approximation characteristics, which is more suitable for image feature expression, and makes the results more close to the real situation.

Shear wave transform has a unique bionic system, and its mathematical expression can be described as:

$$\Psi_{XY}(\psi) = \{\psi_{c,d,e}(a) \\ = |\det X|^{c/2}\psi(Y^d X^c a - e) : c, d \in Z, e \in Z^2\} \tag{16}$$

In formula (16), $\psi \in L^2 R^2$ and L represent the length of the wavelet, R represents the wavelet transform coefficients, c, d, and e represent the shear wave transform parameters, Z represents the set of real numbers, and X and Y can form a $2 \times$ The invertible matrix of 2, $|\det X| = 1$. If $\Psi_{XY}(\psi)$ conforms to the Parseval framework, i.e. a compact framework, then the elements of $\Psi_{XY}(\psi)$ are considered as synthesized wavelets.

Assuming the continuous Shearlet transformation of function $f \in L^2 R^2$ is set to:

$$SH_\psi f(c, d, e) = <f, \psi_{(c,d,e)}> \tag{17}$$

According to formula (16) and (17), the texture image is divided into two parts: low frequency information and high frequency information. Among them, the low frequency component is the image irradiation component, namely brightness, contrast, gray value, etc. The high frequency component is the texture and detail features of the image, i.e. the average gradient, spatial frequency, etc.

On this basis, the multi-scale Retinex method was used to solve the low frequency coefficient after separation, and the best estimated radiation component was obtained. However, the result of low frequency coefficient may be positive or negative. If it is negative, the enhancement effect of the image will be affected to some extent. In this way, linear mapping method is used to map the coefficient $F(x, y)$ to the interval $[0,125]$, which is also an enhancement processing of the texture image.

$$F'(x, y) = \frac{F(x, y) - F_{\min}}{F_{\max} - F_{\min}} \times 125 \tag{18}$$

In formula (18), $F_{\max}$ and $F_{\min}$ represent the two extreme values of the low-frequency subband coefficients, while $F'(x, y)$ represents the normalized result, which is treated as an input image and then enhanced according to the multiscale Retinex formula. In order to better reconstruct this part of the image, the multi-scale Retinex processed coefficients are mapped to the interval of $[0, F_{\max} - F_{\min}]$.

Since the details and noise of the texture image are basically retained in the high-frequency subband after the shear wave transformation, the separation of noise and image can be solved by selecting the appropriate threshold value. The scale will become more refined with the increase of decomposition level, and the shear wave allows different sizes and directions of each scale [14]. In order to solve the problem of noise amplification in the process of fuzzy image enhancement, the threshold method is used to suppress the high frequency subband coefficient of shear wave, so as to remove the noise formed in the process of decomposition and the noise of texture image itself. There is a direct correlation between image enhancement effect and threshold value, so the following formula is used to determine the threshold value, namely:

$$T = \lambda \varepsilon \tag{19}$$

In formula (19), λ represents a constant quantity, and its value satisfies $[0, 2]$ between $[0,2]$; ε represents the minimum value of the variance of shear wave noise in different directions at the minimum scale, namely:

$$\varepsilon = \min\left(\varepsilon_{x,y}\right) \tag{20}$$

The noise variance of the high-frequency subband is obtained using the robust median value, i.e.

$$\varepsilon_{x,y} = mediam|D_{x,y}| \tag{21}$$

In formula (21), $D_{x,y}$ represents the coefficients of each high-frequency subband. In this paper, the decomposition layer of the shear wave transformation is 3, and the number of directions is 2, 5, and 7. According to the hard threshold T shrinkage method, denoise the high-frequency coefficients, and the calculation process is as follows.

$$D'_{x,y} = \begin{cases} D_{x,y}, & D_{x,y} \geq T \\ 0, & D_{x,y} < T \end{cases} \tag{22}$$

The texture image processed by the above steps basically removes the interference of light and noise, but the contrast of the image is low and the sense of hierarchy is not good. Therefore, the fuzzy contrast method is used in this paper to process the multi-scale Retinex reconstructed image. The fuzzy contrast formula is set as:

$$F = |\mu_{x,y} - \overline{\mu}_{x,y}|/|\mu_{x,y} - \overline{\mu}_{x,y}| \tag{23}$$

In formula (23), $\mu_{x,y}$ represents the grayscale value of the pixel; $\overline{\mu}_{x,y}$ represents 2 × 2 windows remove the average value of the central domain. This calculation process is similar to the low-frequency component, and the grayscale value of the reconstructed image is transformed to the range of [0125] through formula (18). Due to the fast processing speed of this process, the image enhancement effect is average and the contrast is low. Taking into account the above factors, a linear membership function is used to further improve it. The reconstructed image is transformed into a fuzzy domain, and $x_{a',b'}$ describes the grayscale situation of any pixel, which is represented as follows:

$$\mu_{x,y} = \frac{x_{a',b'}}{L+1} \tag{24}$$

Then, the fuzzy contrast of the reconstructed image is obtained according to the threshold formula, and the nonlinear transformation of F can be obtained:

$$F' = \psi(F) \tag{25}$$

After transforming $\psi(F)$, make $\psi(0) = 0$, $\psi(1) = 1$, and the selected nonlinear enhancement function is represented as:

$$\psi(a') = \frac{1 - e^{-ka'}}{1 - e^{-k}} \tag{26}$$

The membership function adjusted by F' can be described in the following form:

$$\mu_{a',b'} = \begin{cases} \bar{\mu}_{a',b'}(1 - F'), & \mu_{a,b} \leq \bar{\mu}_{a',b'} \\ 1 - (1 - \bar{\mu}_{a',b'})(1 - F'), & \mu_{a,b} > \bar{\mu}_{a',b'} \end{cases} \tag{27}$$

Finally, the fuzzy domain is converted into the space domain, and then the image with enhanced contrast is obtained, that is, the enhanced texture image. The calculation process is as follows:

$$a'_{a',b'} = \mu'_{a',b'} \times (L+1) \tag{28}$$

In formula (28), $a'_{a',b'}$ is the enhanced texture image.

In summary, the design of a texture image feature enhancement processing method based on visual saliency model has been completed, and the framework structure of this method is summarized, as shown in Fig. 3.

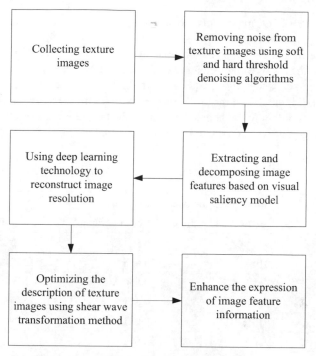

Fig. 3. Framework structure diagram of the method

7 Experiment and Analysis

To verify the feasibility of the texture image feature enhancement processing method based on visual saliency model designed above, the following experiments are designed.

The method proposed in this article will be used as the experimental group, with traditional visual communication based image enhancement methods and global and local multi feature based image enhancement methods as control group A and control group B, respectively. Use three processing methods to enhance the same set of texture images.

The obtained images are all taken in low light and shaded areas. Number the 10 texture images obtained as JPG-#1, JPG-#2, JPG-#3, JPG-#4, JPG-#5, JPG-#6, JPG-#7, JPG-#8, JPG-#9, and JPG-#10. Strictly follow the implementation steps of the three processing methods to complete the image enhancement processing.

To objectively evaluate the three processing methods, firstly, the distortion of the processed image is selected as the evaluation indicator, which can be calculated using the following formula:

$$LOE = \frac{1}{m} \sum_{x=1}^{m} RD(x) \tag{29}$$

In Formula (29), *LOE* represents the distortion coefficient of the texture image after enhancement; m represents the number of pixels in the image; $RD(x)$ represents the relative order difference between the original image and the enhanced image. According to this formula, the distortion coefficient *LOE* of the texture image processed by the three groups of processing methods is calculated. In terms of image enhancement effect, if the value of *LOE* is smaller, it means that its features are more natural and the distortion rate is lower. According to the above discussion, the brightness distortion results of the three processing methods are counted, as shown in Table 1.

Table 1. Distortion coefficients of images processed by three methods

Image number	AImage distortion coefficient *LOE*		
	Experimental group	Control group A	Control group B
JPG-#1	0.325	0.624	0.742
JPG-#2	0.314	0.664	0.756
JPG-#3	0.315	0.650	0.826
JPG-#4	0.327	0.671	0.845
JPG-#5	0.315	0.681	0.825
JPG-#6	0.326	0.692	0.791
JPG-#7	0.313	0.674	0.783
JPG-#8	0.326	0.682	0.925
JPG-#9	0.317	0.634	0.921
JPG-#10	0.325	0.665	0.846

From the data in Table 1, it can be seen that after enhancing the five images, the image distortion coefficients of the experimental group were controlled between 0.313 and 0.327, while the image distortion coefficients of the control group A fluctuated between 0.624 and 0.692, while the image distortion coefficients of the control group B all exceeded 0.700. Through the comparison of the experimental results obtained above, it can be concluded that the distortion coefficient of the texture image is the lowest in the experimental group after enhancement processing, indicating that after the experimental group method processing, the natural characteristics of the image are maintained well, and the distortion rate is effectively controlled.

To further compare the clarity of texture images processed by three different processing methods, the average gradient was chosen as the evaluation index to reflect the changes in clarity of the original image after enhancement processing. The calculation formula for the average gradient is:

$$G(x, y) = I(dxi + dyj) \tag{30}$$

In Formula (30), $G(x, y)$ represents the average gradient of the image after enhanced processing; I represents pixel value; (i, j) stands for pixel coordinates. The higher the average gradient $G(x, y)$, the clearer the image.

According to the above formula (30), taking JPG-#3 of the above five pedestrian action images as an example, the average gradient of the images processed by the three enhancement processing methods was compared with the original image, and the experimental results as shown in Fig. 4 were obtained.

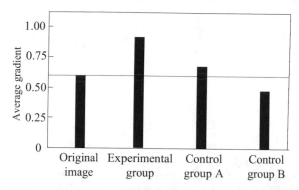

Fig. 4. Comparison of processing effects of three enhancement processing methods

From the bar chart shown in Fig. 4, it can be seen that the average gradient of the experimental group and the control group A treatment method has been slightly improved compared to the original figure, while the average gradient of the control group B treatment method has actually decreased compared to the original figure. Therefore, the above experiments can further prove that the experimental group method can effectively improve the clarity of texture images.

Based on the above experimental results, it can be concluded that the method proposed in this paper not only maintains good feature naturalness in the processed texture image, but also improves the clarity of the image. This indicates that the method proposed in this paper effectively achieves comprehensive optimization of visual effects on texture images.

8 Conclusion

In this paper, a feature enhancement method of texture image based on visual saliency model is designed. Firstly, after the texture image is collected, the texture image is denoised. Then, the features are extracted and decomposed based on the visual saliency model, and the resolution of the texture image is reconstructed using deep learning technology. Finally, shear wave transform is used to describe the texture image to enhance the expression of feature information.

The experimental results show that the enhanced texture image has lower distortion coefficient and higher clarity, which indicates that the proposed method achieves the design expectation effectively.

In practical applications, this method can enhance the features of texture images by utilizing visual saliency models, thereby improving the visibility and discrimination of texture features. By highlighting the most significant texture regions in the image, this method can help analysts better understand and recognize texture images. For example, in texture image classification tasks, using this method can make texture features more vivid and clear, reduce background noise interference, and thus improve the accuracy and stability of classification. This is of great significance for many application scenarios such as texture recognition, material classification, medical image analysis, etc.

Aknowledgement. 2022 Hubei Provincial Department of Education Science Research Program Guiding Project: Application of Gesture Interaction Technology in Public Art Design in Smart Cities (B2022422).

References

1. Xue, Y., Zhang, X., Zhao, J.: Study on low Illumination image enhancement based on quantum behaved particle swarm optimization. Optical Technique **47**(04), 500–506 (2021)
2. Yu, T., Li, Y., Lan, C.: Bionic image enhancement algorithm based on top-bottom hat transformation. J. Computer Appl. **40**(05), 1440–1445 (2020)
3. Gan, S., Qiu, L.: Research on low illumination image enhancement based on visual communication. Laser Journal **42**(9), 114–118 (2021)
4. Liu, Y., Zhu, S.: Image enhancement algorithm based on global and local multi features. Chinese Journal of Liquid Crystals and Displays **35**(5), 508–512 (2020)
5. Xu, R., Wang, Z., Zong, T.: Edge enhancement of medical image based on improved Gaussian filter. Information Technology **44**(4), 75–78 (2020)
6. Ji, C., Wang, D., Huang, X., et al.: Saliency calculation based on the fusion of enhanced contour features and spatial semantic information. J. Computer-Aided Design & Computer Graphics **32**(11), 1813–1821 (2020)
7. Liang, Y., Ma, N., Liu, H.: Deep learning based salient region detection. J. Data Acquisition & Processing **35**(03), 474–482 (2020)
8. Qian, Y., Lu, J., et al.: A lightweight low illumination image enhancement method based on information multiple distillation. J. Shanxi University (Natural Science Edition) **44**(5), 887–896 (2021)
9. Liu, M., Tang, L., Xiong, D., et al.: Research on image enhancement model based on adaptive fractional anisotropic diffusion. J. Hubei Minzu University: Natural Science Edition **40**(01), 58–66+109 (2022)
10. Jiang, Z., Wu, X., Zhang, S.: Low-illumination image enhancement based on MR-VAE. Chinese J. Computers **43**(7), 1328–1339 (2020)
11. Lin, Z.: Multi-scale detail enhancement method for two-dimensional animated images based on bilateral filtering. J. Qiqihar University(Natural Science Edition) **37**(01), 56–61 (2021)
12. Tian, Z., Wang, M., Zhang, Y.: Image enhancement algorithm based on dual domain decomposition. Acta Electronica Sinica **48**(07), 1311–1320 (2020)
13. Song, G., Du, H., Wang, P., Liu, X., Han, H.: Texture detail preserving image interpolation algorithm. Computer Science **46**(S1), 169–176 (2019)
14. Fan, Z., Liu, B.: Research on adaptive enhancement technology of low illumination image based on improved Retinex. Industry and Mine Automation **47**(S1), 126–130 (2021)

Author Index

© ICST Institute for Computer Sciences, Social Informatics and Telecommunications Engineering 2024
Published by Springer Nature Switzerland AG 2024. All Rights Reserved
L. Yun et al. (Eds.): ADHIP 2023, LNICST 548, p. 477, 2024.
https://doi.org/10.1007/978-3-031-50546-1

Printed in the United States
by Baker & Taylor Publisher Services